Eta Carinae and the Supernova Impostors

Astrophysics and Space Science Library

For further volumes:
http://www.springer.com/series/5664

Kris Davidson • Roberta M. Humphreys
Editors

Eta Carinae and the Supernova Impostors

Springer

Editors
Kris Davidson
Minnesota Institute for Astrophysics
University of Minnesota
116 Church Street SE
Minneapolis, MN 55455
USA

Roberta M. Humphreys
Minnesota Institute for Astrophysics
University of Minnesota
116 Church Street SE
Minneapolis, MN 55455
USA

ISSN 0067-0057
ISBN 978-1-4614-2274-7 e-ISBN 978-1-4614-2275-4
DOI 10.1007/978-1-4614-2275-4
Springer New York Dordrecht Heidelberg London

Library of Congress Control Number: 2012930990

Cover illustration: Color-composite image of the Carina Nebula, revealing exquisite details in the stars and dust in the nebula. Eta Carinae is at the center with the Keyhole Nebula just to its right. North is up and East is to the left. **Credit:** ESO.

Printed on acid-free paper

Springer is part of Springer Science+Business Media (www.springer.com)

To the observers of the Southern Sky.

Preface

Eta Carinae is well known to astronomers for its many superlatives. It is the most massive, most luminous star in our region of the Milky Way. During its "Great Eruption" 170 years ago it ejected 10 $M_{\odot}$ or more, creating its famous Homunculus Nebula. We now know that its Great Eruption was the energetic, non-terminal explosion of a very massive star nearing the end of its short life. But η Car and its Great Eruption may not be unique, some weird event, only observed once.

In 1965 Fritz Zwicky proposed a class of supernova events he called "Type V", relatively faint at maximum with long durations. There were only two members of his Type V group, η Car and SN1961v. These stars were not true supernovae and today we recognize several stars that had giant eruptions in which they increase their total luminosity and become bright enough to be mistaken for underluminous supernovae. In addition to η Car, examples include SN1961v, SN1954j, P Cyg 400 years ago, V1 in N2363 and probably the "Pistol star" near the Galactic Center. The modern supernova surveys are producing a growing list of sub-luminous outbursts from objects in other galaxies. These "supernova impostors" are apparently very massive stars undergoing eruptions similar to η Car and other unstable stars such as the Luminous Blue Variables (LBVs). Hence the title of this book. These stars are very rare, however, and most have been observed only sparsely.

The most massive stars are at the intersection of several critical areas of modern astrophysics including the chemical evolution of galaxies via their high mass loss episodes, as the likely progenitors of gamma-ray bursters and the most luminous supernovae, and as candidates for the "first stars" in the Universe. Extremely luminous supernovae, like SN2006gy, may be the terminal explosions of very massive stars surrounded by extensive circumstellar material from previous eruptions. The role of high mass loss events like that of η Car on the pre-supernova state of these stars is clearly of increasing interest in astrophysics.

Our understanding of this remarkable object has benefited from the work and insight of numerous astronomers over the past century many of whom are now deceased. Due to the limitations on the size of this volume, it was not possible to invite all current researchers for a contribution. The editors therefore gratefully acknowledge the contributions of Patricia Whitelock and Michael Feast,

Chris Sterken and Antonin Van Genderen, Roberto Viotti, Augusto Damineli, Ted Gull, John Meaburn, Otmar Stahl, Noam Soker, and Bob Gehrz.

The contributed chapters in this volume provide a comprehensive review of η Car, including its Great Eruption, historical evolution and recovery, its wind and ejecta and its relation to other very massive unstable stars, the origin of their instabilities and eventual fate. The contributions are written by experts in their fields and hence are intended for professional astronomers, graduate students in astrophysics, and physicists interested in the final stages of stellar evolution. These reviews are based on the most recent observations from space and ground-based telescopes and the current state of our theoretical understanding, nevertheless, we must emphasize that η Car never ceases to surprise us.

Minneapolis, MN

Kris Davidson
Roberta M. Humphreys

Contents

Chapter 1
Eta Carinae: From 1600 to the Present

Roberta M. Humphreys and John C. Martin

Abstract In this introductory chapter we review η Car's historical photometric and spectroscopic record with emphasis on its recovery from its Great Eruption and important recent advances in our understanding of the most massive, most luminous star in our corner of the Milky Way.

1.1 Introduction

Eta Carinae is famous for its many superlatives as the most luminous, most massive star in our region of the Milky Way and for its "Great Eruption" from 1837 to 1858 that formed its bipolar "Homunculus" nebula. η Car is also our closest example of a supernova impostor.[1] The Great Eruption and continuing instability during η Car's recovery provide us with a remarkable opportunity to study close-up one of the most massive stars known during the final stages in its evolution; a star that may be in a pre-supernova state. Throughout this volume, however, we emphasize that η Car is not some bizarre, unique, or extreme case in stellar evolution. There are numerous very massive stars in our galaxy, some rivalling η Car in brilliance, even in its own neighborhood, and many that show evidence for instabilities. In Fig. 1.1, we show a schematic HR Diagram illustrating η Car's relation to the massive star population in general, to other extremely luminous stars, and to stars famous for their instabilities.

[1] See Chap. 11.

R.M. Humphreys (✉)
Minnesota Institute for Astrophysics, University of Minnesota, Minneapolis, MN, USA
e-mail: roberta@umn.edu

J.C. Martin
University of Illinois, Springfield, IL, USA
e-mail: jmart5@uis.edu

K. Davidson and R.M. Humphreys (eds.), *Eta Carinae and the Supernova Impostors*, Astrophysics and Space Science Library 384, DOI 10.1007/978-1-4614-2275-4_1,

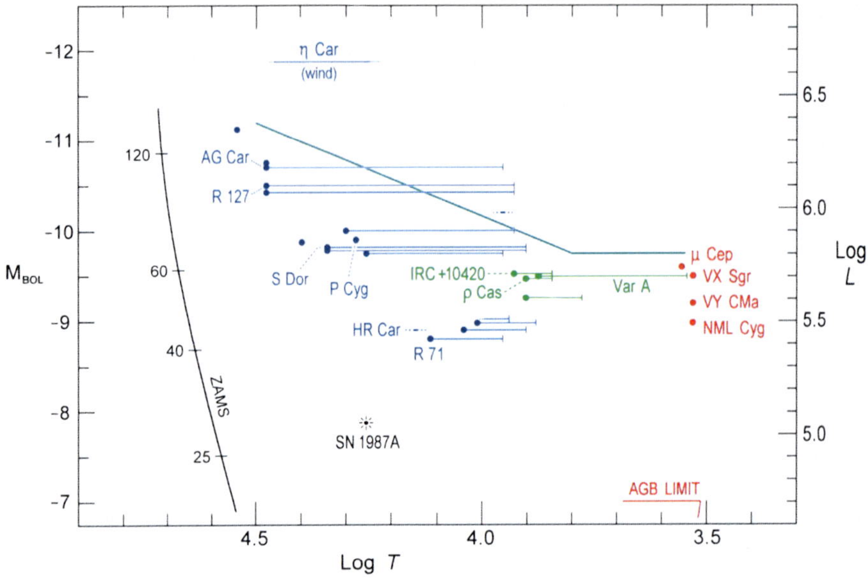

Fig. 1.1 A schematic HR Diagram showing the zero-age main sequence, the empirical upper luminosity boundary with the locations of η Car (its wind), selected LBVs (*blue*), luminous red supergiants (*red*) and intermediate-type hypergiants (*green*). The horizontal lines represent the apparent transits in the HRD during their optically thick wind state or "eruptions"

1.2 The Great Eruption and Recovery 1837–1955

Because of its location in the Southern sky, the first historical record of η Car doesn't appear until about 1600 (Bayer's Atlas (1603), Halley (1677), Noel (1685–1689), see Innes [53]). In those days it was known as η Argus and up until the early 1800s it was usually reported as a fourth or second magnitude star, although Frew [37] has suggested that instead of fluctuations between these two states, the star was gradually brightening from 1600 to ~1800. The observational record shows that it had definitely reached second magnitude by about 1800. Between 1820 and 1830 it began to show more variability, and may have been in an enhanced mass loss state when it oscillated between second and first magnitude for 10 years or more before the Great Eruption. Figure 1.2 shows the light curve from 1600 to 1900.[2] Based on what we now know about η Car and the Luminous Blue Variables (LBVs), Davidson and Humphreys [24] suggested that the pre-eruption variability may have been due to shifts in apparent temperature of the star's wind or envelope, typical of

[2]The pre-1900 light curves included in this paper are based, for the most part, on the compilation by Innes [53] converted to modern magnitudes as described by de Vaucouleurs and Eggen [31]. Also see the recent discussion by Frew [37] and Smith and Frew [89].

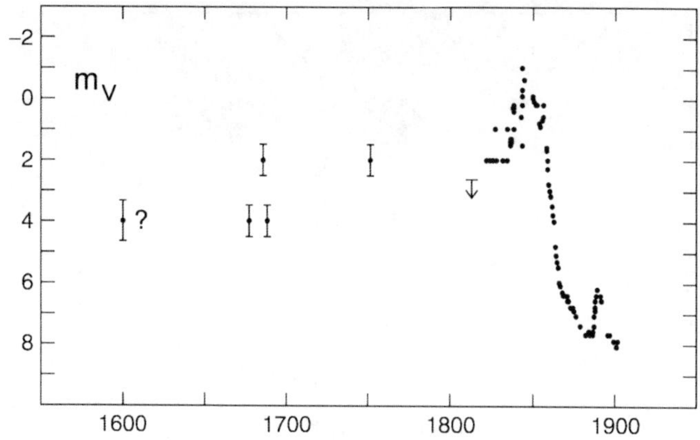

Fig. 1.2 The historical light curve from 1600 to 1900 (From Humphreys et al. [51])

LBV or S Doradus-type variability,[3] not to actual changes in luminosity. Beginning in late 1837, η Car rapidly brightened much more, to zero apparent magnitude and even brighter.

1.2.1 The Great Eruption 1837–1858

In 1838 Herschel [45] reported that η Car rapidly brightened from near first magnitude to brighter than zero apparent magnitude. It reached apparent magnitude ~ -1 in 1843 and again in late 1844 – early 1845 (see Fig. 1.3), and was the second brightest star in the sky. It then oscillated near zero magnitude for several more years until by 1858 it had faded back to a first magnitude star. Frew [37, 89] has recently identified additional photometry from this period for a more complete record of its variability during the eruption. The expanded light curve at the time of the Great Eruption (Fig. 1.3) shows evidence for considerable instability with oscillations on the order of a magnitude or more in only a few weeks. It is also remarkable that it stayed at or near zero apparent magnitude for 20 years and that it actually increased its total luminosity.

Eta Car's present luminosity of $10^{6.7}\,L_\odot$ or $M_{bol} = -12$ mag is well determined from its spectral energy distribution that peaks in the thermal infrared (see Sect. 1.3.1) and its distance of 2,300 pc.[4] When it was zero apparent magnitude during the eruption, its absolute visual brightness was $M_v \approx -13.5$ mag, assuming a total visual extinction A_v of 1.7 mag from its membership in the cluster Tr 16 and no significant circumstellar extinction at that time. Davidson [19] showed that with

[3] See Chap. 10.

[4] See Chap. 2 for a review of its distance and foreground extinction.

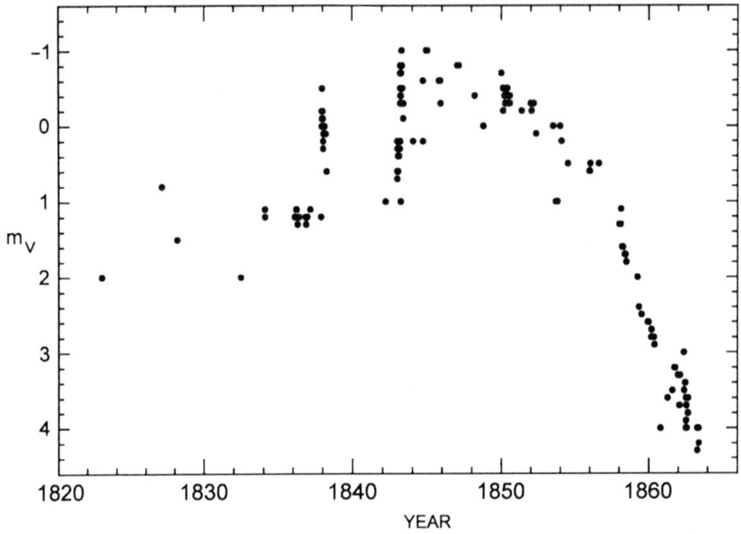

Fig. 1.3 The Great Eruption circa 1838–1858 based on magnitudes in Frew [37] and Smith and Frew [89]. There are no recorded magnitudes from 1838 to 1841

extremely high mass loss, the temperature of the optically thick wind formed during an eruption would approach a minimum at ~6,500 K. By comparison with LBVs in eruption, η Car's apparent temperature was probably cooler than normal LBVs and its bolometric correction near zero. Rest at al. [78] have recently reported the discovery of the 'light echo' from η Car's Great Eruption with an absorption line spectrum similar to an early G-type star.

During the eruption, its total luminosity was thus $M_{bol} \approx -13.5$ mag or $10^{7.3}\,L_{\odot}$, and when η Car briefly reached apparent magnitude -1 in 1843 and in 1844–1845, it would have been -14.5 mag. *It thus increased its total luminosity by a factor of 10 during the Great Eruption.* With this luminosity and temperature, η Car's apparent photosphere or optically thick wind would have been ≈10 A.U, or more in radius. It total luminous energy during the 20 year eruption was on the order of $10^{49.7}$ ergs and for comparison, its kinetic energy was also $\sim 10^{49.7}$ ergs (Sect. 1.3.3).

Eta Car began to slowly fade after 1850, and beginning in 1858 it declined more rapidly reaching a minimum near 7.5 apparent magnitude by 1880. We now know that it formed dust probably during or soon after its eruption, but we do not know if the initial fading was due to dust or if the dust formed after the cessation of the outburst. The dust formation zone for η Car is ≈200–600 A.U. With an expansion velocity of $\sim 600\,\mathrm{km\,s^{-1}}$ [12, 70, 83], the ejecta would have reached this distance in ≈10 years. Thus circumstellar dust should have begun to form prior to the observed decline, and the eruption may have actually continued after 1858. The exact time when the eruption ended is not clear from the light curve. Humphreys et al. [51] however point out a brief period in the early 1860s when its apparent brightness changed very slowly, if at all, at about third magnitude, resembling a shoulder or plateau on its decline, which may signal the end of the actual eruption. It had probably ended, though, by 1869 when Le Sueur [58] described an emission line spectrum.

1.2.2 Post Eruption

We know very little about the state of the star soon after the Great Eruption. For this reason, Le Sueur's [58] visual observation about 10 years later is especially intriguing. He described five emission lines, the Fraunhofer lines C (Hα), D, "b group" and F (Hβ) plus the "principal green nitrogen line". The Fraunhofer b group, due to Mg I absorption in the Sun (5167–5184 Å), is very likely [Fe II] 5158 Å and Fe II 5169 Å emission in η Car [105]. The "D" line is especially interesting, because although Na I D is present today in η Car as a complex mix of absorption and emission, it is not a strong emission line. Walborn and Liller [105] suggested that it was actually He I 5876 Å, and if so, it had to be at least as strong then as it is now. In current groundbased spectra, it is comparable in strength to the [N II] 5755 Å line which apparently was not noticed by Le Sueur. They also suggested that the "principal green nitrogen line" was due to [Fe II] λ 5018 because of Le Sueur's statement that is was not just an extension of the nebular emission line, presumably [O III] λ 5007, across the spectrum of η Car. It is clear from the text that this was a difficult observation, near the limit of what could be seen, so it is possible that He I 5876 Å and other emission lines were much stronger then. As this was only about 10 years after the Great Eruption during which η Car had thrown off its entire outer envelope, we can only speculate about the origin of He I and the other emission lines. They could have originated in a hot wind or even from what was then the star's photosphere.

1.2.3 The Second Eruption 1887–1895

Eta Car's second or lesser eruption lasted about 7 years (1887–1895) with a maximum at $m_V \approx 6.2$, about 1.5 magnitudes above its minimum. Agnes Clerke [8] reported an absorption line spectrum with no "dark" i.e. emission lines, from a visual observation in 1889. The first photographic spectra of η Car were obtained during the second eruption as part of the Harvard Observatory objective prism survey. The spectrum from June 1893 is of especially good quality and shows an early F-type supergiant absorption line spectrum with emission lines of hydrogen, Fe II, and [Fe II] that has been described by many authors [4, 6, 47, 105, 115]. Walborn and Liller [105] provide a good reproduction of the original photographic spectrum, and a digitized scan and extracted linear spectrum can be seen in Humphreys et al. [52].

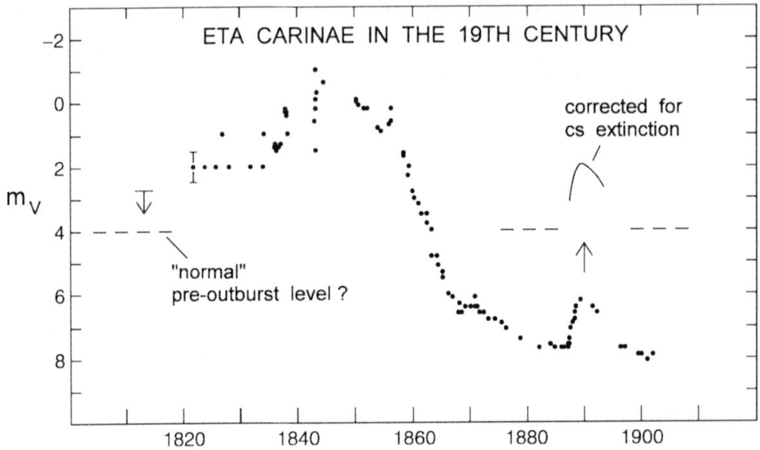

Fig. 1.4 The second or lesser eruption corrected for circumstellar extinction (From Humphreys et al. [51])

Its visual brightening of ∼2 mag, a duration of several years, and an F-type supergiant spectrum at maximum light from a slow, cool opaque wind or pseudo-photosphere are typical characteristics of an S Doradus or classical LBV–type eruption [50]. The absorption lines had a velocity of $-180\,\mathrm{km\,s^{-1}}$ [115] relative to the hydrogen emission lines indicative of an expanding envelope and consistent with the comparatively slow winds in LBV eruptions. The hydrogen P Cygni absorptions have a velocity of $-300\,\mathrm{km\,s^{-1}}$ significantly slower than η Car's current polar wind [24, 86], but comparable to the polar expansion velocity in the "Little Homunculus" [54, 87], which was probably ejected in the 1890s eruption. The second eruption was very similar to what we now describe as LBV eruptions or instabilities.[5]

The second eruption was not as minor as it may appear on the historical light curve (Fig. 1.2). When corrected for a realistic estimate of the circumstellar extinction at that time [51], the star's apparent brightness would have been about second magnitude (Fig. 1.4) and its expanded, cool envelope would have reached ∼6 A.U. We now know that the star also ejected about $0.2\,\mathrm{M}_\odot$ of relatively slow moving material in both the equatorial and the polar directions [28, 54, 87]. Thus in both eruptions, η Car ejected mass with a bipolar structure, but the kinetic energy in the lesser one was only $10^{46.9}$ ergs. Assuming that the star maintained a near constant bolometric luminosity in the second eruption, like LBVs, its radiated energy was $10^{48.6}$ ergs, significantly greater than the kinetic energy, but still an order of magnitude less than in the Great Eruption.

The cause of this second outburst is not known, but the most likely explanations for LBV-type eruptions include the opacity-modified Eddington limit, subphotospheric gravity-mode instabilities, and a super-Eddington continuum-driven wind.[6]

---

[5]See Chap. 10.

[6]See Chaps. 10, 12, and reviews by Humphreys and Davidson [50] and Glatzel [42].

The significant difference in the kinetic and radiant energies in the two eruptions indicates that they were definitely on a different scale and may have had different physical causes.

The F-type absorption lines had vanished by 1895 when the second eruption had ended. Beginning with the observations in 1902, the spectra begin to show the prominent emission line spectrum of hydrogen, [Fe II], and Fe II that is familiar in all subsequent groundbased spectra of η Car.

1.2.4 Quiescence ~1900–1941

The period from about 1900–1941 was one of relative quiesence for η Car. Its apparent brightness was basically constant at ~8th magnitude, and the spectra from this period are remarkably alike [52]. But most importantly, the relatively strong He I emission lines and high excitation emission lines of [Fe III] and [Ne III], conspicuous in spectra today, were absent or very weak prior to 1944. Feast et al. [33] noted that He I emission was not present in pre-1920 spectra. A quantitative analysis of digitized scans of the Harvard objective prism spectra from 1902 to 1941 by Humphreys et al. [52] showed that the He I λ 4471 emission line was either very weak or not present during that time. He I emission and the high excitation lines of [Ne III] and [Fe III] were first reported in Gaviola's spectral series [39] obtained from 1944 to 1951. The lack or weakness of these lines implies a scarcity of UV radiation prior to 1944 which could not be explained either by extinction due to dust or an optically thick wind from η Car [52]. See the Chap. 3 for more discussion of a possible explanation.

1.2.5 The Spectroscopic/Photometric Transition 1941–1952

Between the end of 1939 and mid 1941, η Car brightened about 0.8 mag on the photographic scale [72] . Only 3 years later, the star's spectrum first showed prominent He I emission and other high excitation emission lines not previously observed [33, 52]. It continued to brighten for several years with small oscillations when in 1952 de Vaucouleurs and Eggen [31] reported a rapid increase from 6.9 to 6.5 mag in apparent visual brightness, measured photoelectrically, in only a few weeks from February to March 1952. It is important to realize that these measurements refer to the integrated light of the ejecta i.e. the Homunculus nebula (see Sect. 1.3.3), not just to the central object.[7] The object's rapid brightening during this time was most likely due to dust destruction which might also be caused by

[7] Neither of the two brightenings correspond with a spectroscopic event (see Sect. 1.3.2). There are also two gaps in the photometric record during the twentieth century, 1915–1935 and 1953–1970. We suspect that a brightening of 1 magnitude or more would have been noticed especially in the 1950–1960s when Thackeray was observing η Car frequently.

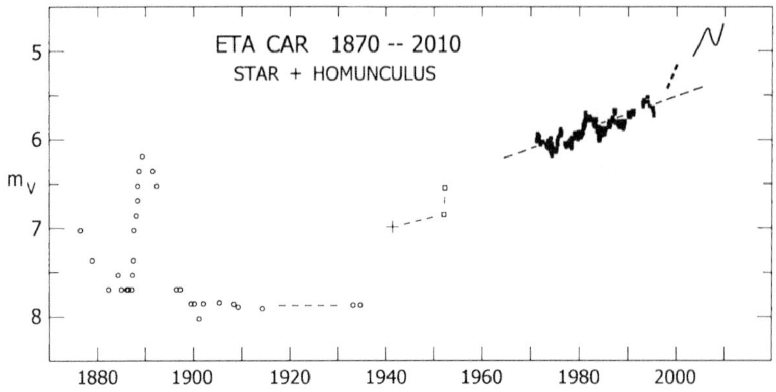

Fig. 1.5 The integrated light curve during the twentieth century shows the rapid brightening in mid-century and the subsequent gradual brightening (Adapted from Humphreys et al. [52])

the increase in UV radiation required for the onset of the He I and high excitation emission lines. Together these spectroscopic and photometric changes suggest a significant change in the wind of η Car. After 1952, its integrated light continued to brighten slowly at a fairly regular rate, attributed to the expansion of the dust envelope [100–102]. Figure 1.5 shows the integrated light curve, star + Homunculus nebula, from 1890 to 2010. The more rapid brightening beginning in 1998 is discussed in Sect. 1.4.

1.3 The Nature of the Central Object: Multiwavelength Observations 1960–2000

By the beginning of the second half of the twentieth century, η Car had once again attracted the attention of astronomers, but its true nature was not understood. In 1963 Gratton [43] wrote the first comprehensive review of η Car's characteristics and behavior and suggested it might be a massive star in formation. Other suggestions ranged from a nova [76], a new class of supernova [118], a slow supernova powered by a pulsar [73] and an unstable F-type supergiant [5]. The latter suggestion, which was closest to its true nature, was based on its spectral appearance in the second eruption. In this section we highlight the major developments that led to our current understanding of η Car as a very massive, evolved star and its instability.

Observational techniques had improved considerably with photoelectric photometry and spectrophotometry. In South Africa, Thackeray was monitoring η Car [95–97] and like Gaviola [39] had noticed an unexpected change in its emission line spectrum [98]. Aller and Dunham [2] and Rodgers and Searle [82] had obtained the first quantitative spectrophotometry.

The most important observation of η Car in the twentieth century, was the crucial discovery by Neugebauer and Westphal [71, 111] that it radiated its huge luminosity in the thermal infrared at 10 μm and longer wavelengths. Combined with an improved distance [34, 35], this led to a realistic estimate of its extremely high luminosity and the realization that its flux in the infrared was comparable to what was observed in the Great Eruption (Sect. 1.2.1). At about the same time, Pagel [74, 75] showed that considerable circumstellar extinction was necessary to account for the reddening of the emission lines. However, to explain the high luminosity, based on the observed energy distribution, he concluded that the central source was a non-thermal, compact object. Davidson's [18] analysis convincingly showed that *η Car was in fact a very luminous, unstable hot star of 100 $M_\odot$ or more.*

With improved spatial and wavelength resolution in the infrared, η Car was modeled as an enshrouded, unresolved hot source surrounded by cool, dusty ejecta, possibly bipolar [41, 81], identified with the visual Homunculus nebula (see Sect. 1.3.3). Measurement of its polarization [103] confirmed that the Homunculus was a reflection nebula. The basic nature of the central object was now recognized – *an extremely luminous, hot star emitting most of its energy in the UV, but absorbed by dust formed in the Great Eruption and re-radiated in the infrared.*

1.3.1 The Central Star

With the long-wavelength observations, integration of η Car's spectral energy distribution [11] combined with its modern distance (2.3 kpc) determined from the expansion of the ejecta [1, 28, 64, 84], its current luminosity is 1.9×10^{40} ergs s^{-1} or $\sim 10^{6.7}\,L_\odot$ and $M_{Bol} \approx -12.0$ mag.[8] During the Great Eruption its total luminosity did indeed increase significantly exceeding the Eddington Limit for its mass. To avoid violating the Eddington Limit at its present luminosity, η Car's present-day mass must be greater than 90 $M_\odot$ Estimates of its initial mass based on its luminosity in combination with stellar models, suggest a zero-age mass near 150 $M_\odot$ and possibly as high as 200 $M_\odot$.

Unfortunately, we cannot observe the central star directly even with the Hubble Space Telescope. It has an extremely dense, opaque wind [24, 46]. Consequently there is no unambiguous estimate of the star's temperature from stellar models. With its intense emission line spectrum, realistic estimates of the characteristic temperature are uncertain, but are of the order of 20,000 K with an effective radius,

[8]This includes a small correction of ~10% for visual/UV radiation that escapes absorption by the dust longwards of 1 μm [24].

where the UV/visual radiation originates, of <0.8 A.U. The actual embedded star may be both hotter and smaller. See the Chap. 3 for a discussion of the central star.[9]

Measured mass loss rates depend somewhat on the wavelength used. Radio observations [112] yielded an estimated $10^{-3.5}\,M_\odot\,\mathrm{year}^{-1}$ while millimeter mapping [11] of the central region gave $\sim 10^{-2.6}\,M_\odot\,\mathrm{year}^{-1}$. But, both the radio and millimeter observations refer to a region much larger than the stellar wind. Independent mass loss estimates by Davidson et al. [22] and by Hillier et al. [46], based on emission lines in the wind, give a consistent mass loss rate of $\sim 10^{-3}\,M_\odot\,\mathrm{year}^{-1}$.

The discovery from the UV emission-line spectrum that the gas in the outer ejecta was nitrogen and helium rich while carbon and oxygen were less abundant than nitrogen [20, 21] verified that η Car was indeed an evolved, massive star and was most likely a post-main sequence object.

1.3.2 Photometric and Spectroscopic Variability

Since its rapid brightening in the 1940s to early 1950s, the integrated light of the central star plus Homunculus has shown a long-term trend slowly increasing in apparent brightness at visual and near-infrared wavelengths [100–102, 113, 114]. The slow visual increase was largely attributed to the expansion of the nebula resulting in a clearing of obscuring material around the star. HST/STIS spectra in 1998–1999 revealed another rapid brightening, this time of the central star, of ~1 magnitude in only 1 year [26]. Since then the integrated light has continued to brighten more rapidly than the previous long-term trend [59, 61] shown in Fig. 1.5. The central star has brightened even more rapidly, Sect. 1.4 and Fig. 1.8.

1.3.2.1 The 5.5 year Spectroscopic Cycle

Beginning with Gaviola's first recorded spectroscopic change in 1948 and Thackeray's observation in 1964, η Car's spectrum had been observed on various occasions to show a dramatic change with the weakening and disappearance of the high excitation emission lines including He I, [Ne III] and [Fe III]. Zanella, Wolf and Stahl [116] observed a "shell" episode in 1981–1983, surmised that the changes were due to a shell ejection, and suggested a recurrence approximately every 17 years based on the previous observations.[10] Near-infrared photometry by Whitelock and Feast and their collaborators [113, 114] showed a 5–6 year periodicity in the infrared flux, but its interpretation was complicated by the increasing brightness

[9] In Chap. 4, Najarro and Hillier derive much lower effective temperatures assuming the standard $\tau = 2/3$ for stellar atmospheres, but as mentioned there this is unrealistic for an opaque wind.

[10] Later, 17 years turned out to be three full 5.5 year cycles.

of the central star. Meanwhile, "spectroscopic events" occurred in 1986 and in 1992. Damineli [13] was the first to recognize an approximately 5.5 year periodicity based on the periodic weakening of the He I λ 10830 line and successfully predicted the next event at the end of 1997. Subsequent multiwavelength observations from X rays to radio during the 1997–1998 and 2003.5 events confirm the periodicity and weakening of the flux during the events which are now recognized to recur approximately every 2023 days [15]. The periodicity of the events is now acknowledged as due to the presence of a second star in a highly eccentric orbit that brings it into the wind of the massive primary at periastron.[11] The spectroscopic events have been attributed to (1) eclipses of a hot secondary star by the primary wind [14, 77], (2) disturbances in the primary wind triggered by a companion near periastron [60, 86], and/or (3) a thermal/rotational recovery cycle [27, 30, 86, 116]. Suggestions (2) and (3) are not necessarily mutually exclusive. As discussed in Sect. 1.4.1, observations during the 2003.5 event support the wind-disturbance/instability suggestion as opposed to a simple eclipse.

Although most astronomers now agree that η Car is a binary very likely with a hotter, less massive, less luminous companion, the secondary star has remained unobservable, even in the UV. It is assumed to be the source of the UV radiation responsible for the high excitation emission lines in the inner ejecta (See Sect. 1.4). A recent photoionization analysis [65] demonstrated that a 40–50 $M_\odot$ star with L $\sim 4 \times 10^5\ L_\odot$ and $T_{eff} \approx$ 39,000 K fits the observed parameters.

1.3.3 The Homunculus, Equatorial Spray, and the Outer Ejecta

Eta Car's distinguishing visual features are its now famous bipolar lobes and equatorial "skirt" or "spray" as imaged by HST. The "halo" surrounding η Car was first noted by van den Bos [99] in 1938. Thackeray [93, 94] described the nebulosity surrounding η Car and sketched it with a 10 arcsec halo. Gaviola [38] may have been the first to actually photograph the nebula which he named the Homunculus or "little man". Figure 1.6 shows several images of the Homunculus nebula obtained over the past 60 years.[12] The nebula's expansion has been measured by many observers, beginning many years prior to HST observations [38, 40, 80, 93]. The numerous results leave little doubt that the bipolar lobes were ejected in the 1840s Great Eruption. Though most of the groundbased images did not show much detail, the

[11] See Chap. 9 for a detailed discussion of the X ray variability and the secondary star.

[12] η Car's visual appearance in a small telescope has changed dramatically in the past 40 years. In the early 1970s, it looked like the Gehrz and Ney [40] image in Fig. 1.6, a bit like a cross between the Pillsbury "Dough Boy" and the "Michelin Man". By the late 1990s, the two lobes were easily visible in a small telescope. The inner ejecta could be seen as a bright red-orange region surrounding the position of the star in the SE lobe. This change was due to the increasing brightness of the star and contrast with that of the Homunculus.

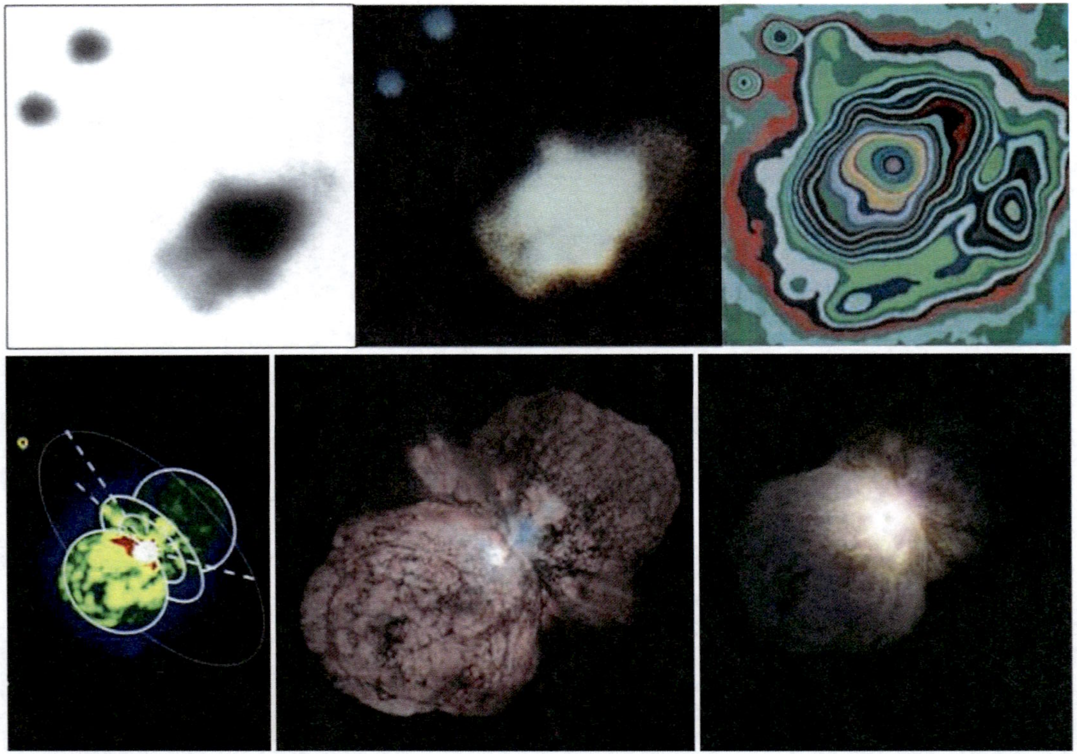

Fig. 1.6 Different views of η Car from 1950 to the present. *Top row* from left to right: the first photograph of the Homunculus circa 1945 [38], groundbased image from 1972 [40], a digital image from 1981 [49], *Bottom row* left to right: 1985 image made from 200 very short exposures [32], 1995 multi-color HST/WFPC2 image, 2009 near-infrared adaptive optics image [3]

term 'bipolar' was applied to the Homunculus long before the HST images. For example, the infrared data [44] was best fit by a pair of hollow lobes, and Duschl's [32] ground-based images obtained with a specialized technique revealed the bipolar lobes and equatorial skirt for the first time, prior to HST imaging.

One of the more surprising developments was the discovery using speckle interferometry of small, bright starlike condensations or "knots" within $0.3''$ of the central star (knot A) [48, 107] , now called the 'Weigelt knots'. Subsequent speckle interferometry [108] and HST imaging and spectroscopy [23]. showed that knots BCD are moving very slowly, compared with the expansion of the Homunculus, at less than $50\,\mathrm{km\,s^{-1}}$, in the equatorial plane, and were apparently ejected in the second eruption.[13]

Although, for the most part, detailed groundbased studies of the Homunculus were hampered by limited spatial resolution, the presence of complex features well outside the Homunculus were first noted by Thackeray [93]. Proper motions of some of the nearer condensations [104, 106] showed that they may be related to the equatorial structure and were probably ejected during or soon after the 1840s eruption, while other outlying nebulosity may be from earlier outbursts depending on the amount of acceleration. Several amazing near-linear rope-like structures have also been identified extending outwards from the Homunculus and apparently emanating from the central star [63, 109]. These "strings" remain unexplained. The strings and other peculiar features are all described in more detail in Chap. 8.

Several studies have revealed the presence of very high velocity gas in the outer ejecta with speeds ranging from $\approx$1,000 $\mathrm{km\,s^{-1}}$ up to 3,200 $\mathrm{km\,s^{-1}}$ [110]. Very recently, Smith [88] has reported ejecta moving at 3,500 $\mathrm{km\,s^{-1}}$ and possibly as high as 6,000 $\mathrm{km\,s^{-1}}$ which he attributes to a blast wave from the 1843 eruption. If correct, the associated kinetic energy would rival that of a true supernova.

The first estimates of the mass lost in the Great Eruption were based on the total dust in the Homunculus from maps of the infrared emission and the resulting extinction in the lobes. They yielded a total ejected mass of $\sim$2–3 $M_\odot$ [24] or more, depending on the gas-to-dust ratio. More recent measurements yield total masses of 12.5–15 $M_\odot$ and possibly as high as 20 $M_\odot$ [69, 85] based on far-infrared emission, with a gas-to-dust ratio of 100.[14] Chapter 8 estimates $\approx$2–4 $M_\odot$ in the outer ejecta. The total mass in the equatorial skirt is much less certain, but most estimates put it below 1 $M_\odot$. The skirt is also a mixture of ejecta from the two eruptions [28]. Assuming that all of this material was ejected in the Great Eruption, *the total mass lost was 15–20* $M_\odot$ *with a corresponding mass loss rate of 0.7–1.0* $M_\odot$ *year*$^{-1}$*!*

The basic parameters that describe what we now know about η Car's physical state and its Great Eruption are summarized in Table 1.1.

[13]See Chap. 6.

[14]See Chap. 7.

Table 1.1 Physical properties of η Carinae

Distance	2,300 pc
Current mass	$\geq 90\,M_\odot$
Current luminosity	$5 \times 10^6\,L_\odot$, $(10^{6.7})$, $(M_{bol} = -12$ mag)
Current mass loss rate	$\sim 10^{-3}\,M_\odot$ year^{-1}
Initial or zero age mass	$\sim$150–200 $M_\odot$
Great eruption 1837–1858	
Duration	20 years
Maximum luminosity	$10^{7.7}\,L_\odot$ $(M_{bol} = -14.5$ mag)
Total luminous energy	$10^{49.7}$ ergs
Total kinetic energy	$10^{49.7}$ ergs
Mass lost	$\approx$15–20 $M_\odot$
Mass loss rate	$\sim 1\,M_\odot$ year^{-1}
Second eruption 1888–1895	
Duration	7 years
Maximum luminosity	$10^{6.7}\,L_\odot$
Total luminous energy	$10^{48.6}$ ergs
Total kinetic energy	$10^{46.9}$ ergs
Mass lost	$\approx 0.2\,M_\odot$
Mass loss rate	$\sim 0.03\,M_\odot$ year^{-1}

1.4 Highlights from the HST Era: 1995–2011

The Hubble Space Telescope first generation imagers and spectrographs provided our first opportunity to separate the light and spectrum of the central star from that of the surrounding ejecta. All groundbased spectra of the central object up until that time were contaminated with light from the nearby ejecta. Davidson et al. [22] used the Faint Object Spectrograph (HST/FOS) to obtain the first spectra of the central star (A) and at positions dominated by light from the Weigelt knots, objects B,C, and D. The spectra were distinctly different (Fig. 1.7). The spectrum of A, the central star, is that of a hot stellar wind while the ejecta, BCD, have a combined spectrum of many forbidden and permitted lines. Indeed, the narrow permitted and forbidden lines and the high excitation emission lines which dominate the groundbased spectra were shown to come from the ejecta. Subsequent observations with the Goddard High Resolution Spectrograph (HST/GHRS) at high spatial and spectral resolution [23] not only confirmed the FOS results, but showed that the Weigelt knots were equatorial and probably ejected after the Great Eruption. The peculiar, remarkably strong Fe II lines at λ 2507 Å and λ 2509 Å that dominate the UV spectrum of η Car were resolved. The lines were believed to be due to some kind of fluorescence, but the intensity ratios were so anomalous that Johansson and Letokhov [55, 56] proposed an astrophysical laser to explain them.

The second generation Space Telescope Imaging Spectrograph (STIS) provided an opportunity to combine high spatial and spectral resolution with wavelength coverage from the UV to 1 μm using the CCD long slit and MAMA echelle detectors. The long slit, covering the central star, its inner ejecta and the Homunculus nebula

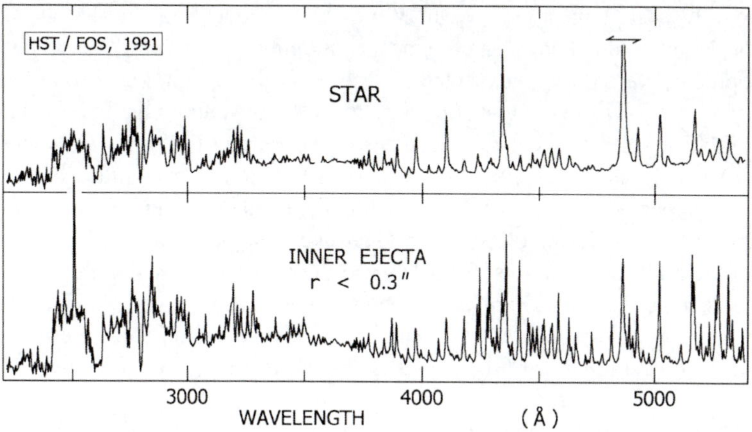

Fig. 1.7 The 1991 HST/FOS spectra from two different postions, on the central star and the inner ejecta, the Weigelt knots, B,C,D, from the near UV to visual (Based on spectra from Davidson et al. [22]). The unusually strong Fe II blend at λ 2507 Å in the ejecta is visible on the left

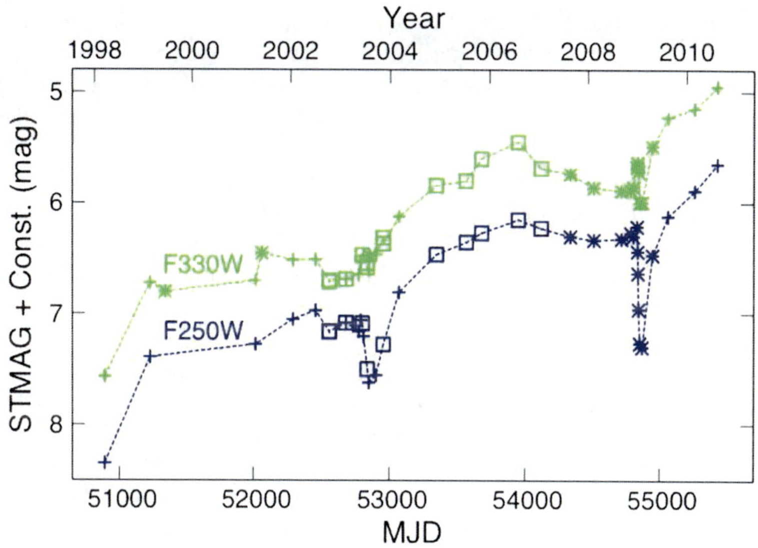

Fig. 1.8 The brightness of the central star in the near-UV with the Hubble Space Telescope from 1998 to 2011 (From Mehner et al. [67])

at a range of slit orientations, led to several important discoveries. But the first discovery from HST/STIS was totally unexpected. Comparison of the continuum in the STIS spectra between 1998 and 1999 showed that η Car, *the central star*, had brightened by a factor of 2 in only a year in the near-UV and visual [26], see Fig. 1.8. Groundbased photometry of the star and the Homunculus confirmed the trend in the visual and near-IR and showed that the nebula had also brightened by about 30%; its largest increase in 30 years. As mentioned above it was known that the Homunculus had been slowing brightening for about 30 years, but these developments were quite unexpected and not easy to explain. Davidson et al. [26] suggested that it was likely that some circumstellar dust along the line of sight, hot grains relatively near the star, had been destroyed perhaps due to a small increase in the star's UV flux. Consequently the grain formation zone moved futher out, causing the star to appear significantly brighter while the nebula brightened a lesser amount.

Subsequent monitoring of the central star using the ACS and WFPC2 images as well as the STIS observations, from 1998 to the present, show that the central star has continued to brighten (Fig. 1.8) [59, 61, 67] with a dip corresponding to the time of the spectroscopic events in 2003.5 and 2009. The very prominent decrease in the near-ultraviolet at the time of the events is due to the greatly increased opacity from the "forest" of Fe II lines

Although the bipolar lobes and equatorial skirt are easily visible in the images, the STIS/CCD long slit spectroscopy revealed evidence from the kinematics of the equatorial debris that it contained material ejected during both eruptions [28]. Spectroscopic mapping of the nebula by Ishibashi et al. [54] exhibited the presence of a previously unknown bipolar emission nebula embedded within the much larger Homunculus and extended along the major axis up to $\pm 2''$ from the central star. It was consequently named the "Little Homunculus". The expansion velocity of the associated gas suggests that it was ejected in the lesser 1890s eruption. Estimates of the mass in the Little Homunculus of 0.1 to 0.2 $M_\odot$ are much less than in the Homunculus and suggest a mass loss rate of $\sim 10^{-2}\,M_\odot\,\mathrm{year}^{-1}$ during the second eruption. *Thus η Car ejected material in the polar and equatorial directions during both eruptions but with very different energies and very likely different physical causes.*

Recent ground based near-infrared imaging with adaptive optics [3, 7] revealed a multi-lobed structure embedded in the SE lobe of the Homunculus. Artigau et al. [3] demonstrated that the knots and filaments were also ejected during the two eruptions and are located behind the Little Homunculus and the SE lobe, 2,000–4,000 A.U. from the star. These filamentary features are thus associated with the equatorial skirt or debris but on the far side of the Homunculus.

The early FOS and GHRS observations along the major axis of the Homunculus identified a position in the SE lobe, now known as FOS4 [117], that reflects the spectrum of the central star from its polar region. The Homunculus reflection nebula thus provides a unique opportunity to observe the spectrum of the star and its spectroscopic events from different directions. Long slit spectra from several different positions in the nebula showed that η Car has an aspherical stellar wind. Smith et al. [86] found that the P Cygni absorption profiles in the Balmer lines were latitude dependent with higher velocities and stronger absorption toward the poles. This was quite surprising implying that the mass loss flux toward the poles was higher than in the equatorial zone. The wind was fast and dense at the poles with a higher degree of ionization than the lower density wind at lower latitudes. Equally important, comparing the hydrogen and He I profiles from 1998 during the spectroscopic event, with observations 2 years later, they found that the structure of the wind changes during the 5.5 year cycle. During the event, the polar wind

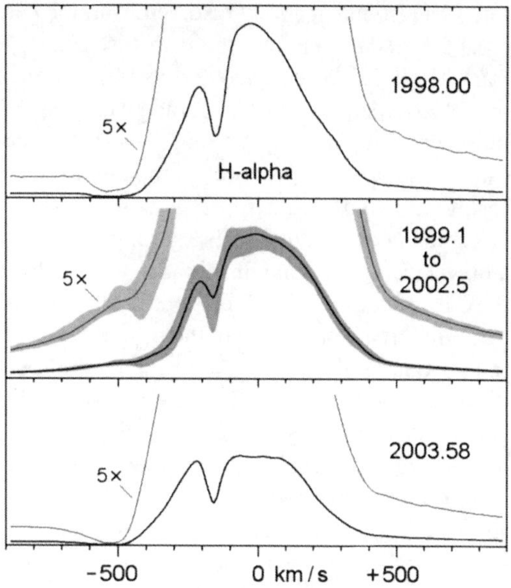

Fig. 1.9 The flat-topped Hα line profile during the 2003.5 event was very different than in the 1998.0 and at intermediate dates (From Davidson et al. [29])

remains largely unchanged while the wind at lower latitudes becomes denser and more closely resembles the polar wind. Comparison of STIS and UVES data from the 2003.5 event [90] later confirmed that the changes in the wind during the event were latitude dependent. This is a crucial result supporting suggestions that during the spectroscopic events, the star ejects a shell in the equatorial region. Detailed latitude structure like this is observable for very few other stars.

1.4.1 The HST Treasury Project: The 2003.5 Event

The first STIS observations were not obtained until Jan 1998, after the 1997–1998 event had already begun. Coverage was therefore limited. The HST Treasury Project on η Car was specifically designed to get thorough spectroscopic and image coverage of the 2003.5 event. One of the more surprising results was the discovery that there were significant changes in the star's wind between the events in 1997–1998 and in 2003. In the 2003.5 event the profiles of the Hα and Hβ lines, the two strongest lines in the stellar wind, were unlike any that had been seen before. Their distinctive flat tops [29] shown in Fig. 1.9, were quite different from their appearance in 1997–1998. Another unexpected discovery was the presence of He II emission first noticed in groundbased spectra by Steiner and Damineli [91]. The HST/STIS spectra provided superior spatial resolution and showed that the He II emission in the star first appeared about 3 months before the event, grew to a brief maximum at the time of the event [60] and then quickly disappeared. It reappeared for a second smaller maximum [91] before weakening again. The He II emission line was also not spatially resolved from the central star and therefore originated in its dense stellar wind, not in the diffuse ejecta. In the colliding wind binary models for η Car's 5.5 year period and its X ray light curve, the shock fronts near η Car would be expected to produce some weak He II emision. The line was far too bright, however, at its maximum strength, to be explained by conventional models for η Car's wind, and the proposed companion star's wind cannot supply enough energy for the required shock fronts. The primary star's wind therefore had to provide a larger than normal but temporary supply of energy. Martin et al. [60] concluded that a mass ejection or wind disturbance was necessary to fuel the observed emission.

The observational evidence, especially from the observations with HST/STIS thus support a disturbance in the primary's wind or mass ejection event, most likely in the equatorial zone, as the origin of the spectroscopic changes observed during the events [9, 16, 25, 86]. The close approach of the less massive secondary during periastron very likely triggers the instability in the massive primary's wind, a star which is already within a few percent of the Eddington limit.[15] But then why were there no apparent events prior to 1948, or perhaps more correctly, why did the star's spectrum look like it was always in an event, the spectroscopic low excitation state, prior to 1944? See the Chap. 3 for more discussion of this question.

1.5 A Change of State

With the demise of STIS in 2004, we continued to monitor the central star photometrically with HST/ACS/HRC, and after its failure, with HST/WFPC2. By 2008, the central star was about 2 magnitudes brighter than it was 10 years before [62]. The brightening from the UV-to-red wavelengths is probably due to a combination of dust destruction and reduced dust formation, requiring a change in the UV radiation and/or the mass loss rate i.e. a change in the wind. With the evidence that the 2003.5 event had been qualitatively different from the preceding 1998.0 event, it was therefore especially important to observe η Car during the 2009.0 event, but without the advantages of STIS. Historically, groundbased spectra of η Car represented an inseparable mixture of the central star and its bright ejecta, but with the dramatic brightening of the star relative to the inner ejecta, the ground-based spectra are now much less contaminated. The multiwavelength observations of the 2009 event, including groundbased spectroscopy [17, 67, 79, 92], produced several surprises. The UV light curve (Fig. 1.8) exhibited a much deeper minimum at both wavelengths than in 2003.5. A much smaller difference between 2003.5 and 2009.0 was observed in the visual [36]. The X ray light curve[16] showed an unexpected early recovery from the X ray minimum compared to the previous events most likely due to a decrease in the wind density of the primary [57]. More complete

[15] See Chaps. 3 and 9.

[16] See Fig. 9.7 in Chap. 9

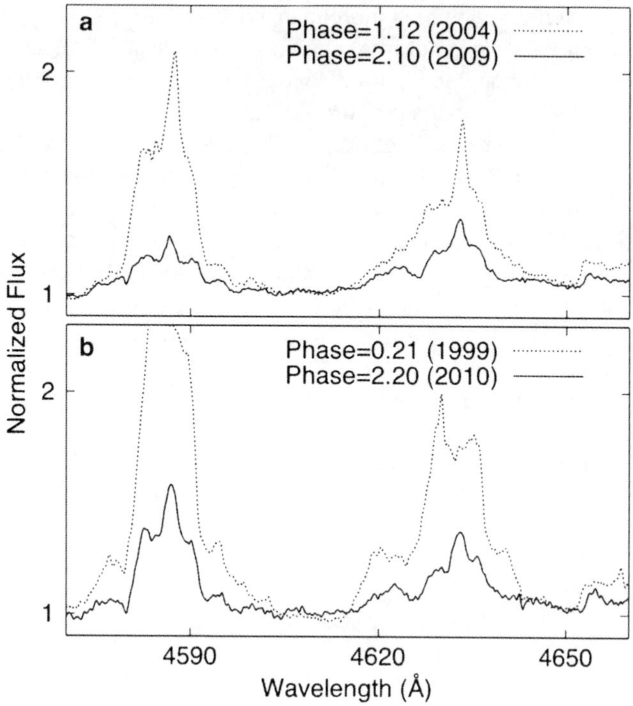

Fig. 1.10 A comparison of line profiles of blends of Fe II, [Fe II], Cr II, and [Cr II] near 4600A. The two panels show HST/STIS data at the same phase in successive spectroscopic cycles (Adpated from Mehner et al. [66])

temporal coverage of the behavior of the peculiar He II emission exhibited a close correlation with the X ray flux [67].

The accelerated brightening of the central star beginning in the late 1990s together with major differences among the 1998.0, 2003.5, and 2009.0 events, suggested that the stellar wind of the massive primary was changing much more rapidly than it had in the previous 40 years or so. Very likely the mass loss rate and consequently the density of the wind was decreasing. But the spectrum of the central star, from HST spectroscopy from 1991 to 2004, showed no major changes except during the events. Thus it was extremely important when observations with the post-servicing mission (SM4) STIS, after the 2009 event, revealed important changes in the spectrum of the star. Mehner et al. [66] found that major stellar wind emission features (Fig. 1.10) had decreased by a factor of 2 or more! The most likely explanation is a significant decrease in η Car's primary wind density and therefore a decrease in the mass loss rate by a factor of 2–3 times [10, 57, 68]. Looking back, at the groundbased Gemini/GMOS spectra from 2007 to 2010, weakening of the wind features and changes in other lines are apparent before the 2009 event [62, 66], and Mehner et al. [68] have subsequently found a gradual change in the hydrogen and wind emission lines beginning more than a decade ago. The change of state in η Car thus began with the rapid brightening, but significant changes in the wind emission lines were not apparent until recently; an indication of the strength of the wind which has dominated its spectrum since 1900.

Eta Car's recovery from the Great Eruption has been highlighted by three periods of rapid photometric and spectroscopic change; the second eruption in the 1890s, the rapid transition in the 1940s, and the accelerated brightening beginning in the late 1990s accompanied by a major decrease in the density of the wind. These transition periods are interestingly spaced approximately 50 years apart. As of Spring 2011, the strength of the wind emission lines have not continued to decrease [68]. The wind density appears to have stabilized in this state, for now. But, will η Car now continue in this state for several decades? Will the wind density continue to decline revealing the underlying star, or will it increase and revert to its former appearance?

Acknowledgements It is a pleasure to thank our colleagues Kris Davidson and Andrea Mehner for careful reading of this chapter.

References

1. D.A. Allen, D.J. Hillier, The shape of the Homunculus nebula around Eta Carinae. Publ. Astron. Soc. Aust. **10**, 338–341 (1993)
2. L.H. Aller, T. Dunham Jr., The spectrum of Eta Carinae in 1961. ApJ. **146**, 126–141 (1966)
3. É. Artigau, J.C. Martin, R.M. Humphreys, K. Davidson, O. Chesneau, N. Smith, Penetrating the homunculus – near-infrared adaptive optics images of Eta Carinae. AJ. **141**, 202–215 (2011)
4. B.J. Bok, The spectrum of η Carinae . Pop. Astron. **38**, 399 (1930)
5. G.R. Burbidge, A speculation concerning the evolutionary state of Eta Carinae. ApJ. **136**, 304–307 (1962)
6. A.J. Cannon, Classification of spectra. Harv. Ann. **28**, 175 (1901)
7. O. Chesneau et al., The Sub-arcsecond dusty environment of Eta Carinae. A&A. **435**, 1043–1061 (2005)
8. A.M. Clerke, Southern star spectra. Observatory **11**, 429–432 (1888)
9. M.F. Corcoran, K. Ishibashi, J.H. Swank, R. Petre, The X-ray light curve of η Carinae: refinement of the orbit and evidence for phase-dependent mass loss. ApJ. **547**, 1034–1039 (2001)
10. M.F. Corcoran, K. Hamaguchi, J.M. Pittard, C.M.P. Russell, S.P. Owocki, E.R. Parkin, A. Okazaki, Recent X-ray variability of Carinae: the quick road to recovery. ApJ. **725**, 1528 (2010)
11. P. Cox, P.G. Mezger, A. Sievers, F. Najarro, L. Bronfman, E. Kreysa, G. Haslam, Millimeter emission of Eta Carinae and its surroundings. A&A. **217**, 168–174 (1995)
12. D.G. Currie, D.M. Dowling, E.J. Shaya, J. Hester, P. Scowen, E.J. Groth, R. Lynds, E.J. O'Neil Jr., The wide field/planetary camera instrument definition team: astrometric analysis of the homunculus of Eta Carinae with the hubble space telescope. AJ. **112**, 1115 (1996)
13. A. Damineli, The 5.52 year cycle of Eta Carinae. ApJ. Lett. **460**, L49+ (1996)
14. A. Damineli, A. Kaufer, B. Wolf, O. Stahl, D.F. Lopes, F.X. de Araújo, η Carinae: binarity confirmed. ApJ. Lett. **528**, L101–L104 (2000)
15. A. Damineli et al., The periodicity of the η Carinae events. MNRAS. **384**, 1649–1656 (2008)

16. A. Damineli et al., A multispectral view of the periodic events in η Carinae. MNRAS. **386**, 2330–2344 (2008)
17. A. Damineli et al., He II 4686A in Eta Car: the data and modeling, in *Eta Carinae in the Context of the Most Massive Stars*, ed. by T.R. Gull, A. Damineli (2009), pp. 7–8. arXiv:0910.3158
18. K. Davidson, On the nature of Eta Carinae. MNRAS. **154**, 415–427 (1971)
19. K. Davidson, The relation between apparent temperature and mass-loss rate in hypergiant eruptions. ApJ. **317**, 760–764 (1987)
20. K. Davidson, N.R. Walborn, T.R. Gull, The remarkable spectrum of some material ejected by Eta Carinae. ApJ. **254**, L47–L51 (1982)
21. K. Davidson, R.J. Dufour, N.R. Walborn, T.R. Gull, Ultraviolet and visual wavelength spectroscopy of gas around Eta Carinae. ApJ. **305**, 867–876 (1986)
22. K. Davidson, D. Ebbets, G. Weigelt, R.M. Humphreys, A.R. Hajian, N.R. Walborn, M. Rosa, HST/FOS spectroscopy of Eta Carinae: the star itself, and ejecta within 0.3 arcsec. AJ. **109**, 1784–1796 (1995)
23. K. Davidson, D. Ebbets, S. Johansson, J.A. Morse, F.W. Hamann, HST/GHRS observations of the compact slow wjecta of Eta Carinae. AJ. **113**, 335–345 (1997)
24. K. Davidson, R.M. Humphreys, Eta Carinae and its environment. Ann. Rev. Astron. Astrophys. **35**, 1–32 (1997)
25. K. Davidson, Why the binary hypothesis isn't a pancea. ASP Conf. Ser. **179**, 304–315 (1999)
26. K. Davidson, T.R. Gull, R.M. Humphreys, K. Ishibashi, P. Whitelock, L. Berdnikov, P.J. McGregor, T.S. Metcalfe, E. Polomski, M. Hamuy, An unusual brightening of Eta Carinae. AJ. **118**, 1777–1783 (1999)
27. K. Davidson, K. Ishibashi, T.R. Gull, R.M. Humphreys, N. Smith, η Carinae: testing a binary orbit model with the hubble space telescope/space telescope imaging spectrograph. ApJ. Lett. **530**, L107–L110 (2000)
28. K. Davidson, N. Smith, T.R. Gull, K. Ishibashi, D.J. Hillier, The shape and orientation of the Homunculus Nebula based on spectroscopic velocities. AJ. **121**, 1569–1577 (2001)
29. K. Davidson et al., A change in the physical state of η Carinae? AJ. **129**, 900–906 (2005)
30. K. Davidson, The physical nature of Carinae, in *The Fate of the Most Massive Stars*, ed. by R. Humphreys, K. Stanek. ASP Conference Series, vol. 332 (Astronomical Society of the Pacific, San Francisco, 2005), pp. 101–110
31. G. de Vaucouleurs, O.C. Eggen, The brightening of η Carinae. PASP. **64**, 185–190 (1952)
32. W.J. Duschl, K.-H. Hofmann, F. Rigaut, G. Weigelt, Morphology and kinematics of Eta Carinae. RevMexAA SdC **2**, 17–22 (1995)
33. M. Feast, P. Whitelock, F. Marang, Variability of η Carinae – III. MNRAS. **322**, 741–748 (2001)
34. A. Feinstein, η Carinae and the trumpler 16 cluster. PASP. **75**, 492–497 (1963)
35. A. Feinstein, The OB stars in Carina-Centaurus. MNRAS. **143**, 273–287 (1969)
36. E. Fernández-Lajús et al., Optical photometry of the 2009.0 event of Eta Car. NewA **15**, 108 (2010)
37. D.J. Frew, The historical record of Carinae I. The visual light curve, 1595–2000. J. Astron. Data **10**(6), 1–76 (2004)
38. E. Gaviola, Eta Carinae I. The nebulosity. ApJ. **111**, 408–413 (1950)
39. E. Gaviola, Eta Carinae II. The spectrum. ApJ. 118, 234–251 (1953)
40. R.D. Gehrz, E.P. Ney, The core of Eta Carinae. Sky Telesc. **44**, 4–6 (1972)
41. R.D. Gehrz, E.P. Ney, E.E. Becklin, G. Neugebauer, The infrared spectrum and angular size of Eta Carinae. Astrophys. Lett. **13**, 89–93 (1973)
42. W. Glatzel, Instabilities in the most massive evolved stars, in *The Fate of the Most Massive Stars*, ed. by R.M. Humphreys, K.Z. Stanek. ASP Conference Series, vol. 332 (Astronomical Society of the Pacific, San Francisco, 2005), pp. 22–31
43. L. Gratton, The problem of Eta Carinae, in *Star Evolution*, ed. by L. Graton (Academic, New York, 1963), pp. 297–311

44. J.A. Hackwell, R.D. Gehrz, G.L. Grasdalen, The internal structure of the dust shell of Eta Carinae deduced from six channel 8–13 micron mapping. ApJ. **311**, 380–399 (1986)
45. J. Herschel, *Results of Astronomical Observations made During the Years 1834–1888 at the Cape of Good Hope* (Smith Elder, London, 1847), pp. 32–47
46. D.J. Hillier, K. Davidson, K. Ishibashi, T.R. Gull, On the nature of the central source in η Carinae. ApJ. **553**, 837–860 (2001)
47. D. Hoffleit, Note on the spectrum of Eta Carinae. Harv. Bull. **893**, 11–14 (1933)
48. K.-H. Hofmann, G. Weigelt, Speckle masking observation of Eta Carinae. A&A. **203**, L21–L22 (1988)
49. R.M. Humphreys, K. Davidson, The most Luminous stars. Cover picture. Science **223**, 243 (1984)
50. R.M. Humphreys, K. Davidson, The Luminous blue variables: astrophysical geysers. PASP. **106**, 1025–1051 (1994)
51. R.M. Humphreys, K. Davidson, N. Smith, Eta Carinae's second eruption and the light curves of the Eta Carinae variables. PASP. **111**, 1124–1131 (1999)
52. R.M. Humphreys, K. Davidson, M. Koppelman, The early spectra of Eta Carinae 1892–1941 and the onset of its high excitation emission spectrum. AJ. **135**, 1249–1263 (2008)
53. R.T.A. Innes, Observations of variable stars, η Argus. Cape Ann. **9**, 75B–78B (1903)
54. K. Ishibashi, T.R. Gull, K. Davidson, N. Smith et al., Discovery of a little Homunculus within the Homunculus Nebula of Carinae. AJ. **125**, 3222–3236 (2003)
55. S. Johansson, V.S. Letokhov, Anomalous Fe II spectral effects and high H I Lyα temperature in gas blobs near η Carinae. Astron. Lett. **30**, 58–63 (2004)
56. S. Johansson, V.S. Letokhov, Astrophysical lasers operating in optical Fe II lines in stellar ejecta of η Carinae. A&A, **428**, 497–509 (2004)
57. A. Kashi, N. Soker, Explaining the early exit of Eta Carinae from its 2009 X-Ray minimum with the accretion model. ApJ. Lett. **701**, L59–L62 (2009)
58. A. Le Sueur, On the Nebulae of Argo and Orion. Proc. R. Soc. **18**, 245–250 (1870)
59. J.C. Martin, M.D. Koppelman, η Carinae's brightness variations since 1998: Hubble Space telescope observations of the central star. AJ. **127**, 2352–2361 (2004)
60. J.C. Martin, K. Davidson, R.M. Humphreys, D.J. Hillier, K. Ishibashi, On the He II emission in η Carinae and the Origin of its Spectroscopic events. ApJ. **640**, 474–490 (2006)
61. J.C. Martin, K. Davidson, M.D. Koppelman, The Chrysalis opens? photometry from the η Carinae Hubble Space telescope treasury project, 2002–2006. AJ. **132**, 2717–2728 (2006)
62. J.C. Martin, K. Davidson, R.M. Humphreys, A. Mehner, Mid-cycle changes in Eta Carinae. AJ. **139**, 2056–2065 (2010)
63. J. Meaburn, R.D. Wolstencroft, J.R. Walsh, Echelle and spectropolarimetric observations of the Eta Carinae nebulosity. A&A, 181, 333–342 (1987)
64. J. Meaburn, An updated Proper-Motion/Spectropolarimetric distance to η Carinae, in *Eta Carinae at the Millennium*, ed. by J.A. Morse, R.M. Humphreys, A. Damineli. ASP Conference Series, vol. 179 (Astronomical Society of the Pacific, San Francisco, 1999), pp. 89–95
65. A. Mehner, K. Davidson, G.J. Ferland, R.M. Humphreys, High excitation emission lines near Eta Carinae and its likely companion star. ApJ. **710**, 727–742 (2010)
66. A. Mehner, K. Davidson, R.M. Humphreys, J.C. Martin, K. Ishibashi, G.J. Ferland, N.R. Walborn, A sea change in Eta Carinae. ApJ. Lett. **717**, L22–L25 (2010)
67. A. Mehner, K. Davidson, J.C. Martin, R.M. Humphreys, K. Ishibashi, G. Ferland, Critical differences and clues in η Car's 2009 event. ApJ. **740**, 2 (2011)
68. A. Mehner, K. Davidson, R.M. Humphreys, K. Ishibashi, J.C. Martin, M.-T. Ruiz, Secular changes in Eta Carinae's wind 1998–2011. ApJ in press (2012)
69. P.W. Morris, L.B.F.M. Waters, M.J. Barlow, T. Lim, A. de Koter, R.H.M. Voors, P. Cox, Th de Graauw, Th Henning, S. Hony et al., Discovery of a massive equatorial torus in the Carinae stellar system. Nature **402**, 502–504 (1999)
70. J.A. Morse, J.R. Kellogg, J. Bally, K. Davidson, B. Balick, D. Ebbets, Hubble Space telescope proper-motion measurements of the Carinae Nebula. ApJ. Lett. **548**, 207 (2001)

71. G. Neugebauer, J.A. Westphal, Infrared observations of Eta Carinae. ApJ. **152**, L89–L94 (1968)
72. D.J.K. O'Connell, Photographic light curve of Eta Carinae. Vistas Astron. **2**, 1165–1171 (1956)
73. J.P. Ostriker, J.E. Gunn, Do pulsars make supernovae? ApJ. **164**, L95–L104 (1971)
74. B.E.J. Pagel, Intrinsic reddening of Eta Carinae. Nature **221**, 325–327 (1969)
75. B.E.J. Pagel, Energy budget for the infrared radiation from Eta Carinae. ApJ. Lett. **4**, 221–224 (1969)
76. C. Payne-Gaposchkin, *The Galactic Novae* (Dover, New York, 1957)
77. J.M. Pittard, M.F. Corcoran, In hot pursuit of the hidden companion of Eta Carinae: an X-ray determination of the wind parameters. A&A. **383**, 636–647 (2002)
78. A. Rest et al., Light echoes reveal an unexpectedly cool eta Carinae during its 19th century Great Eruption. Nature, in press (2012)
79. N.D. Richardson, D.R. Gies, T.J. Henry, E. Fernández-Lajús, A.T. Okazaki, The Hα variability of η Carinae during the 2009.0 Spectroscopic event. AJ. **139**, 1534–1541 (2010)
80. A.E. Ringuelet, Note on the Nebulosity around Eta Carinae. Z. f. Astrophys. **46S**, 276–278 (1958)
81. G. Robinson, A.R. Hyland, J.A. Thomas, Observation and interpretation of the infra-red spectrum of Eta Carinae. MNRAS. **161**, 281–292 (1973)
82. A.W. Rodgers, L. Searle, Spectrophotometry of the object η Carinae. MNRAS, **135**, 99–119 (1967)
83. N. Smith, R.D. Gehrz, Proper motions in the ejecta of Eta Carinae with a 50 Year baseline. AJ. **116**, 823 (1998)
84. N. Smith, Dissecting the Homunculus Nebula around Eta Carinae with spatially resolved near-infrared Spectroscopy. MNRAS. **337**, 1252–1268 (2002)
85. N. Smith, R.D. Gehrz, P.M. Hinz, W.F. Hoffmann, J.L. Hora, E.E. Mamajek, M.R. Meyer, Mass and kinetic energy of the Homunculus Nebula around η Carinae. AJ. **125**, 1458–1466 (2003)
86. N. Smith, K. Davidson, T.R. Gull, K. Ishibashi, D.J. Hillier, Latitude-dependent effects in the stellar wind of Eta Carinae. ApJ. **586**, 432–450 (2003)
87. N. Smith, Doppler Tomography of the little Homunculus: high-resolution spectra of [FeII]16435 around Eta Carinae. MNRAS. **357**, 1330–1336 (2005)
88. N. Smith, A blast from the 1843 Eruption of η Carinae. Nature **455**, 201–203 (2008)
89. N. Smith, D.J. Frew, A revised historical light curve of Eta Carinae and the timing of close Periastron encounters. MNRAS. **415**, 2009 (2011)
90. O. Stahl, K. Weis, D.J. Bomans, K. Davidson, T.R. Gull, R.M. Humphreys, A Spectroscopic event of η Car viewed from different directions: the data and first results. A&A. **435**, 303–312 (2005)
91. J.E. Steiner, A. Damineli, Detection of He II λ 4686 in η Carinae. ApJ. **612**, L133–L136 (2004)
92. M. Teodoro et al., He II 4686 in Eta Carinae and the nature of the 5.538-year cycle. arXiv:1104.2276
93. A.D. Thackeray, Nebulosity surrounding Eta Carinae. Observatory **69**, 31–33 (1949)
94. A.D. Thackeray, Some southern stars involved in Nebulosity. MNRAS. **110**, 524–530 (1950)
95. A.D. Thackeray, Identifications in the spectra of Eta Carinae and RR Telescopii. MNRAS. **113**, 211–236 (1953)
96. A.D. Thackeray, Spectra of the Polarized Halo around Eta Carinae. Observatory **81**, 99–102 (1961)
97. A.D. Thackeray, The infra-red spectrum of Eta Carinae. Observatory **81**, 102–104 (1961)
98. A.D. Thackeray, Coude spectra of Eta Carinae and the strongest lines of [Fe II] and [Ni II]. MNRAS. **135**, 51–81 (1967)
99. W.H. van den Bos, Measures of double stars – 23rd series. Union Obs. Circ. **100**, 522 (1938)
100. A.M. van Genderen, M. de Groot, P.S. The, The evolutionary status of Eta Car based on historical and modern optical and infrared photometry. Space Sci. Rev. **66**, 219–223 (1993)

101. A.M. van Genderen, M.J.H. de Groot, P.S. The, The optical and near infrared variability of Eta Carinae: a binary Luminous blue variable? A&A. **283**, 89–110 (1994)
102. A.M. van Genderen, C. Sterken, M. de Groot, O. Stahl, J. Andersen et al., uvby-Hβ photometry of Eta Car 1992–94. A&A. **304**, 415–430 (1995)
103. N. Visvanathan, Polarization in η Carinae. MNRAS. **135**, 275–86 (1967)
104. N.R. Walborn, The complex outer shell of Eta Carinae. ApJ. **204**, L17–L19 (1976)
105. N.R. Walborn, M.H. Liller, The earliest spectroscopic observations of Eta Carinae and its interaction with the Carina nebula. ApJ. **211**, 181–186 (1977)
106. N.R. Walborn, B.M. Blanco, A.D. Thackeray, Proper motions in the outer shell of Eta Carinae. Apj. **211**, 181–186 (1978)
107. G. Weigelt, J. Ebersberger, Eta Carinae resolved by speckle interferometry. A&A. **63**, L5–L6 (1986)
108. G. Weigelt et al., Revista Mexicana de Astronomia y Astrofisica Serie de Conferencias. **2**, 11–16 (1995)
109. K. Weis, W.J. Duschl, Y.-H. Chu, The nature of strings in the Nebula around η Carinae. A&A. **349**, 467–474 (1999)
110. K. Weis, W.J. Duschl, D.J. Bomans, High velocity structures in, and the X-ray emission from the LBV Nebula around Carinae. A&A. **367**, 566–576 (2001)
111. J.A. Westphal, G. Neugebauer, Infrared observations of Eta Carinae to 20 Microns. ApJ. **156**, L45–L48 (1969)
112. S.M. White, R.A. Duncan, J. Lim, G.J. Nelson, S.A. Drake, M.R. Kundu, The radio source around Eta Carinae. ApJ. **429**, 380–384 (1994)
113. P.A. Whitelock, M.W. Feast, B.S. Carter, G. Roberts, I.S. Glass, The infrared spectrum and variability of Eta Carinae. MNRAS. **203**, 385–392 (1983)
114. P.A. Whitelock, M.W. Feast, C. Koen, G. Roberts, B.S. Carter, Variability of Eta-Carinae. MNRAS. **270**, 364–372 (1994)
115. C.A. Whitney, The spectrum of Eta Carinae in 1893. Harv. Bull. **921**, 8–14 (1952)
116. R. Zanella, B. Wolf, O. Stahl, Spectroscopy of the shell episode of Eta Car (1981–1983). A&A. **137**, 79–84 (1984)
117. T. Zethson, S. Johansson, K. Davidson, R.M. Humphreys, K. Ishibashi, D. Ebbets, Strange velocities in the equatorial ejecta of Eta Carinae. A&A. **334**, 211–220 (1999)
118. F. Zwicky, in *Supernovae*, ed. by L.H. Aller, D.B. McLaughlin. Stars and Stellar Systems, vol. VIII (University of Chicago Press, Chicago, 1965), pp. 367–423

Chapter 2
The Company Eta Carinae Keeps: Stellar and Interstellar Content of the Carina Nebula

Nolan R. Walborn

Abstract The remarkable, interrelated stellar and nebular constituents of η Carinae's habitat, the giant Carina Nebula (NGC 3372), are reviewed. They are relevant to the interpretation of η Car itself. They include some of the hottest and most massive O stars known, frequently in close multiple systems; mysterious nebular structures; recently recognized ongoing star formation; the most extreme high-velocity interstellar absorption-line profiles in the Galaxy; and both stellar and diffuse X-ray sources.

2.1 Introduction

The fame and brilliance of η Carinae sometimes overwhelm its equally fascinating stellar and nebular environment. But the context of associated phenomena can provide vital clues to the nature of what may otherwise appear to be a bizarre peculiar object, for instance, its initial mass and age. In fact, the giant H II region known as the Carina Nebula, or NGC 3372, is one of the most powerful laboratories for investigation of the early evolution of the most massive stars, because of its relative proximity and moderate extinction, independently of η Car itself. The hottest spectral class, type O3, (now expanded to O2 through O3.5) was first recognized there. The massive stellar content of the Carina Nebula has been reviewed in detail by Walborn [77] and Smith [58]. Current star formation, likely triggered by the first-generation O stars, a ubiquitous characteristic of giant H II regions, has only recently been recognized in the Carina Nebula as a result of new infrared observations. This region also harbors extreme and in some respects unique

N.R. Walborn (✉)
Space Telescope Science Institute, 3700 San Martin Drive, Baltimore, MD 21218, USA
e-mail: walborn@stsci.edu

K. Davidson and R.M. Humphreys (eds.), *Eta Carinae and the Supernova Impostors*, Astrophysics and Space Science Library 384, DOI 10.1007/978-1-4614-2275-4_2,

Fig. 2.1 The entire Carina Nebula (Image made from UK Schmidt plates by David Malin)

high-energy phenomena, revealed by the interstellar absorption lines and X-rays. The Carina Nebula is a dynamic, transitory cauldron of astrophysical interactions from which we have yet much to learn (Fig. 2.1).

The distance to the Carina Nebula was reviewed by Walborn [77] and by Davidson and Humphreys [19]. Using the luminosities of the O-type stars in the ionizing cluster Trumpler 16, Walborn estimated its distance as $D \approx 0.8^{(R-4)} \times$ 2,250 pc with a small formal error, while Davidson and Humphreys adopted $2{,}300 \pm 200$ pc based upon photometric distances to Tr 16 plus the expansion of the η Car ejecta measured by Allen and Hillier [4] ($\sim$2,200 pc). The spectroscopic/photometric parallaxes of the OB stars are uncertain because of anomalous reddening laws that may vary from star to star, and which remain to be determined from future ultraviolet through infrared analyses of their energy distributions. Color-magnitude diagram distance determinations are also uncertain, because of superimposed subclusterings with a range of ages, and subluminous, zero-age OB main sequences at progressively lower masses with increasing age. Overluminous pre-main sequences at still lower masses also complicate the analysis of color-magnitude diagrams. Moreover, in this direction we are looking down a spiral arm, so there are numerous background, as well as a few foreground, objects in the field. Subsequently, Meaburn [38], Davidson et al. [18], and Smith [57] independently derived distances of 2,300, 2,250, and 2,300 pc, respectively, from the motions of the ejected nebulosity. Systematic errors dominate, but the quoted uncertainties are in the range of ± 100 to ± 200 pc. The four nebular parallaxes agree remarkably well. These results imply that the typical ratio of total to selective extinction must be $R \sim 4$ toward the ionizing stars [77]. *The distance of η Carinae, and by implication of the Carina Nebula with which it has been shown to be associated, is thus now well established at 2,300 pc.* The diameter of the H II region is about 40 pc, and the total mass of ionized hydrogen is of order $10^4 M_\odot$, a thousand times that of the Orion Nebula, but one to two orders of magnitude below that of 30 Doradus in the Large Magellanic Cloud.

2.2 Stellar Content

The principal ionizing clusters of the Carina Nebula are Trumpler 16 (Tr 16), which contains η Car, and the very compact Trumpler 14 (Tr 14) in the northern part (Fig. 2.2); and Collinder 228 (Cr 228) in the southern part of the Nebula. The O3 subclass was introduced by Walborn [73, 74] to describe six stars in Tr 14 and Tr 16 with spectral types earlier than the earliest (O4) MK standards. One of them, HD 93129A, the brightest star in Tr 14, presented a qualitatively new kind of O-type spectrum, with a narrow N IV λ4058 emission line stronger than the normal Of N III $\lambda\lambda$4634–4640–4642 lines, which was denoted as Of*, and strong N V $\lambda\lambda$ 4604–4620 absorption lines. These N IV and N V features are present in O4f spectra, but much more weakly and with the N IV weaker than N III; that is, the O3f* spectrum is a higher-ionization analogue of the O4f. There are also three WN-A, or WNL (i.e., narrow-line, late-type, high-luminosity WN), stars present, which bear strong spectroscopic relationships to the O3f*, in terms of progressive envelope or wind development ([75,81]; Fig. 2.3). One of the WN-A stars is in Tr 16, while the other two are at the western and southern edges of the Nebula. This stellar population corresponds to an evolutionary age of $\leq$2 Myr (recently reduced to that value by the recognition that this kind of WN star is likely still burning H in the core) and upper masses of $\geq 100\,M_\odot$. Evidence that Tr 14 is somewhat younger ($\leq$1 Myr), and Cr 228 somewhat older, was discussed by Walborn [77], as was the speculative possibility that the two errant WN-A stars are runaways from Tr 16. Because of the smooth spectroscopic progression from the associated types O3 V through O3 If* to WN-A, it is likely that η Car is a *post*-WN-A star, if its current state is essentially a product of very massive single-star evolution, as opposed to a binary interaction. Langer et al. [33] advocated the same progression from a theoretical standpoint, and Smith and Conti [60] from a rediscussion of WN spectra.

In a review of the O3 subclass [83], it was subdivided into 3 including the new types O2 and O3.5. HD 93129A is now the prototype of the O2 If* category on the basis of its large N IV/N III emission-line ratio (Fig. 2.4), while the other Carina

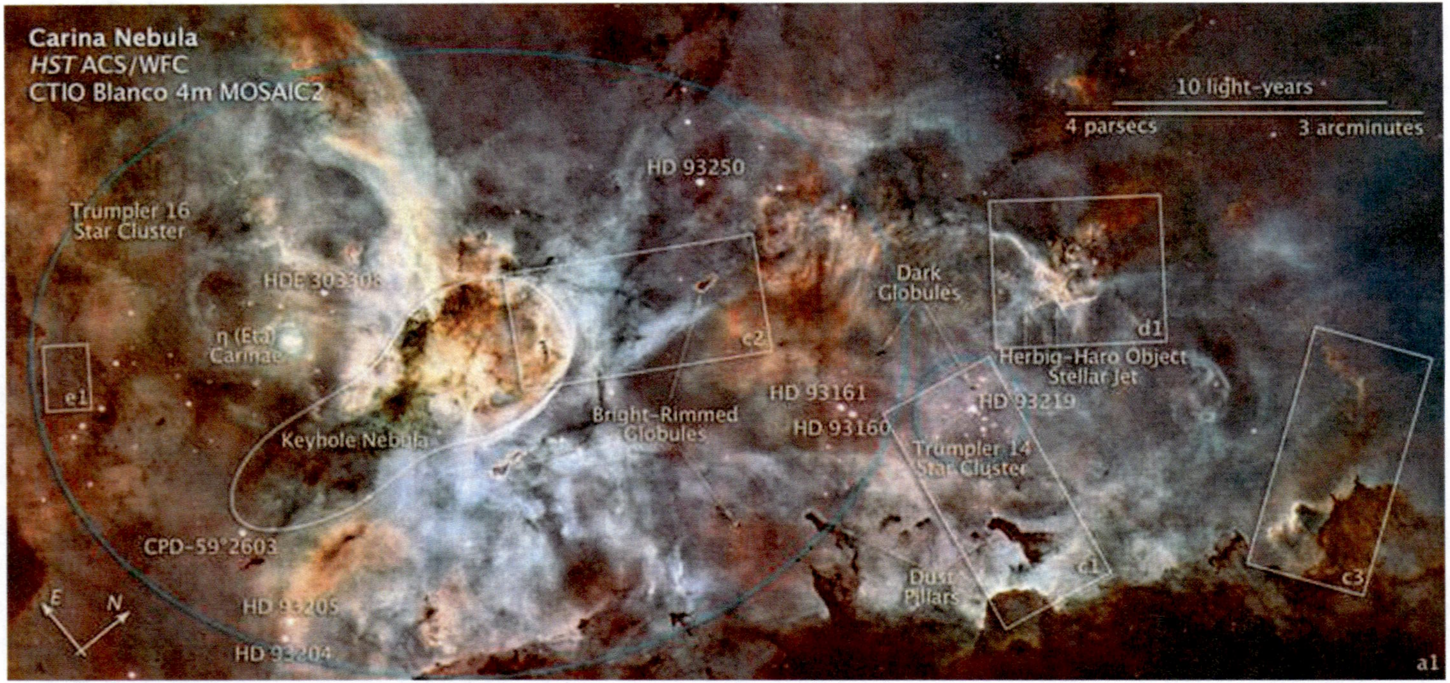

Fig. 2.2 The northern part of the Carina Nebula as imaged in Hα by the advanced camera for surveys on the *Hubble Space Telescope*. Color has been added by means of groundbased images in other nebular lines from the Cerro Tololo Inter-American Observatory 4m telescope. The outlines and identifications of the principal features discussed in the text were added by Zolt Levay (STScI) (Image credit: NASA, ESA, Nathan Smith, and The Hubble Heritage Team (STScI/AURA))

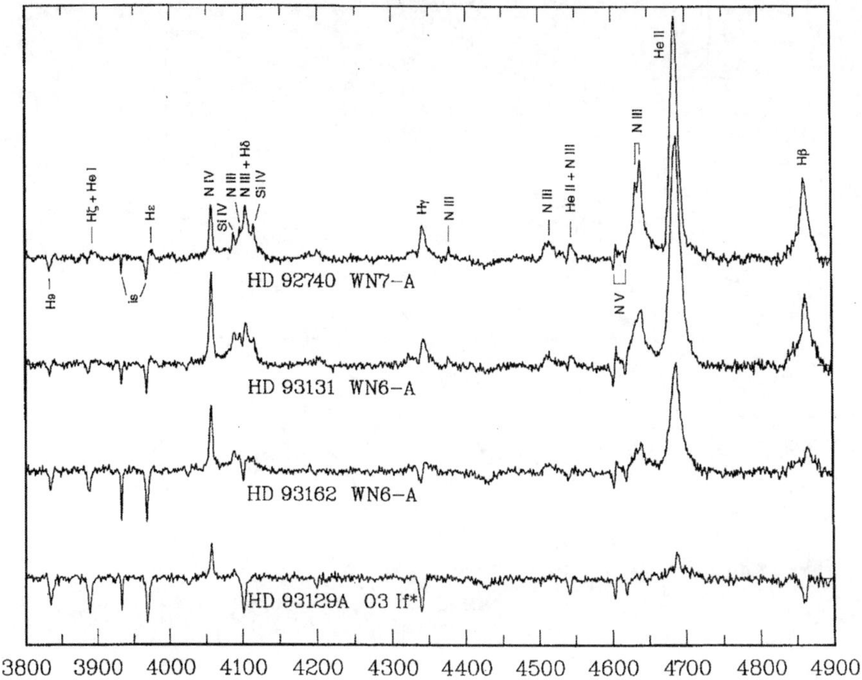

Fig. 2.3 Rectified, digital blue-violet spectrograms of HD 93129A (now O2 If*) and the three WN-A stars in the Carina Nebula (From [81])

Nebula members have been reclassified as O3.5 V((f+)), except one (HDE 303308) back to O4 V((f+)) based on the higher quality digital data now available (Fig. 2.5). (The ((f)) notation signifies strong He II λ4686 *absorption* combined with weak N III emission; while the "+" qualifier denotes the additional presence of Si IV λλ4089–4116 emission lines, which is subsumed by f*.)

A large percentage of the most massive stars are found in spectroscopic binary and compact multiple systems; the Carina Nebula population is no exception. Outstanding examples are the O3.5 V((f)) + O8 V system HD 93205 [42] and the WN7-A + O8-9 HD 92740 [53]. Another interesting case is FO 15 (O5.5 Vz + O9.5 V; [45]; "Vz" means that the He II λ4686 absorption line is stronger relative to other He lines than in normal class V spectra, indicating possible subluminosity and zero-age-main-sequence status [80], which is supported by the physical parameters derived for this star. Triple lines have been found in the spectra of CPD −59° 2603 (O7 V + O9.5 V + B0.2 IV; [50]) and CPD −59° 2636 (O7 V + O8 V + O9 V; [1]). Even more extreme is the double spectroscopic binary HD 93206 in Cr 228 (O9.7 Ib:(n) + O9 III), which has two sets of lines with different periods [34, 43], but has yet to be resolved spatially despite attempts with groundbased speckle and the *HST* Fine Guidance Sensor interferometer.

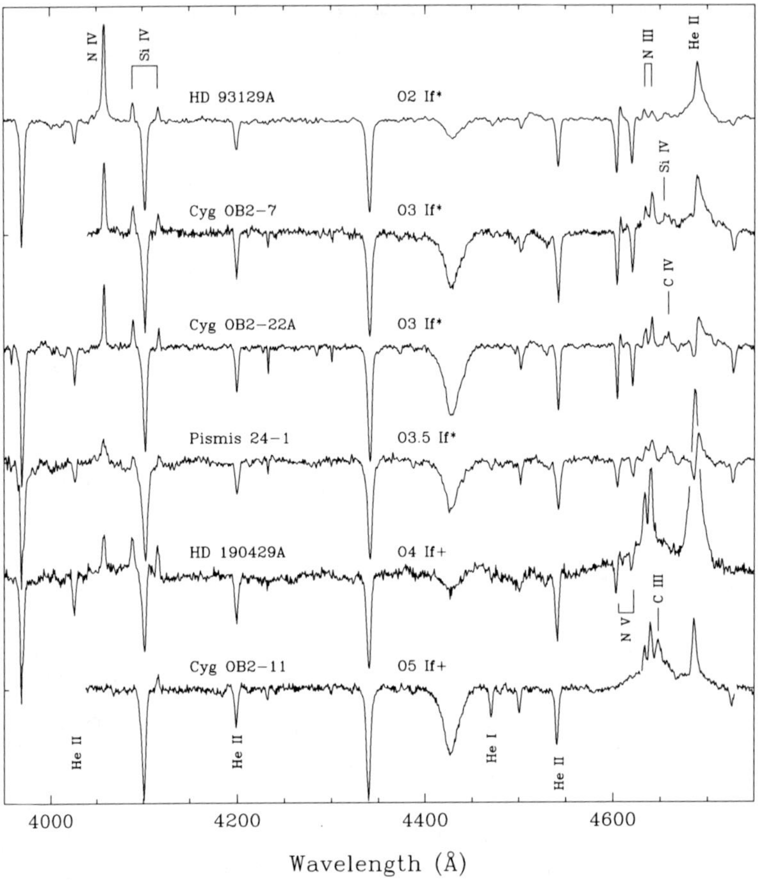

Fig. 2.4 An early Of supergiant spectral sequence, showing the extreme N IV/N III emission-line ratio in HD 93129A, leading to its earliest spectral type (Courtesy of Ian Howarth, from [79])

The Carina Nebula presents an attractive opportunity for a systematic investigation of multiplicity in an OB population, and such has been undertaken with *HST*/FGS [44]. Five new binaries were found in a preliminary subsample, including most remarkably HD 93129A with a separation of 55 milliarcsec and Δm_V of 0.9 (Fig. 2.6), likely implying yet another very early O main-sequence member of Tr 14. This pair was subsequently resolved in direct images with the *HST* Advanced Camera for Surveys (Fig. 2.7), and even more remarkably, relative motion of the components has been detected with both instruments; this is the most massive visual/astrometric binary known [36]. Likely related high-energy observations of the system are discussed later. HD 93129A (now Aab) has provided an anchor point for both the classification and quantitative analysis of the most massive stars (e.g., [70]). These results for relatively nearby stars emphasize anew the fundamental issues introduced by the endemic multiplicity of the O stars, frequently at

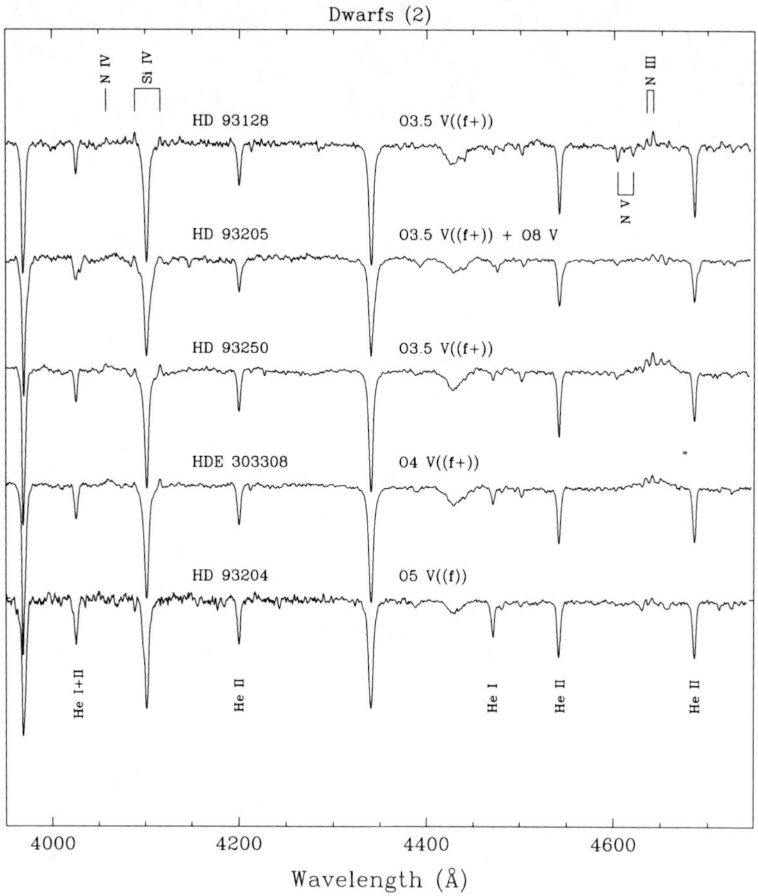

Fig. 2.5 An early O dwarf spectral sequence in the Carina Nebula (From [83])

or beyond the detection capabilities of current techniques, and the importance of continually addressing them at the state of the art. A binary nature of η Car would not be surprising, in view of the characteristics of its associated population.

Extensive spectroscopic, optical and/or infrared photometric, and astrometric studies of the Carina Nebula clusters [6,20,37,68,72] have provided useful data and interesting discussions. Nevertheless, the large disparities in reddening, distance, and age results document the uncertainties summarized in the Introduction above. It is likely that they will remain unresolved until UV–IR energy distributions and space interferometric astrometry become available for hundreds of individual stars. Smith [58] has made a complete census of the known OB stellar content of the Carina Nebula and calculated their radiative and mechanical energy inputs to the Nebula, which are relevant to the following sections.

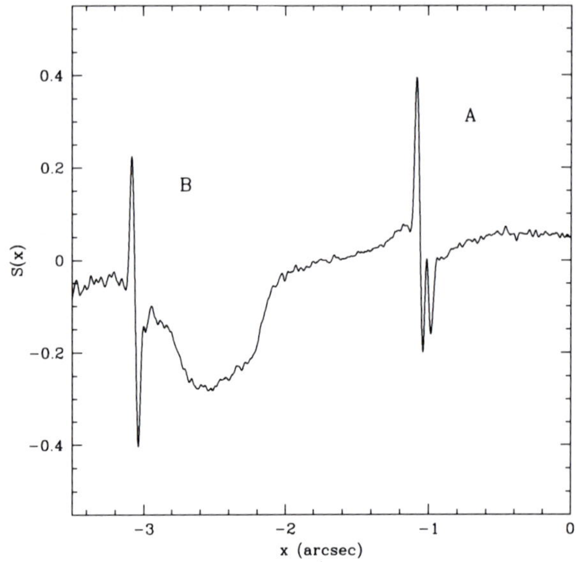

Fig. 2.6 *HST* Fine Guidance Sensor scan of HD 93129Aab and B, showing the duplicity of the former (From [44] (The broad dip between the two stellar traces is an artifact))

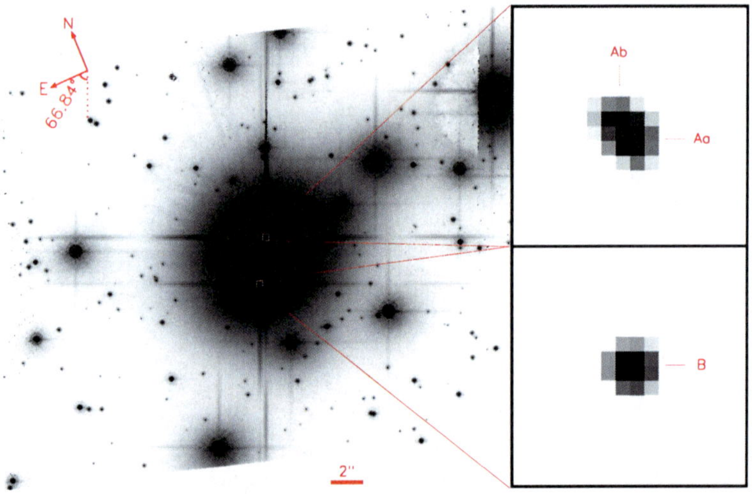

Fig. 2.7 *HST* Advanced Camera for surveys High Resolution Camera images of the same two stars. (*left*) F850LP image of the HD 93129 system. (*right*) Enlargements of Aab (*top*) and B (*bottom*) from the F220W image. The separation of the former pair measured from this image is 51 milliarcsec in the N-S direction. An HRC pixel is 0.025″ square and each of the right panels is 0.4″ square (Courtesy of Jesús Maíz Apellániz)

2.3 Nebular Structures

The early radio-continuum observations of the Carina Nebula revealed two strong peaks in the northern part of the nebula, called Carina I and II [27]. They coincide with optical structures, the first with a semicircular dust incursion into the H II near the compact cluster Tr 14 (Fig. 2.2), lower corner of "c1" box, and the second with a curious apparent bright ring a few arcminutes northwest of η Car (Fig. 2.2, top of "Keyhole Nebula", further discussed below). Subsequent radio observations with higher spatial resolution showed greater complexity in these structures [12, 31, 51, 86].

Deharveng and Maucherat [21] and Walborn [76] showed that the Car II ring is a physical structure best defined in the light of [S II]. There are other strange [S II] structures in the vicinity, and the general appearance of the northern part of the Carina Nebula is completely different in the light of [O III]. The gas kinematics of the region was further investigated by Meaburn et al. [39], and Kennicutt et al. [32] have discussed the implications of this inhomogeneity for standard nebular analyses, finding that the oxygen abundance is unaffected but there are problems with nitrogen and sulfur. Images of the Car II ring with the *HST* Wide Field Planetary Camera 2 revealed further intricate details, including barely resolved dark globules and the astounding adjacent "Finger Nebula", evidently sculpted by the WN star HD 93162 (Fig. 2.8; [78]). The ring may have been created by η Car's polar wind [56], and the apparently helical structure at its western edge might be a "twisted trunk" [15]. The Finger Nebula was analyzed in detail by Smith et al. [65] and is further mentioned below in the context of ongoing, triggered star formation. Smith et al. [67] also

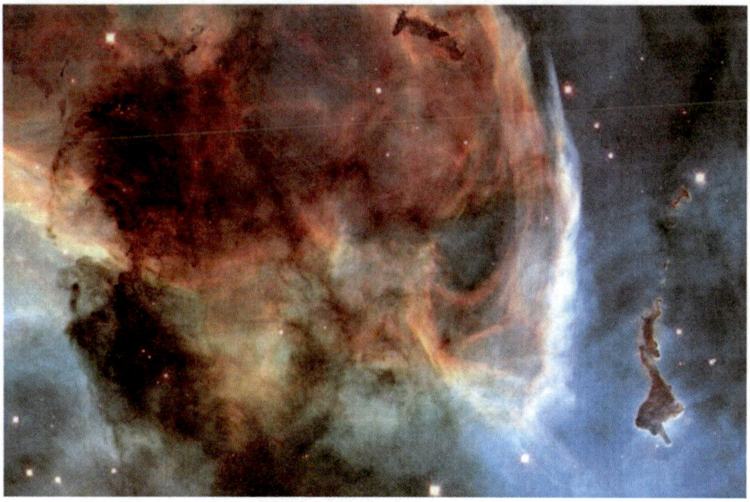

Fig. 2.8 *HST* Wide Field Planetary Camera 2 image of the Keyhole/Carina II ring and the Finger Nebula (Credit: NASA, ESA, the author, Rodolfo Barbá, and The Hubble Heritage Team)

discussed further aspects of the possible interactions of η Car's winds with the immediately surrounding region of the Nebula.

The association of η Car with the Carina Nebula has been demonstrated twice, first by Walborn and Liller [82] who discovered nearby dust clouds reflecting the peculiar spectrum of the star, and then by Allen [3] who showed that the redward component of the double nebular emission lines is occulted by the Homunculus. López and Meaburn [35] found that the reflected Hα profile is different from that observed directly toward η Car and suggested a lagged time variation; however, [10, 62] reinterpreted the difference as caused by the spatial anisotropy of the wind emission, with the reflection revealing the polar profile. The presence of reflection has important implications for the drawing of the Carina Nebula by John Herschel at the time of η Car's Great Eruption during the nineteenth century. He drew the bright nebulosity immediately to the west of the variable as far brighter and more extensive than it appears today, defining his "Keyhole Nebula" (the upper part of which is the Car II ring). The accuracy of Herschel's drawing was questioned by Bok [9] on the assumption that the nebulosity was entirely ionized; η Car at maximum would not have been hot enough to excite it, and the recombination time would have been far longer in any event. Since we now know that this nebulosity has a reflection component, it is likely that the latter was considerably brighter when the variable was at maximum, and that it faded with a short lag during the subsequent decline, so that Herschel's drawing is correct.

Molecular observations toward the Carina Nebula provide a different, complementary perspective. Early observations in the H_2CO and OH lines showed two strong peaks, coincident not with the radio-continuum maxima but with the dense V-shaped dust lanes that divide the northern and southern parts of the Nebula [22, 23, 28]. More extensive CO observations revealed an elongated giant molecular cloud with a mass of order $10^5 M_\odot$, the Carina Nebula being located near one end [29]. CO observations with higher spatial resolution discovered discrete cloudlets with masses of order $10 M_\odot$ in the vicinity of Car II [16], which are undoubtedly related to the subject of the following section. Infrared H_2 and PAH [11] and mm continuum [14] observations have been made in the same region; while far-IR line, CO, and CS have been observed throughout the Nebula [13, 41, 88].

Smith and Brooks [59] presented a thorough discussion of the global properties and energetics of the Carina Nebula, based on multiwavelength observations and a census of the inputs from the current stellar content. They conclude that the latter energy is more than sufficient to account for the IR luminosity, ionization, and kinematics of the Nebula, so there is no need to invoke an SNR in the region on energetic grounds.

2.4 Current Star Formation

Until recently, the Carina Nebula was generally regarded as an evolved H II region, with no particular evidence of ongoing star formation. However, more powerful IR (and optical) instrumentation in the Southern Hemisphere and in space has

Fig. 2.9 Enlargement from Fig. 2.2 of pillar and jet structures signaling new star formation in the northern Carina Nebula (Credit: NASA, ESA, Nathan Smith, and The Hubble Heritage Team)

drastically revised that perspective. Megeath et al. [40] discovered a small IR cluster within a bright-rimmed globule southeast of and facing η Car and Tr 16. A larger scale view from the *Midcourse Space Experiment* revealed a number of similarly oriented dust pillars and IR sources somewhat further to the southeast [61]. There are some additional cloudlets near Car II, as well as three possible ultracompact H II regions and a young stellar object with a disk in the vicinity of Tr 14 [48]. Numerous proplyd candidates were found in groundbased images [63], and many more will undoubtedly be reported from the ACS images (Fig. 2.2). It is quite possible that the "Finger Nebula" noted above will give birth to a small cluster of low-mass stars, if the condensations responsible for its protuberances are sufficiently dense [65]. A giant Herbig-Haro flow was discovered in the southern part of the Nebula [64], and another small, newborn cluster nearby [30, 49, 66]. Several CO cores associated with IR sources and outflows, and further young objects in the vicinity of Tr 14 and Car I, have been presented [69, 87], respectively. The unprecedented pillar plus jet structures in Fig. 2.9, from the ACS data, appear to have been generated by the O6 stars HD 93160 and HD 93161AB (Fig. 2.2); they will enlighten us further about the physics of triggered star formation not only in Carina, but in general. Thus, the Carina Nebula also displays the ubiquitous evidence for triggered, second-generation star formation in H II regions [78]. Even more spectacular massive star formation will likely be triggered soon around the periphery of the current stellar-wind bubble in the Nebula, when η Car and its massive siblings end their lives as core-collapse supernovae [59].

2.5 Interstellar Absorption Lines

The interstellar absorption lines toward stars in the Carina Nebula have been studied by the author and collaborators for many years with progressively more powerful instrumentation, up to the state of the art with the *HST* Space Telescope Imaging Spectrograph ([17, 84, 85]; and references therein; Fig. 2.10). They present the most extreme profiles known in the Galaxy, similar in fact to some QSO narrow-line systems. Radial velocities in the strong, low-ionization lines range from nearly -400 to $+200\,\mathrm{km\,s^{-1}}$, with up to 26 components resolved in a single line of sight. The rapid spatial variations in the profiles over small angular distances, higher frequency of complex profiles toward more massive stars, and large temporal variations in several components toward one star observed twice with STIS, all suggest that the high-velocity features arise in the near foregrounds of the stars toward which they are observed, i.e. within the Nebula. Although a model does not yet exist, interaction between the stellar winds and dense surrounding material is a possible origin of the phenomenon. Alternatively, Walborn et al. [85] emphasized the ubiquity of positive- as well as negative-velocity components, which may indicate expanding "shells" predominantly in the foreground of the stars, and thus possibly related to star-formation activity or an SNR on the near side of the Nebula. If this kinetic energy is being converted to thermal energy by collisions, temperatures of a few million degrees would result, which is likely relevant to the following section.

In contrast, the dominant components in the high-ionization and excited-state interstellar lines have velocities similar to the blueward components of the double nebular emission lines, that is, they arise in the near side of the globally expanding H II region. Thus, they provide valuable absorption diagnostics of the ionized material.

2.6 X-Rays

The *Einstein Observatory* discovered that all of the bright stars in the Carina Nebula, including η Car, are X-ray sources, but more surprisingly, that the entire Nebula is a source of diffuse soft X-rays [54, 55]. The more advanced capabilities of the *Chandra Observatory* and *XMM-Newton* have greatly expanded those early results [2, 5, 24, 25, 52, 71]. In particular, X-ray emission from tens of additional OB stars as well as hundreds of likely T Tauri stars has been found. The latter contributed to the original "diffuse" emission, but a true diffuse component remains. In view of the energy correspondence, the diffuse X-rays are very likely related to the high-velocity motions revealed by the interstellar absorption lines, as discussed in the previous section. An extensive new survey of the Carina Nebula undertaken by *Chandra* during 2007 (PI L. Townsley) will provide even more detailed information.

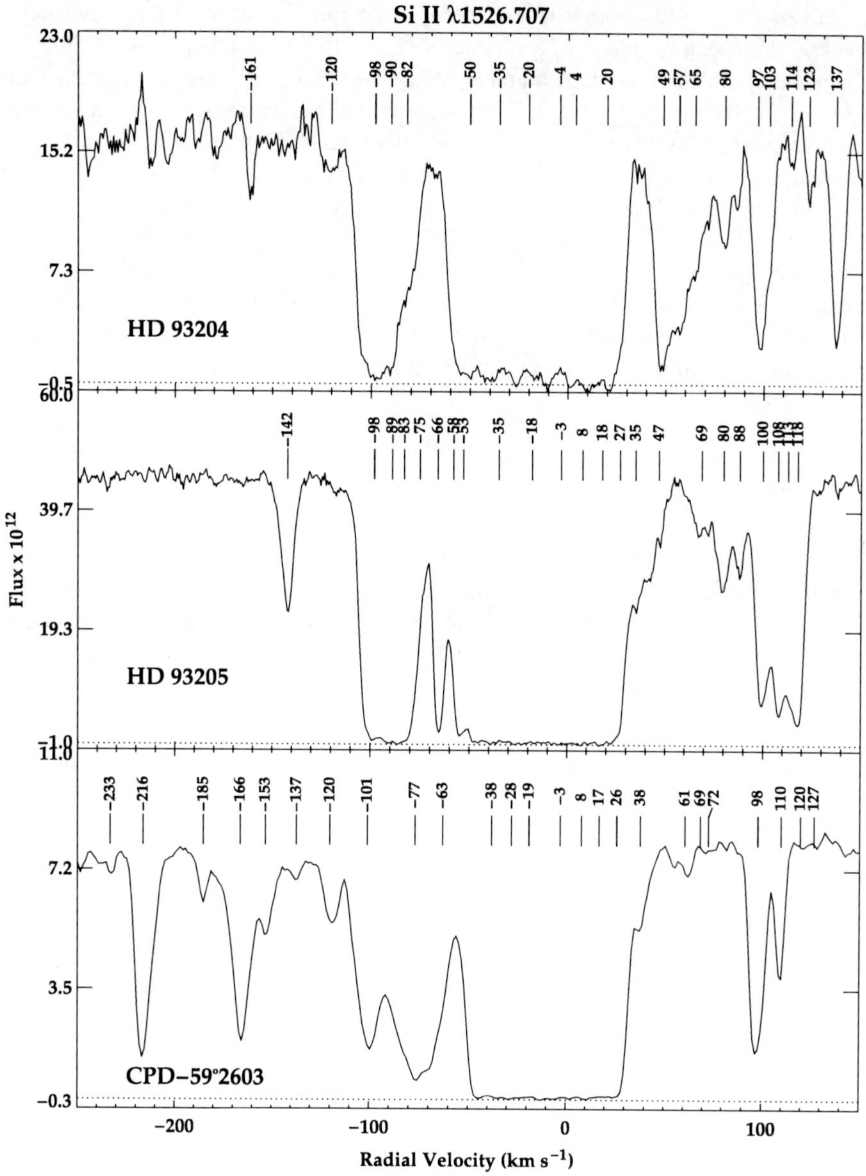

Fig. 2.10 Examples of extreme high-velocity interstellar absorption-line structure in the Carina Nebula. HD 93204 and HD 93205 are just 20″ apart, or 0.22 pc in projection (From [85])

Several of the OB and WR members of the ionizing clusters have unusually hard X-ray spectra, indicating possible binary colliding-wind emission. Three of them are of outstanding interest and have related UV/optical/radio properties. One of them is HD 93129A, the O2 If* prototype and massive astrometric binary that dominates Tr 14, as discussed above. It has also been found to be a nonthermal radio source [7, 8]. Another is the WN6-A member of Tr 16 HD 93162, the sculptor of the "Finger Nebula", which has one of the the highest X-ray luminosities known for a Wolf-Rayet star. It may be a colliding-wind binary on that basis [46, 47], and it has subsequently been found to be a long-period spectroscopic binary [26]. Finally, the remarkable X-ray and related properties of the likely binary η Car itself are discussed elsewhere in this volume. The primaries of these systems may well represent an evolutionary sequence of stars near the upper stellar mass limit.

2.7 Summary

Numerous studies have revealed the remarkable population of very massive stars and the complexity of the environment associated with η Car. But, as has been emphasized, we still need the multiwavelength spectral energy distributions of many of these stars to accurately determine their total luminosities and initial masses for comparison with η Car.

One of the outstanding problems in our present understanding of star formation, and of the Carina Nebula in particular, is why so many extremely massive stars form in such close proximity and at essentially the same time. What is the formation mechanism of such very massive stars?

References

1. J.F. Albacete Colombo, N.I. Morrell, G. Rauw, M.F. Corcoran, V.S. Niemela, H. Sana, Optical spectroscopy of X-Mega targets–IV. CPD $-59°$ 2636: a new O-type multiple system in the Carina Nebula. MNRAS. **336**, 1099–1108 (2002)
2. J.F. Albacete Colombo, M. Méndez, N.I. Morrell, XMM-Newton X-ray observations of the Carina nebula. MNRAS. **346**, 704–718 (2003)
3. D.A. Allen, The location of Eta Carinae in NGC 3372. MNRAS. **189**, 1P–4P (1979)
4. D.A. Allen, D.J. Hillier, The shape of the Homunculus Nebula around Eta Carinae. PASA. **10**, 338–341 (1993)
5. I.I. Antokhin, G. Rauw, J.M. Vreux, K.A. van der Hucht, J.C. Brown, XMM-Newton X-ray study of early type stars in the Carina OB1 association. A&A. **477**, 593–609 (2008)
6. J. Ascenso, J. Alves, S. Vicente, Lago MTVT: NTT and VLT diffraction limited imaging of Trumpler 14: revealing a massive core-halo cluster. A&A. **476**, 199–215 (2007)
7. P. Benaglia, B. Koribalski, Radio observations of HD 93129A: the earliest O star with the highest mass loss? A&A. **416**, 171–178 (2004)
8. P. Benaglia, B. Koribalski, J.F. Albacete Colombo, Radio detection of colliding wind binaries. PASA. **23**, 50–63 (2006)

9. B. Bok, *A Study of the η Carinae Region*. Harvard Reprints Ser 1 No 77 (Harvard, Cambridge, MA, 1932), pp. 1–90
10. P. Boumis, J. Meaburn, M. Bryce, J.A. López, The Hα echoes of η Carinae from 1985 to 1997. MNRAS. **294**, 61–68 (1998)
11. K.J. Brooks, M.G. Burton, J.M. Rathborne, M.C.B. Ashley, J.W.V. Storey, Unlocking the Keyhole: H_2 and PAH emission from molecular clumps in the Keyhole Nebula. MNRAS. **319**, 95–102 (2000)
12. K.J. Brooks, J.W.V. Storey, J.B. Whiteoak, H110α recombination-line emission and 4.8 GHz continuum emission in the Carina nebula. MNRAS. **327**, 46–54 (2001)
13. K.J. Brooks, P. Cox, N. Schneider, J.W.V. Storey, A. Poglitsch, N. Geis, L. Bronfman, The Trumpler 14 photodissociation region in the Carina Nebula. A&A. **412**, 751–765 (2003)
14. K.J. Brooks, G. Garay, M. Nielbock, N. Smith, P. Cox, SIMBA observations of the Keyhole Nebula. ApJ. **634**, 436–441 (2005)
15. P. Carlqvist, G.F. Gahm, H. Kristen, Theory of twisted trunks. A&A. **403**, 399–412 (2003)
16. P. Cox, L. Bronfman, The molecular gas content of the Keyhole nebula. A&A. **299**, 583–590 (1995)
17. A.C. Danks, N.R. Walborn, G. Vieira, W.B. Landsman, J. Gales, B. García, Rapid temporal variations of interstellar absorption lines in the Carina Nebula. ApJ. **547**, L155–L159 (2001)
18. K. Davidson, N. Smith, T.R. Gull, K. Ishibashi, D.J. Hillier, The shape and orientation of the Homunculus Nebula based on spectrscopic velocities. AJ. **121**, 1569–1577 (2001)
19. K. Davidson, R.M. Humphreys, Eta Carinae and its environment. ARA&A. **35**, 1–32 (1997)
20. K. DeGioia-Eastwood, H. Throop, G. Walker, K.M. Cudworth, The star formation history of Trumpler 14 and Trumpler 16. ApJ. **549**, 578–589 (2001)
21. L. Deharveng, M. Maucherat, Optical study of the Carina Nebula. A&A. **41**, 27–36 (1975)
22. H.R. Dickel, Carina Nebula: a possible interpretation of the molecular observations. A&A. **31**, 11–16, (1974)
23. H.R. Dickel, J.V. Wall, The OH absorption against the Carina Nebula. A&A. **31**, 5–10 (1974)
24. N.R. Evans, F.D. Seward, M.I. Krauss, T. Isobe, J. Nichols, E.M. Schlegel, S.J. Wolk, Chandra observations of associates of η Carinae. I. Luminosities. ApJ. **589**, 509–525 (2003)
25. N.R. Evans, E.M. Schlegel, W.L. Waldron, F.D. Seward, M.I. Krauss, J. Nichols, S.J. Wolk, Chandra observations of associates of η Carinae. II. Spectra. ApJ. **612**, 1065–1080 (2004)
26. R. Gamen, E. Gosset, N. Morrell, V. Niemela, H. Sana, Y. Nazé, G. Rauw, R. Barbá, G. Solivella, The first orbital solution for the massive colliding-wind binary HD 93162 (=WR 25). A&A. **460**, 777–782 (2006)
27. F.F. Gardner, D.K. Milne, P.G. Mezger, T.L. Wilson, The Carina nebula at 6 cm. A&A. **7**, 349–358 (1970)
28. F.F. Gardner, H.R. Dickel, J.B. Whiteoak, The H_2CO absorption against the Carina Nebula. A&A. **23**, 51–54 (1973)
29. D.A. Grabelsky, R.S. Cohen, L. Bronfman, P. Thaddeus, Molecular clouds in the Carina arm–The largest objects, associated regions of star formation, and the Carina arm in the galaxy. ApJ. **331**, 181–196 (1988)
30. G.F. Hägele, J.F. Albacete Colombo, R.H. Barbá, G.L. Bosch, G287.84–0.82: an infrared star cluster in the Carina nebula. MNRAS. **355**, 1237–1243 (2004)
31. W.K. Huchtmeier, G.A. Day, The Carina Nebula at 3.4 and 6 cm. A&A. **41**, 153–164 (1975)
32. R.C. Kennicutt Jr., F. Bresolin, H. French, P. Martin, An empirical test and calibration of H II region diagnostics. ApJ. **537**, 589–612 (2000)
33. N. Langer, W.R. Hamann, M. Lennon, F. Najarro, A.W.A. Pauldrach, J. Puls, Towards an understanding of very massive stars. A new evolutionary scenario relating O stars, LBVs, and Wolf-Rayet stars. A&A. **290**, 819–833 (1994)
34. K.C. Leung, A.F.J. Moffat, W. Seggewiss, The massive multiple system HD 93206 (QZ Carinae) in the great Carina Nebula. ApJ. **231**, 742–750 (1979)
35. J.A. López, J. Meaburn, The broad Balmer profiles from η Carinae scattered by dust over the Car II region. RMxAA. **13**, 27–32 (1986)

36. J. Maíz Apellániz, L. Úbeda, N.R. Walborn, E.P. Nelan, IMF biases and how to correct them, in *Resolved Stellar Populations*, ed. by D. Valls-Gavaud, M. Chávez. arXiv:astro-ph/0506283 (2005 in press)
37. P. Massey, J. Johnson, Massive stars near Eta Carinae–The stellar content of Tr 14 and Tr 16. AJ. **105**, 1225–1233 (1993)
38. J. Meaburn, An updated proper-motion/spectropolarimetric distance to η Carinae, in *Eta Carinae at the Millennium*, ed. by J.A. Morse, R.M. Humphreys, A. Damineli. ASP Conference Series, vol 179 (ASP, San Francisco, 1999), pp. 89–95
39. J. Meaburn, J.A. López, D. Keir, Insect-eye, Fabry-Perot observations of the large-scale motions within the Carina nebula (NGC 3372, RCW 53). MNRAS. **211**, 267–276 (1984)
40. S.T. Megeath, P. Cox, L. Bronfman, P.R. Roelfsema, Evidence for ongoing star formation in the Carina nebula. A&A. **305**, 296–307 (1996)
41. M. Mizutani, T. Onaka, H. Shibai, Origin of diffuse C II 158 micron and Si II 35 micron emission in the Carina nebula. A&A. **423**, 579–592 (2004)
42. N.I. Morrell, et al., Optical spectroscopy of X-Mega targets–II. The massive double-lined O-type binary HD 93205. MNRAS. **326**, 85–94 (2001)
43. N.D. Morrison, P.S. Conti, Spectroscopic studies of O-type binaries. VI–The quadruple system QZ Carinae (HD 93206). ApJ. **239**, 212–219 (1980)
44. E.P. Nelan, N.R. Walborn, D.J. Wallace, A.F.J. Moffat, R.B. Makidon, D.R. Gies, N. Panagia, Resolving OB systems in the Carina nebula with the Hubble space telescope fine guidance sensor. AJ. **128**, 323–329 (2004)
45. V.S. Niemela, N.I. Morrell, E. Fernández Lajús, R. Barbá, J.F. Albacete Colombo, M. Orellana, Optical spectroscopy of X-Mega targets in the Carina nebula–VI. FO15: a new O-type double-lined eclipsing binary. MNRAS. **367**, 1450–1456 (2006)
46. A.M.T. Pollock, M.F. Corcoran, Evidence for colliding winds in WR 25 from XMM-Newton observations of X-ray variability. A&A. **445**, 1093–1097 (2006)
47. A.J.J. Raassen, K.A. van der Hucht, R. Mewe, I.I. Antokhin, G. Rauw, J.M. Vreux, W. Schmutz, M. Güdel, XMM-Newton high-resolution X-ray spectroscopy of the Wolf-Rayet object WR 25 in the Carina OB1 association. A&A. **402**, 653–666 (2003)
48. J.M. Rathborne, M.G. Burton, K.J. Brooks, M. Cohen, M.C.B. Ashley, J.W.V. Storey, Photodissociation regions and star formation in the Carina nebula. MNRAS. **331**, 85–97 (2002)
49. J.M. Rathborne, K.J. Brooks, M.G. Burton, M. Cohen, S. Bontemps, The giant pillars of the Carina Nebula. A&A. **418**, 563–576 (2004)
50. G. Rauw, H. Sana, I.I. Antokhin, N.I. Morrell, V.S. Niemela, J.F. Albacete Colombo, E. Gosset, J.M. Vreux, Optical spectroscopy of XMEGA targets in the Carina Nebula–III. The multiple system Tr 16-104 (=CPD -59° 2603). MNRAS. **326**, 1149–1160 (2001)
51. D.S. Retallack, Radio emission from Eta Carinae and the surrounding nebula. MNRAS. **204**, 669–674 (1983)
52. K. Sanchawala, W.P. Chen, H.T. Lee, Y.H. Chu, Y. Nakajima, M. Tamura, D. Baba, S. Sato, An X-ray and near-infrared study of young stars in the Carina nebula. ApJ. **656**, 462–473 (2007)
53. J. Schweickhardt, W. Schmutz, O. Stahl, Th. Szeifert, B. Wolf, Revised mass determination of the super massive Wolf-Rayet star WR 22. A&A. **347**, 127–136 (1999)
54. F.D. Seward, T. Chlebowski, X-ray emission from the Carina Nebula and the associated early stars. ApJ. **256**, 530–542 (1982)
55. F.D. Seward, W.R. Forman, R. Giacconi, R.E. Griffiths, F.R. Harnden Jr., C. Jones, J.P. Pye, X-rays from Eta Carinae and the surrounding nebula. ApJ. **234**, L55–L58 (1979)
56. N. Smith, Near-infrared and optical emission-line structure of the Keyhole Nebula in NGC 3372. MNRAS. **331**, 7–12 (2002a)
57. N. Smith, Dissecting the Homunculus nebula around Eta Carinae with spatially resolved near-infrared spectroscopy. MNRAS. **337**, 1252–1268 (2002b)
58. N. Smith, A census of the Carina Nebula–I. Cumulative energy input from massive stars. MNRAS. **367**, 763–772 (2006) [err **368**, 1983–1984]
59. N. Smith, K.J. Brooks, A census of the Carina Nebula–II. Energy budget and global properties of the nebulosity. MNRAS. **379**, 1279–1292 (2007)

60. N. Smith, P.S. Conti, On the role of the WNH phase in the evolution of very massive stars: enabling the LBV instability with feedback. ApJ. **679**, 1467–1477 (2008)
61. N. Smith, M.P. Egan, S. Carey, S.D. Price, J.A. Morse, P.A. Price, Large-scale structure of the Carina nebula. ApJ. **532**, L145–L148 (2000)
62. N. Smith, K. Davidson, T.R. Gull, K. Ishibashi, D.J. Hillier, Latitude-dependent effects in the stellar wind of η Carinae. ApJ. **586**, 432–450 (2003a)
63. N. Smith, J. Bally, J.A. Morse, Numerous proplyd candidates in the harsh environment of the Carina nebula. ApJ. **587**, L105–L108 (2003b)
64. N. Smith, J. Bally, K.J. Brooks, HH 666: the axis of evil in the Carina nebula. AJ. **127**, 2793–2808 (2004a)
65. N. Smith, R.H. Barbá, N.R. Walborn, Carina's defiant Finger: HST observations of a photoevaporating globule in NGC 3372. MNRAS. **351**, 1457–1470 (2004b)
66. N. Smith, K.G. Stassun, J. Bally, Opening the treasure chest: a newborn star cluster emerges from Its dust pillar in Carina. AJ. **129**, 888–899 (2005a)
67. N. Smith, J.A. Morse, J. Bally, The [O III] Veil: astropause of η Carinae's wind? AJ. **130**, 1778–1783 (2005b)
68. M. Tapia, M. Roth, R.A. Vázquez, A. Feinstein, Imaging study of NGC 3372, the Carina nebula–I. UBVRIJHK photometry of Tr 14, Tr 15, Tr 16, and Car I. MNRAS. **339**, 44–62 (2003)
69. M. Tapia, P. Persi, J. Bohigas, M. Roth, M. Gómez, Imaging study of NGC 3372, the Carina nebula–II. Evidence of activity in the complex Trumpler 14/Car I photodissociation region. MNRAS. **367**, 513–526 (2006)
70. G. Taresch, R.P. Kudritzki, M. Hurwitz, S. Bowyer, A.W.A. Pauldrach, J. Puls, K. Butler, D.J. Lennon, S.M. Haser, Quantitative analysis of the FUV, UV, and optical spectrum of the O3 star HD 93129A. A&A. **321**, 531–548 (1997)
71. L.K. Townsley, P.S. Broos, E.D. Feigelson, G.P. Garmire, Parsec-scale X-ray flows in high-mass star-forming regions, in *Massive Star Birth: A crossroads of Astrophysics*, ed. by R. Cesaroni, M. Felli, E. Churchwell, M. Walmsley. IAU Symposium, vol. 227 (CUP, Cambridge, 2005), pp. 297–302
72. R.A. Vázquez, G. Baume, A. Feinstein, P. Prado, Investigation on the region of the open cluster Tr 14. A&AS. **116**, 75–94 (1996)
73. N.R. Walborn, Some extremely early O stars near eta Carinae. ApJ. **167**, L31–L33, (1971)
74. N.R. Walborn, Some characteristics of the Eta Carinae complex. ApJ. **179**, 517–525 (1973)
75. N.R. Walborn, Some morphological properties of WN spectra. ApJ. **189**, 269–271 (1974)
76. N.R. Walborn, Forbidden Sulfur II structures in the Carina nebula. ApJ. **202**, L129–L130 (1975)
77. N.R. Walborn, The stellar content of the Carina nebula. Rev. Mexicana Astron. Astrof. Ser. Conf. **2**, 51–55 (1995)
78. N.R. Walborn, The pillars of the second generation, in *Hot Star Workshop III: The Earliest Stages of Massive Star Birth*, ed. by P.A. Crowther. ASP Conference Series, vol. 267 (ASP, San Francisco, 2002), pp. 111–125
79. N.R. Walborn, The earliest O-yype stars, in *A Massive Star Odyssey: From Main Sequence to Supernova*, ed. by K.A. van der Hucht, A. Herrero, C. Esteban. IAU Symposium, vol. 212 (ASP, San Francisco, 2003), pp. 13–21
80. N.R. Walborn, Optically observable zero-age main-sequence O stars, in *Massive Stars: From Pop III and GRBs to the Milky Way*, ed. by M. Livio, E. Villaver. STScI Symposium Series, vol. 20 (CUP, Cambridge, 2009), pp. 167–177
81. N.R. Walborn, E.L. Fitzpatrick, The OB Zoo: a digital atlas of peculiar spectra. PASP. **112**, 50–64 (2000)
82. N.R. Walborn, M.H. Liller, The earliest spectroscopic observations of Eta Carinae and its interaction with the Carina nebula. ApJ. **211**, 181–183 (1977)
83. N.R. Walborn, et al., A new spectral classification system for the earliest O stars: definition of Type O2. AJ. **123**, 2754–2771 (2002a)

84. N.R. Walborn, A.C. Danks, G. Vieira, W.B. Landsman, Space telescope imaging spectrograph observations of high-velocity interstellar absorption-line profiles in the Carina nebula. ApJS. **140**, 407–456 (2002b)
85. N.R. Walborn, N. Smith, I.D. Howarth, G. Vieira Kober, T.R. Gull, J.A. Morse, Interstellar absorption-line evidence for high-velocity expanding structures in the Carina nebula foreground. PASP. **119**, 156–169 (2007)
86. J.B.Z. Whiteoak, High-resolution images of the dust and ionized gas distributions in the Carina Nebula. ApJ. **429**, 225–232 (1994)
87. Y. Yonekura, S. Asayama, K. Kimura, H. Ogawa, Y. Kanai, N. Yamaguchi, P.J. Barnes, Y. Fukui, High-mass cloud cores in the η Carinae giant molecular cloud. ApJ. **634**, 476–494 (2005)
88. X. Zhang, Y. Lee, A. Bolatto, A.A. Stark, CO ($J = 4$–3) and [C I] observations of the Carina molecular cloud complex. ApJ. **553**, 274–287 (2001)

Chapter 3
The Central Star: Instability and Recovery

Kris Davidson

Abstract Many clues to η Carinae's structure and physics can be found in its observed behavioral timescales, ranging from a few days to a few centuries. Here these are reviewed in three categories: the 5.5-year spectroscopic cycle with its spectroscopic events, the Great Eruption of 1830–1860, and the subsequent unsteady recovery. Unfortunately, the underlying thread shared by these stories is a scarcity of theoretical effort. Eta Car may or may not be an exception to general rules; but either way, as the most observable very massive star, it reveals major gaps in existing theory. Recent observers have found many unexpected, highly suggestive discoveries in this topic, including phenomena that cannot be observed in any other known object. A variety of specific unsolved theoretical problems are noted here.

3.1 A Counter-Paradigm

Here is a tale of Unfinished Business and Unexplored Territory. "Eta Carinae" amounts to a topic, not just an object; observers have established its outlines but theorists have not yet done their part. The 1830–1860 Great Eruption remains almost as mysterious as it was 20 years ago. It may prove freakish due to the companion star; or, on the other hand, maybe not. *Until these and related questions have been explored better, no account of mass loss and evolution at the top of the H-R diagram should be considered realistic.*

This object has revealed more unpredicted phenomena than all the other known very massive stars ($M \gtrsim 60\,M_{\odot}$) combined. For two centuries it has been in an almost unbelievably rare physical state. Survivor of the most famous supernova

K. Davidson (✉)
Minnesota Institute for Astrophysics, University of Minnesota, Minneapolis, MN, USA
e-mail: kd@astro.umn.edu

K. Davidson and R.M. Humphreys (eds.), *Eta Carinae and the Supernova Impostors*, Astrophysics and Space Science Library 384, DOI 10.1007/978-1-4614-2275-4_3,

impostor event, it is now recovering on a timescale suitable for observation. A companion star recurrently perturbs it, varying some of the parameters like a lab experiment. We know the luminosity unusually well, we know the direction of the polar axis, and we can observe the star from a range of directions via circumstellar reflection. The mass outflow, which is in some respects the most extreme known stellar wind, has changed dramatically in the past decade. Meanwhile the wind exhibits spectacular gasdynamic instabilities, while outlying ejecta show spectroscopic phenomena seen almost nowhere else. Even if we refer only to the luminosity and the eruption ejecta, nothing else like η Car has been found in any nearby galaxy. Yet it is only 2.3 kpc away, an amazingly improbable fact.

But a disadvantage of "combining more points of interest" than any other extrasolar star [30] is a centrifugal tendency in research. Choosing papers at random, first we may read that η Car is chiefly an X-ray binary. Then others explain, instead, that infrared studies are paramount; or the dense wind, or exotic nebular spectra, or periodic fluctuations, etc. These are all good problems, but the deepest and broadest for astrophysics concern *the nature of the star itself, and why it erupted.* For more than 30 years there have been empirical hints that sporadic events, rather than conventional line-driven winds, may account for much or even most of the mass loss in sufficiently massive stars [23, 33, 34, 77]. Theorists neglected to foresee how precarious is a state within, say, 30% of the Eddington limit.

This point deserves amplification. Old-fashioned idealized stars behaved gracefully near the classical Eddington limit, simply refusing to cross it. But real stars sometimes erupt when L/M becomes comparable to that value [34]. Soon after this fact was recognized, temperature-dependent opacity seemed a likely culprit. Rotation attracted attention after 1990, though its possible synergism with opacity was usually ignored (see refs. in Sect. 3.4 below). Eta Car's bipolar Homunculus ejecta-nebula became a famous hint in those years. Nevertheless, theoretical discussions continued to employ euphemisms such as "enhanced mass loss" – as though the destabilizing factors would merely increase the quasi-steady wind [48]. One reason, of course, was that no one could simulate a real eruption. Published evolution tracks in the HR diagram relied on simple empirical prescriptions for the long-term mass loss rates. Those accounts blurred the distinction between structural effects of rotation, etc., as analyzed there, vs. the poorly known instabilities that actually cause outbursts. Stellar-evolution papers thus give an exaggerated – and, one hopes, unintended – impression of the extent to which mass loss is known and/or understood.

LBV's are usually cited as the classic massive eruptors. But η Car's giant outburst, though atypical, involved the same physics and offers the greatest wealth and diversity of observational clues. Regarding cosmology, instabilities near the Eddington Limit may affect Population III stars, and might even prevent such objects from having the high masses commonly attributed to them. A basic question is whether or not low metallicity immunizes a star against the mysterious giant-eruption instability [34, 77]. So long as η Car remains ill-understood, all discussions of evolutionary tracks in the upper HR diagram, as well as Pop III, rest on shaky ground. Crucial phenomena may have been omitted!

This article concerns the physics of η Car, basic or conjectural, with regard to its timescales. Diverse observations indicate characteristic times of several days, a few weeks or months, 5.5, ~20, ~50, and ~200 years, in addition to the current evolutionary timescale which may have almost any value up to 10^5 years. This is a review of the situation, not of the literature; many of the citations really mean "and other references therein."[1] In addition to timescales, characteristic energies will also play their roles here. Some points noted below have received little attention in recent years, and a few have not yet appeared elsewhere.

3.2 Stellar Parameters

First consider the vital quantities. Eta's total luminosity, $L_{tot} \approx 5 \times 10^6 L_\odot$, can be observed as thermal IR re-emission by dust in the Homunculus ejecta-nebula, plus a correction for UV-to-near-IR light that diffuses out [12, 79]. L_{tot} is more robust than the UV extrapolations employed for most other very massive stars. It's most likely within $\pm 15\%$ of the truth, unless substantial UV light escapes through gaps in the Homunculus. Unfortunately there have been no careful attempts to detect long-term changes in the integrated IR flux. The secondary star probably supplies 4–20% of the total (see below), so $L_1 \approx 4.5 \times 10^6 L_\odot$ seems a good working assumption for the primary. The corresponding Eddington Limit minimum mass is 85–105 $M_\odot$, depending on the surface abundance and ionization of helium.

We can't place the primary star in an H-R diagram, because it has been hidden in an opaque wind for two centuries. Observables in the wind are insensitive to the star's physical radius, usually defined as the sonic point in the outflow [7, 31]. The historical record allows an interesting but questionable deduction. Suppose that η Car's apparent temperature varies over the long term between a "normal" hot quiescent state and a cooler-looking eruptive state [34]. Further suppose that Halley's magnitude estimate in 1677 represented the former: $m_V \approx 3.3$ [22]. This almost entirely represents the primary star, see below. Correcting for interstellar extinction [12] we find $m_{V,0} \approx 1.6$, compared to $m_{bol,0} \approx -0.1$ based on the star's luminosity. Thus we deduce a bolometric correction close to -1.7 magnitude, indicating $T_{eff} \sim 20{,}000$ K for the "normal" state, much cooler than the ZAMS. This result seems mildly surprising because (L, T_{eff}) lies well above the empirical upper boundary in the HRD as we usually draw it [34] – though, admittedly, that boundary has never been well defined at its high-temperature end. On the other hand, 20,000 K is a good temperature for bistability [68]. The corresponding radius is about $180 R_\odot \approx 0.8$ AU.

Several obvious forms of error may lurk in the above reasoning, but it's the best we have. If Frew's calibration of Halley's magnitude [22] was wrong and $m_V \approx 4.0$,

[1]So many papers have been published since 2000 that a "complete" list now amounts to an academic display, tending by weight of numbers to obscure the works that really altered the topic.

for example, then $T_{\rm eff} \approx 25{,}000$ K which does not greatly alter the case. If the wind was opaque even before 1700, and the star itself was hotter than 25,000 K, then there is no evident way to measure it, except, perhaps, via the theoretical mass-loss vs. temperature relation, a modernized version of Fig. 1 in [7]. The wind may become transparent in the foreseeable future, Sect. 3.5 below.

We know that η Car is somewhat evolved, because its ejecta have a substantially higher-than-normal helium abundance and the CNO is mostly nitrogen [13, 18]. A reasonable guess for helium in the star's outer layers is $Y \approx 0.5$ [31]. It belongs to a cluster whose estimated age approaches the generic 3 Myr lifetime of a very massive star [12, 88]. Its luminosity and evolved state together imply an initial ZAMS mass in the range 160–260 $M_\odot$, depending on rotation and other assumptions [61]. Combining this statement, the classical Eddington Limit, and the fact that substantial mass has already been lost [76, 79], the primary star's current mass is probably in the range 120–170 $M_\odot$, with surface Eddington factor $\Gamma \sim 0.6$–0.8 or possibly higher. But the un-robust character of this assertion is obvious!

Apart from one or two doubtful claims, the companion star has not been observed directly (Sect. 3.3 below); so we must guess its parameters from indirect evidence. Three clues are (1) its far-UV photon output must photoionize enough helium in outlying ejecta to account for emission lines seen there, (2) the X-ray spectrum of the colliding-wind shocked region indicates a wind speed close to 3,000 km s^{-1}, and (3) the object cannot be so luminous that we would detect it in the overall spectrum. These constraints lead to $T_{\rm eff} \approx 37{,}000$–$43{,}000$ K and L between 10^5 and 10^6 $L_\odot$, see Fig. 12 in [57]. The low end of this luminosity range, however, seems inadequate to explain the He I lines formed in the primary wind (again dependent on photoionization by the secondary); so we can reasonably adopt $L_2 \sim 10^{5.8} L_\odot$ within a factor of 2. Such a star would be an O4–O5 supergiant with mass $\sim$60 $M_\odot$ [54, 57, 61]. If we regard the suspected age of the primary star as another indicator for the secondary, then $\sim$40,000 K at 2 Myr would normally imply $M_2 \approx 50 M_\odot$ with $L_2 \approx 10^{5.6} L_\odot$. At 3 Myr no star should be that hot [54], unless it has been altered by accretion [43]. In any case, if this object were not concealed by the primary star and ejecta it would be one of the best-known massive stars, with $m_V \sim 7.5$. Its probable 5.5-year orbit is noted below.

3.3 The 5.5-Year Spectroscopic Cycle

Eta Car's most fundamental problems involve the Great Eruption c.1840 and its aftermath, with characteristic timescales of 10–200 years. Before examining them, though, we must acknowledge the 5.5-year spectroscopic cycle which has been a focus of research since 1997. This cycle may be essential for understanding the long-term behavior – or, on the other hand, maybe not.

As noted throughout this volume, since 1948 observers of η Car have occasionally recorded "spectroscopic events." High-excitation emission lines fade or disappear for a few weeks or months, the 2–10 keV X-rays become unsteady and then disappear temporarily, and other changes occur [37, 38, 60, 95]. Some of the spectral changes occur in ejecta located almost 1,000 AU from the star, probably signaling a weakened UV output. Eventually a recurrence period of 5.54 years was noticed for these events [4–6, 93], good evidence for a binary companion.[2] If the mass of the system is between 150 and 240 $M_{\odot}$ (Sect. 3.2 above), then a 5.54-year orbital period implies a semimajor axis in the range 16.5–19.5 AU. Since the dramatic part of each event occurs within $\Delta t < 120$ days, most authors agree that it must happen near periastron in a highly eccentric orbit. If this means within, say, $\pm 90°$ of periastron, it implies eccentricity $\varepsilon > 0.78$ and periastron separation $D_{\mathrm{min}} < 4$ AU. Most recent authors assume $\varepsilon \approx 0.9$ and $D_{\mathrm{min}} \approx 1.8$ AU. The controversial orientation of the orbit doesn't need to be discussed here; see references and caveats in Sect. 6.2 of [60].

Another probable fact should be mentioned. The hot secondary star is thought to supply most of the helium-ionizing photons, $h\nu > 24.6$ eV (Sect. 3.2 above). Thus it indirectly causes the high-excitation He I, [Ne III], etc., lines, which arise in He^{+} zones. These are precisely the features that fade or disappear in a spectroscopic event, and they carry far more energy than the hard X-rays. One can almost define a spectroscopic event as a temporary failure of the secondary star's far UV output [95]. The He^{+} zones exist in certain parts of the primary wind, in outlying ejecta, and possibly in dense condensations within the shocked structure. (The secondary wind can be ignored because its densities are low.) Concerning the emission physics, see [36, 57, 59, 60] and references cited therein.

Incidentally, there is no reason to think that dust formation plays a significant role in the spectroscopic events. At a distance of 30 AU from the primary star, for example – roughly the apoastron separation – the radiation density temperature is about 2,400 K, which implies grain temperatures close to 3,000 K if we assume reasonable absorption and emission efficiencies. This is too hot for normal grain formation, and 30 AU is far larger than the active region near periastron. Invoking dust therefore violates Ockham's razor by "multiplying the hypotheses" compared to other proposed models. It should only be regarded as a last, desperate resort if other explanations fail – which is not the case, see below.

The observable 2–10 keV X-rays presumably originate in a pseudo-hyperboloidal shocked region where the two stellar winds collide; see Fig. 3.1, Fig. 2 in [69], Fig. 3 in [66], and Chap. 9. From 1997 until sometime after 2003, many or most authors assumed that an event occurs when and because the shocked region passes behind the primary star's wind. This opinion arose from a perceived similarity of

[2]This development merits two historical comments. (1) In retrospect the 5.5-year period should have been identified when the 1986 event occurred, since [95] had noticed the 16.5-year spacing of the 1948, 1964, and 1981 events. (Others had passed unnoted in 1953, 1959, 1970, 1975.) But the 1986 and 1992 events were only minimally reported, see Refs. in [4, 5]. (2) The periodicity was good evidence for a binary, but not conclusive as most authors asserted. A single-star model remained conceivable [10, 78]. Even today, a degree of skepticism is justified by the checkered history of this question. Various points cited as incontrovertible evidence in 1997–2004 were soon controverted, and turned out to be fallacious. See the Elephant Parable in [10].

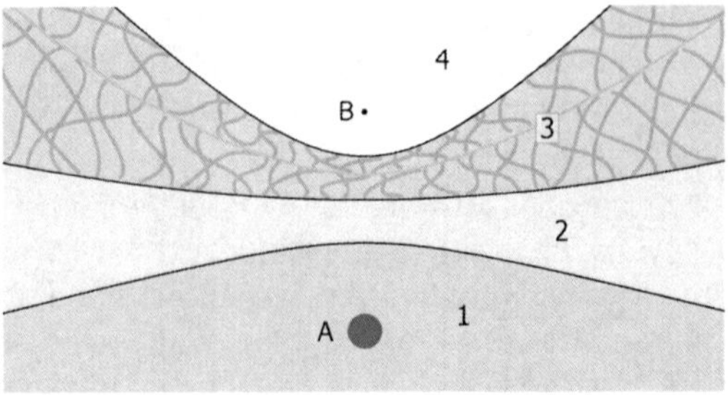

Fig. 3.1 Schematic wind zones during most of η Car's 5.5-year orbit. *A* and *B* are the two stars; region *1* is the undisturbed primary wind; *2* is the primary wind zone where the secondary star photoionizes helium; *3* is the wind-wind shocked region; and *4* is the low-density secondary wind. In reality the shapes are poorly known and the shocks *2*|*3* and *3*|*4* have chaotic irregularities, cf. Fig. 4 in [87] and Fig. 2 in [69]. Zones *4* and probably *3* are transparent and produce very little near-UV, visual, or IR radiation. Near periastron the overall structure becomes violently unstable

the X-ray flux behavior to known X-ray eclipsing binaries [37, 69, 86]. But in fact the resemblance isn't really very good; the UV and visual spectral features are more important to explain because they carry far more energy flux; and eclipse models failed in other respects [9, 10, 51, 66]. The secondary star and shocked region may pass behind part of the primary wind during or after an event, but the main phenomena have other causes [51, 60].

At present three proposed effects seem likely culprits. One, two, or all of them might occur together. They are far more interesting than eclipse effects, because they may have longer-term implications for the primary star.

1. *Enhanced mass loss?* The primary star's mass outflow may increase when the secondary approaches periastron [9, 51]. If so this would cause some of the "shell ejection" effects proposed for a single star in the best early account of a spectroscopic event [95]. It may reduce the 4–16 eV UV output which determines the strengths of some varying emission lines. This idea appears credible in terms of tidal forces. The basic tidal perturbation parameter is

$$\chi = \frac{M_2}{(1-\Gamma)M_1}\left(\frac{R}{D}\right)^3, \tag{3.1}$$

where R is the radius of either the primary star or its wind acceleration zone, D is the separation between stars, and $\Gamma = |g_{\rm rad}/g_{\rm tot}|$ approaches unity at the Eddington Limit. Most likely M_2 is about the same size as $(1-\Gamma)M_1$, and R is of the order of 1 AU (Sect. 3.2 above). Moreover, the primary's rotation rate may be comparable to its companion's orbital rate $d\phi/dt$ near periastron. Altogether we should not be surprised if the wind changes when $D \lesssim 3$ AU. (See also [3].)

Indeed the situation may be synergistic: tidal distortion can lower the local temperatures, thereby increasing the opacity, so Γ rises, amplifying the response.[3] A milder version of this idea is that tidal forces may alter the latitude structure of the wind [78]. Several clues in the 2003 and 2009 spectroscopic events appear to support enhanced mass outflow, but they aren't clear enough to prove it [51, 60].

2. *Shock instabilities?* The colliding-wind shock structure is expected to become very unstable near periastron [10, 51, 81, 84]. When that happens, the structure disintegrates and collapses in a few days or weeks. At most times the slow primary wind (roughly 500 km s^{-1}) and the fast secondary wind (3,000 km s^{-1}) each have their own initial shock fronts (Fig. 3.1); the former is always unstable, while the latter usually stabilizes the overall structure and produces the observable hard X-rays [66,69,81]. The 2–10 keV X-ray emission zone will disappear if the secondary shock becomes unstable. Equally interesting, the breakup process can produce a flood of very soft X-rays, due to a complex tangle of subshocks and oblique flows.

Two specific instabilities have been proposed for this case. First, several authors [10, 51, 81] emphasized the generic fact that strong radiative cooling makes a shock unstable [87]. High densities and comparatively low temperatures (i.e., low shock speeds) favor radiative cooling, and indeed η Car's primary-wind shock is probably always unstable for this reason. Near periastron the same may happen to the secondary-wind shock. Then Soker [84] drew attention to another, even more promising phenomenon. Near periastron the soft X-rays can increase the secondary wind's ionization [85]. If so then the conventional line-driving acceleration weakens and the secondary wind slows, an effect sometimes called "radiative inhibition." The shocks move closer to the secondary star because the wind-wind momentum balance has been altered. Hard X-rays disappear as the highest temperatures decrease, soft X-rays dominate, and in an extreme case the primary wind can overwhelm and suppress the secondary wind.[4] The two instabilities just outlined differ physically but have broadly similar consequences. They might even occur together.

Observed timescales seem reasonable. The main part of each event occurs in typically 30 days, and the most rapid part in about 5 days (e.g. [60]). The pre-collapse shock structure has a size scale 1–3 AU and local relative flow speeds 100–500 km s^{-1}, indicating characteristic flow times 3–50 days. The secondary star's orbital speed is of the order of 300–400 km s^{-1} near periastron.

---

[3]Similar behavior was proposed long ago for rotation, with a conceptual relation to the equally hypothetical "Modified Eddington Limit" instability. See Sect. 5 in [34], Sect. 5 in [96], and [47, 48].

[4]Meanwhile the soft X-rays and the secondary star's far UV presumably increase the ionization in some parts of the primary wind. Tentatively, we can assume that this is a lesser effect because the primary wind is far denser than the secondary wind.

During η Car's 2003 and 2009 events, the dramatic behavior of He II λ4687 emission (induced by soft X-rays) vs. 2–10 keV X-rays, was beautifully consistent with the shock breakup idea; see [51] and especially [60]. Shock modeling provides other reasons to favor it [66]. Hence we can regard the concept as nearly proven.

3. *Accretion by the companion star?* If the primary wind quashes the secondary wind during a spectroscopic event (see above), then matter from the primary wind may fall onto the secondary star [1, 41–43, 81, 83, 84]. Only a tiny fraction of the primary outflow is accreted, but that can be enough to slightly expand the secondary star's photosphere, reducing T_{eff}. This fact can greatly reduce the supply of helium-ionizing photons, which are indirectly responsible for the high-excitation emission lines that fade or disappear during an event [36, 57, 60, 83, 95]. This idea has implications far beyond the spectroscopic event problem, see Sect. 3.5 below.

Altogether these three phenomena appear capable of explaining η Car's spectroscopic events so well that there is no immediate need to seek alternatives. The highest priority for the time being is to test whether these ideas are as good as they seem to be. Observations in 2003.5 and 2009.0 strongly support process 2, process 3 is a believable consequence, and process 1 appears helpful for other clues [51, 60]. The primary wind appears to have changed between 2003 and 2009, so differences between those two events can be likened to varying parameters in a lab experiment [59, 60]. Concerning broader and deeper issues, *the probable effects listed above apply also to η Car's historical development,* see Sects. 3.4 and 3.5 below.

But we obviously need realistic simulations of processes 1–3, with emphasis on "realistic." In principle they might turn out to be quantitatively fallacious. Most of the above assertions about soft X-ray production during shock breakup depend on intuition and semi-quantitative reasoning; we need simulations of radiative output during the disintegration/collapse, and – harder to conceptualize – during the later re-formation of the large-scale shock structure. Trustworthy models are not easy to achieve, and a range of parameter space must be explored. Spherical or axial symmetry, neglect of inhomogeneities in each wind, oversimplified radiative processes, etc. can lead to misleading results.

Eta Car's observable 2–10 keV X-rays have been featured in many discussions of this topic – arguably at the expense of better indicators – but few authors have acknowledged how difficult they are to calculate realistically. These hard X-rays originate near the fast secondary-wind shock, 3|4 in Fig. 3.1. Only a small fraction of the wind kinetic energy is converted to photons above 1 keV, and the conversion efficiency depends on inhomogeneities in both winds ("clumping"). The relevant parameters are extremely uncertain and may involve size scales several orders of magnitude smaller than the overall shock structure. With strong inhomogeneities, the observed 2–10 keV flux may be explainable with a secondary-star mass loss rate much smaller than values assumed in published models such as [69] and [66]. But this, in turn, would alter the estimated ratio of wind momenta, which affects the large-scale shock structure, etc. Moreover, the details vary along the secondary star's orbit. In other words, the 2–10 keV flux is *not* a robust component in any model of the 5.5-year cycle and spectroscopic events.

Brief X-ray maxima or flares tended to recur at 85-day intervals in 1997–1998, possibly related to the primary star's rotation and/or pulsation [15]. Since the confidence level was unsatisfying, this phenomenon received little attention and is now practically forgotten – but it has not been disproven. Detecting such a pattern is non-trivial, because models predict systematic drifts in the recurrence interval, not a strict periodicity. The sequence of X-ray flare times preceding the 1997, 2003, and 2009 events may be worth analyzing. This task requires a sophisticated mathematical treatment allowing for quasi-periodicity and differences between the events.

Some spectral changes persist for many months after an event, albeit usually at an undramatic level [95]. Likely reasons involve the varying geometry of gas densities and ionization as the hot secondary star moves along its orbit; see, e.g., [57, 60] and many references therein.

3.4 The Great Eruption

Experts on stellar evolution have been slow to acknowledge disquieting symptoms, hints of major gaps in existing theory that no one has tried hard enough to close. A recent example has been the controversy about steady vs. episodic mass loss (e.g., [23, 70] and older Refs. in [34]). If history is any guide, the present-day consensus is influenced by a collective desire to preserve the validity of dozens of heavily cited publications. A related but much less discussed case has been η Car's Great Eruption of 1830–1860, the prototype supernova impostor event (Fig. 3.2). Apart from a few generalized, qualitative speculations, *stellar astrophysicists have not yet explained the basic facts which were known decades ago* [12, 34].

Unusual problems like these receive little theoretical attention today, not because they are unimportant but because they seem unsuited to existing computer codes! Meanwhile a few promising insights mentioned below have a tantalizingly unfinished air about them. They need and deserve further effort.

As usual let's begin with some quantities. The observed infrared spectrum of the Homunculus, 3–200 μm, indicates a dust mass which should imply the ejecta mass. With plausible assumptions about grain parameters and the dust/gas ratio, one finds that 10–40 $M_{\odot}$ of material was ejected in the Great Eruption [75, 76, 79].[5] Note, however, that the quantitative assumptions are less secure than some authors suggest. Recent accounts note that the gas/dust ratio may be higher than "normal" because oxygen and carbon are scarce in this case; but they give less attention to

[5] Earlier estimates [12] were too small chiefly because they neglected the coolest, far-IR dust.

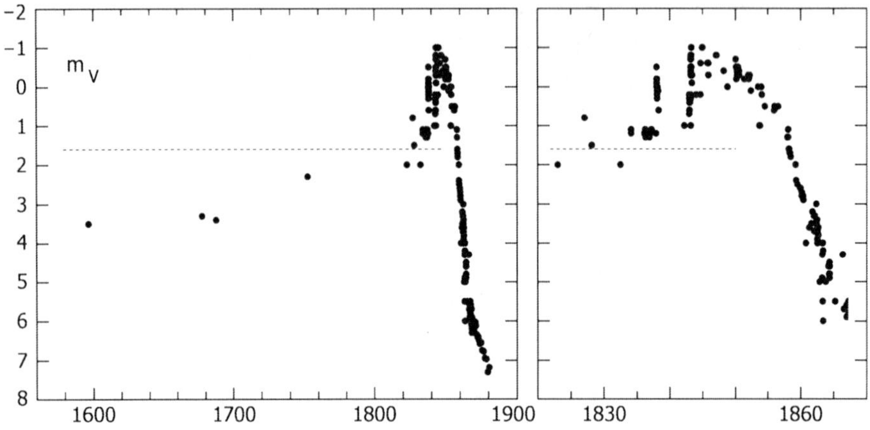

Fig. 3.2 Apparent visual brightness of η Car before 1890, using modern calibrations of the records [22]. Values during the eruption are based on specific comparisons with well-known bright stars. The *horizontal dashed line* shows the maximum that is possible if $L_1 \approx 4.5$ million $L_\odot$

uncertainties in the IR emission efficiency, which may also be abnormal if the grains have unusual compositions and sizes. Theoretical work on grain formation, sizes, and radiative efficiencies in CNO-processed material might be useful here. In one respect the difference between 10 and $40\,M_\odot$ is not great: either value would fill roughly the outer 50–75% of the pre-outburst star in terms of linear radius.

What about energy? Suppose for example that $\Delta M = 15\,M_\odot$ was ejected from a $160\,M_\odot$ star. Then the kinetic energy at $600\,\mathrm{km\,s^{-1}}$ is about 5×10^{49} ergs, and its pre-outburst binding energy, gravitational minus thermal, was very roughly 6×10^{49} ergs. (Since this estimate refers to outer layers, it depends only mildly on the star's internal structure.) The total is roughly proportional to ΔM. To some extent we can also guess the radiative energy release. Using magnitudes listed in [22] and assuming a bolometric correction near zero [7], the record suggests roughly 4×10^{49} ergs of extra radiation between 1835 and 1858, in addition to the star's normal luminosity. This is probably an underestimate, because we've ignored extinction by dust forming in the ejecta. (Is it significant that each of the above three contributions has more or less the same size?) Summary: We should contemplate an energy release of the order of 2×10^{50} ergs in a duration of order 10–20 years. This was roughly 20 times as large as the radiation that would have normally emerged in that time.

The 1596–1870 historical record [12, 22] appears to show three characteristic timescales (Fig. 3.2). First, the star very likely grew in brightness after 1700. Admittedly this claim rests on only one or two reports around 1750, but one of those observations was rather good [22], and, independently, some other supernova impostors or related objects brightened in a qualitatively similar way before erupting [35]. Such brightening indicates a progressively denser wind, but for η Car we cannot tell whether this occurred steadily or sporadically.[6] Either way, a pre-outburst timescale of the order of a century seems likely though not absolutely proven.

Next, the eruption itself was temporally ill-defined. It became noticeable sometime between 1820 and 1837, growing unsteadily [22]. At later times, probable circumstellar extinction makes the end of the event almost impossible to quantify. Given the luminosity and the outward speed of the Homunculus, we expect dust grains to have formed about 3 years after their material was ejected. In this sense maybe we should be surprised that η Car was still bright in the late 1840s, several years after maximum brightness. It certainly did fade progressively after 1850. We simply do not know how long the eruption continued, obscured by dust; very likely something was still happening after 1855. In summary, the duration was somewhere in the range 10–40 years, depending on assumptions and definitions. The thermal timescale for the outer 10–20 $M_\odot$ of the pre-outburst star would have been of the order of 300 years, judging from the energy budget mentioned earlier.

Fluctuations during the event may indicate another, much shorter timescale seldom emphasized in the literature. On one occasion near the peak brightness in 1843, an experienced observer explicitly recorded a temporary 1.5-magnitude brightness drop within a 2-week interval [22]. Lesser fluctuations in intervals of a few weeks were also reported in 1843–1844. Since the observers recorded specific comparisons with bright, well-known stars, we cannot dismiss these changes merely because they seem strangely rapid. Eta Car's opaque-wind photospheric radius must have been 15–30 AU in 1843, implying a characteristic flow time of 40 days or so. Was it possible for the outflow rate, or its morphology or something, to vary fast enough to explain the reported brightness fluctuations? If so, does this tell us anything about the instability? So far this puzzle has gotten practically no attention.

Did the explosion occur in the star's outer layers or in its core? We usually assume the former but a few authors have advocated the latter [12, 29]. In recent years "pulsational pair instability" has attracted attention as a possible mechanism for a pre-SN outburst in a very massive star [94]. Two observed SNae are even known to have had precursor outbursts [24, 67]. Unfortunately the time interval depends on too many factors to be predicted; in this scenario η Car might explode next Thursday, or maybe 1000 years hence. Certain observational facts have been interpreted as evidence for a central explosion. For instance, flow speeds exceeding 2,000 km s^{-1}, possibly up to 6,000 km s^{-1}, occur in ejecta outside the Homunculus lobes [17, 74, 92]; Smith [74] took these to indicate a core-explosion "blast wave." But one senses a likely non sequitur, because *any* outward-moving shock accelerates through the negative density gradient of a star's outer layers and wind. There is no absolute need for it to originate in the core.

Intuitively, at least, three empirical facts appear to disfavor the central-explosion hypothesis. (1) Most of the ejected mass escaped toward fairly high latitudes, in the bipolar Homunculus lobes [15, 73, 79]. This is quite unsurprising for an outer-layer instability which depends both directly and indirectly on surface rotation, but

[6]Some visible outer material may have been ejected before 1800, see Sect. 3.6.

requires special explanation in a central explosion scenario. (Dress rehearsal for a GRB?) (2) As mentioned above, η Car probably brightened in the century preceding the event. This timescale seems reasonable for diffusion and spin-up times related to the emergence of a surface instability (cf. Sect. 3.5 below), but really dramatic instability in an evolved stellar core tends to occur too quickly for the surface to change much beforehand. (3) The same comment also applies to the duration of the eruption. These arguments are not strong enough to prove anything, but they motivate a tentative preference for the outer-layer hypothesis (see, however, Sect. 12.4 in this volume).

A related point concerns the amount of ejected mass. Proposed instabilities in the outer layers generally occur at near-surface depths with much less than $1\,M_\odot$ of overlying material; so how could 10–20$\,M_\odot$ be ejected? A natural answer is the "geyser" analogy: as outer layers successively depart, the instability propagates downward through material that was initially deeper until some structural factor stops it [8, 11, 34, 50].[7] Presumably the removal of each successive layer exposes thermal energy that was formerly trapped, thus causing the surface or sonic point to remain super-Eddington. This old conjecture emphasizes a double question: Why and when does it cease to be true, ending the eruption? The answer probably involves adiabatic expansion, but in any case it is crucial for the overall problem.

The past few years have seen no major additions to the list of near-surface instabilities [26,34,47,68,72]. Leading suspects are the alarmingly complex "strange modes" [25–27]. These do not absolutely depend on opacity, but they tend to be encouraged by opacity peaks such as that which occurs around $T \sim 10^{5.5}$ K. Once the super-Eddington outflow begins, the concept of radiative porosity becomes essential (see [72], and Chap. 12). This phenomenon may itself help to create an opacity-dependent "modified Eddington limit" instability just below the photosphere, qualitatively like an older hypothesis (Sect. 5 in [34]). Alas, even though these ideas are now growing old, *they have not been developed to a state sufficient for a realistic eruption model*. They deserve far more attention than stellar theorists have allowed.

Rotation, obviously pertinent if the bipolar Homunculus is any guide, deserves a few extra comments. The companion star might either stimulate or simulate rotational effects, but that problem is too daunting for now. The simplest initial approach is to pretend that η Car is a single rotating star. Standard accounts of rotation give only minimal guidance for the eruption problem, because they concentrate on steady mass-loss with few predictions of expected rates [20, 47–50, 65]. An eruption, with rapid internal shifting of specific angular momentum, is quite a different proposition. Here are a few puzzles too conspicuous to ignore:

1. *Was rotation essential for the Great Eruption?* (Temporarily ignore the companion star.) The oft-cited "Omega limit" or "$\Omega\Gamma$ limit" merely amounts to supplementing the Eddington fraction Γ with a $R^3\omega^2/GM$ term; this does not

[7] This is the Lagrangian view, of course; members of the Eulerian party say instead that inner layers expand outward through the instability zone. The word "geyser" was first applied in this connection at a meeting in the 1980s – probably [8] – but later was sometimes used as a name for a specific instability model [50]. In fact the analogy is fairly general.

by itself imply a catastrophic event any more than the classical Eddington limit does [34, 49]. If, however, the latitude variations of surface temperature and opacity play a role, then one can imagine various useful phenomena. At least one speculative example has been called "synergistic" [96].

2. *Why was the mass ejection essentially polar,* rather than, say, equatorial? At first sight this looks like the question discussed for stellar winds in Refs. [20, 65, 78]. But in fact it is more complex, because η Car's eruption rapidly ejected material from most of the volume including zones down to $r \lesssim 0.5R_0$. No doubt radiation was channeled toward the poles because those were the easiest escape routes, von Zeipel style; but does this idea lead not only to an enhanced wind but to eruptive instability? Or, alternatively, is bipolarity a red herring?
3. *How significant is the angular momentum budget?* Mass was ejected mainly toward latitudes within 35° of the poles, less than 18% of the spherical solid angle [16, 73]. Since most of this mass came from initial radii $r \sim 0.5R_0$ in the star, we have a near-paradox or at least a clue. For illustration, imagine a massive star with fairly rapid solid-body rotation. Suppose that it suddenly loses, e.g., the outermost 1% of its mass, which filled roughly the outer 35% of its radius. Material just inside that limit then expands on a dynamical timescale, $\lesssim$30 days, to almost the original radius R_0 – thus reducing the surface rotation speed so rotational effects decrease by a factor of 3 or so. Hence any further mass ejection should be comparatively spherical. (Subsequent spin-up of the outer layers will take a much longer time.) Since η Car continued to shed 5–20% in a highly latitude-dependent fashion, obviously its story must be different. As an opposite case, imagine a loss of polar material carrying little angular momentum. Angular momentum per unit mass left within the star thereby increases, accentuating rotational effects. Since these must have been substantial to begin with (otherwise why should the initial ejection have been polar?), the situation will soon get out of hand. Overall implication: The escaping angular momentum per unit mass must have been regulated so that it became neither too large nor too small. Note, again, that this statement follows from the large amount of rapidly ejected mass ΔM, in contrast to a snapshot of an ordinary latitude-dependent stellar wind. It may be a clue to the initial distribution of angular momentum, the nature of the instability, the resulting internal flow pattern, or all three. Very likely these will eventually suggest a reason why the eruption later stopped.
4. *Did the companion star cause the Great Eruption?* This possibility has been obvious ever since the 5.5-year cycle was discovered [4, 9, 44, 46, 76]. Intuitively it makes sense: one can imagine the primary star expanding and approaching some outer instability as it evolved, while the secondary had little effect at most times but came dangerously near at periastron, etc. Tidal effects apply even more forcefully to a latent stellar instability than they do for recent spectroscopic events (Sect. 3.3 above), and some of their consequences may be analogous to rotation. But all attempts to correlate the eruption time with the 5.5-year cycle are highly speculative, because *we do not have enough information about the mass-loss history between 1840 and 1950 to calculate the orbit evolution.*

Consider, for instance, the Great Eruption itself. With a constant mass-loss rate throughout the cycle, an orbit's semi-major axis a and period P increase so that $a'/a \approx M/M'$ and $P'/P \approx (M/M')^2$ with little change in the eccentricity. (M is the total mass of the system, and post-event quantities are primed.) But if the mass loss is concentrated near periastron in an eccentric orbit, then these effects become far larger and the eccentricity grows.[8] Altogether these facts imply a period less than 5 years, possibly less than 4 years, before 1840; and a substantial fraction of the change may have occurred after the 1843 peak. Therefore we cannot meaningfully "predict" the timing in 1843, say, based on the present orbit. Moreover, since the mass loss and eccentricity interact in a fairly complicated way and we do not know dM/dt as a function of time in 1840–1870, we cannot deduce much from the brightness record in Fig. 3.2. The orbit continued to evolve afterward. Since the total mass lost after 1900 may have been anywhere in the range 0.05–0.5 $M_\odot$ [36], and dM/dt may have varied with phase in the orbit, the period in 1900 may have been as short as 2000 days or as long as 2023 days; which implies a range of more than half a year in the time of periastron passage closest to the 1890 second eruption (Sect. 3.5). This fundamental uncertainty has no bearing on the credibility of a periastron-triggered eruption, but it eliminates the most obvious method of proof. The basic hypothesis remains very appealing but there is no clear way to confirm or disprove it. It obviously deserves attention.

Provocation by the secondary star may help to account for the long duration of η Car's event. Other giant eruptions or supernova impostor events typically end after a year or so, anyway less than 5 years, see [35] and Chap. 11. In the case of η Car the secondary may have triggered the event earlier than would have occurred with a single star, and then may have stirred the instabilities in recurrent periastron visits.[9]

Two more remarks in this vein: (1) The velocity kick given to the system by mass ejection is probably undetectable. With reasonable parameters the center-of-mass velocity may have been changed by 5 km s^{-1} or so. (2) The Great Eruption cannot by itself account for the full eccentricity of the present-day orbit. With a periastron distance less than 3 AU, one naturally suspects that the orbit was once circular. Even with the most favorable assumptions, however, the system would have needed to lose at least 90 $M_\odot$ in order to increase the eccentricity from $\varepsilon \approx 0$ to the current value $\varepsilon \gtrsim 0.85$ (cf. [44]). Estimates of the Homuculus mass don't go that high. Is this a hint of earlier eruptions hundreds or thousands of years ago?

Distinctions between "massive" and "very massive" stars include qualitative physical effects. For instance, stellar activity may play a role in η Car as though it had a cooler photosphere. Convection becomes very easy to induce near the

[8] These statements neglect the small fraction of material that falls onto the secondary star rather than escaping from the system [44, 82].

[9] The basic suggestion, that η Car's companion star caused its eruption to persist longer than normal for a supernova impostor, is due to R. M. Humphreys (unpublished).

Eddington Limit – especially in zones with substantial absorption opacity – while strong interior turbulence arises from shear rotation and possibly tidal effects. This line of thought implies magnetic fields as well. Therefore we should not be very surprised if analogs to solar flares, prominences, etc., exist at the base of the wind, even at temperatures well above 10^4 K. They might be energetic enough to help drive the wind or at least to influence its structure, e.g., regarding inhomogeneities and the asymmetrical flows noted below.

Finally, concerning the Great Eruption, we mustn't forget the Homunculus' famous equatorial skirt, unusual among bipolar nebulae. This isn't merely an example of latitude-dependent bistability [45]. Instead of being circularly symmetric, the equatorial structure has radial streamers, rays, and jets [19, 28, 55]. The ragged, irregular morphology near the Homunculus midplane may signal an undiagnosed aspect of the eruption physics [2]. Asymmetric local flows may result, e.g., from periastron tidal forces, or from stellar quasi-activity as mentioned above; this theoretical territory has scarcely been explored. The same can be said of the large-scale bipolar structure's obvious deviations from axial symmetry [16, 63, 80].[10]

Even more peculiar are some radial "strings" of material outside the Homunculus [56, 91], having Doppler speeds consistent with ejection during or earlier than the Great Eruption; see Chap. 8. They are practically as narrow as possible (age × speed of sound) and they resemble aircraft contrails in the sense that they are not exactly straight. The word "bullets" has repeatedly been invoked as a way to form a string, but it doesn't explain why they exist. This is another astrophysical puzzle in serious danger of being forgotten without having been solved.

Rest et al. [71] have detected a light-echo spectrum of η Car's Great Eruption. They assert that its temperature and other details contradict existing views of the nature of that event. In fact, however, the deduced temperature is ill-defined and the spectrum appears reasonably consistent with what one expects from Fig. 1 in [7]. The arguments are too lengthy to review here.

3.5 Aftermath and Recovery

The observational record suggests that η Car is still recovering *in an unsteady manner,* which ought to tell us something (Fig. 3.3). Elementary considerations give only vague pointers to the expected timescale. If extra energy to power the eruption came from the star's outer surviving layers, then a strong entropy gradient

[10] The two-lobes-plus-skirt morphology even had an effect on popular culture. Soon after color HST pictures of the Homunculus appeared in the mid-1990s, space explosions in the revised *Star Wars* and other Hollywood films suddenly became bipolar *with equatorial skirts;* and fireworks manufacturers followed suit. Ironically, some of the most conspicuous "equatorial" features in images of the Homunculus later turned out to be illusory, but Doppler velocities show that there really is an equatorial structure [16].

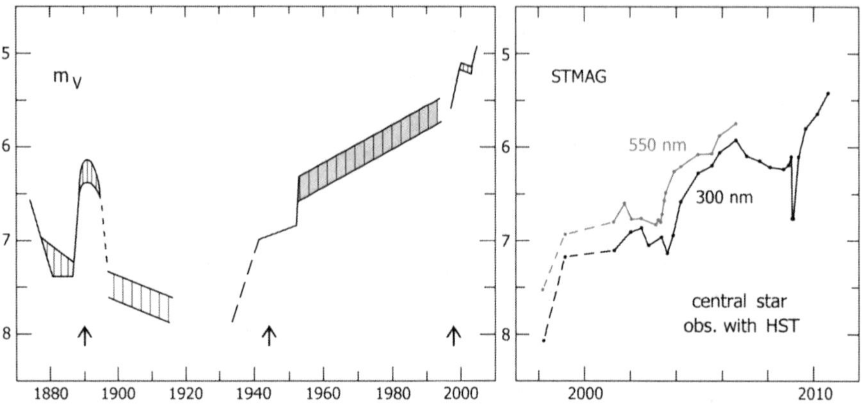

Fig. 3.3 Visual-wavelength photometry of η Car from 1870 to 2010, (Adapted from [36, 52, 60]). See also a similar figure in the Chap. 1. Ground-based data (*left*) include ejecta plus central star, while HST data (*right*) refer only to the central star or rather its opaque wind. *Arrows* point to episodes described in the text

must have resulted, causing extremely vigorous convection. If the "normal" thermal timescale for the most relevant zones was 100–300 years (Sect. 3.4 above), and the post-event convection transported heat much faster than normal, then we can reasonably expect basic thermal recovery to occur in much less than 100 years – say 10–50 years, perhaps. The event should have altered the rotation and magnetic fields, since the outermost surviving layers had to expand by a factor of order 2. Subsequent readjustments shouldn't be much slower than a turbulent diffusion timescale. According to fluid lore, we can very crudely assess this quantity by calculating a sort of equivalent viscosity ν_{eq} which makes a Reynolds number of order 300, and then interpreting ν_{eq} as a diffusion coefficient. Thus

$$\Re = \frac{\ell u}{\nu_{\mathrm{eq}}} \sim 300 \qquad \longrightarrow \qquad t_{\mathrm{diff}} \sim \frac{\ell^2}{\nu_{\mathrm{eq}}} \sim 300\,\frac{\ell}{u}\,, \tag{3.2}$$

where ℓ, u are a local characteristic length and relative speed, e.g. in a rotational shear flow. If we choose $\ell \approx 0.5$ AU and $u \approx 30\,\mathrm{km\,s^{-1}}$ – plausible values for the outer $0.5R_*$ after the eruption – then $t_{\mathrm{diff}} \sim 25$ years. Summary: If forced to predict the recovery timescale with no additional information, one would most likely guess a value appreciably less than the time that has now elapsed since the eruption. But this discrepancy is less than an order of magnitude, almost satisfying in view of the underlying complexity.[11]

[11] If we informally visualize the recovery as something like $1-\exp(-t/t_c)$, then several timescales of t_c may indeed have elapsed since 1860.

Figure 3.3 shows three extraordinary episodes since 1880: an LBV-like outburst around 1890, a sparsely-observed change of state in the 1940s, and another that began in the 1990s. Varying extinction, reflection, and emission in the ejecta make the photometric record hard to interpret, but each episode was also marked by spectroscopic changes. Five distinct eras have spanned 120 years:

1. First, the "little eruption" around 1890. In the absence of circumstellar extinction from the Great Eruption, it would probably have attained $m_V \approx 1.6$ [32, 35]; we know this because the spectrum temporarily resembled an F-type supergiant star, signaling an opaque wind with bolometric correction near zero [34, 36, 89]. The 1890 event produced a bipolar "Little Homunculus" nebula inside the big one, plus equatorial debris and most likely the slow "Weigelt condensations," altogether 0.1–1 $M_\odot$ – see [16, 39] and Chaps. 5–7. It occurred about one thermal or diffusion timescale (in some sense) after the Great Eruption, or $\sim 10^3$ dynamical timescales.
2. In 1900–1940 Eta Car had a rich emission line spectrum. A naive textbook-style estimate of the circumstellar extinction – with the known mass distributed uniformly in a shell, a conventional dust/gas ratio, etc. – would have predicted a visual brightness below 10th magnitude but in fact it was much brighter than that. Modern images of the Homunculus show "granulation" or porous local structures, allowing light to escape preferentially along the easiest paths between density maxima. No one has yet attempted to model this situation quantitatively.
3. Something like a change of state occurred between 1939 and 1953. The brightness rapidly grew, culminating abruptly in 1951–1953, see Refs. in [36]. Meanwhile He I and other high-excitation emission lines first appeared [21]. These are indirectly excited by UV from the secondary star, and their absence before 1940 is not easy to explain [9, 36]. Maybe the primary wind was then dense enough to force accretion onto the secondary star, thereby lowering its $T_{\rm eff}$ and reducing its supply of relevant far-UV photons [60, 83]. If true this account implies a mass loss rate considerably above $10^{-3}\,M_\odot$ year^{-1} before 1940, possibly $\sim 10^{-2}\,M_\odot$ year^{-1}. A conventional line-driven wind cannot attain such high values; continuum driving and perhaps other processes must have played major roles.
4. For about 40 years after 1953, η Car brightened at an average rate of about 0.02 magnitude per year, with temporary excursions smaller than ± 0.3 magnitude. Most of the emergent light was scattered by dust in the Homunculus, though, rather than coming directly from the star. The 1953–1993 brightening is customarily attributed to expansion of the Homunculus, but no one has attempted a realistic model of radiation scattering through the inhomogeneous structure. Note, by the way, that all spectroscopy of η Car before 1991 was *strongly* contaminated by light scattered or emitted in the ejecta. The spectrum of the central star itself – strictly speaking its opaque wind – was not observed until HST became available [14].

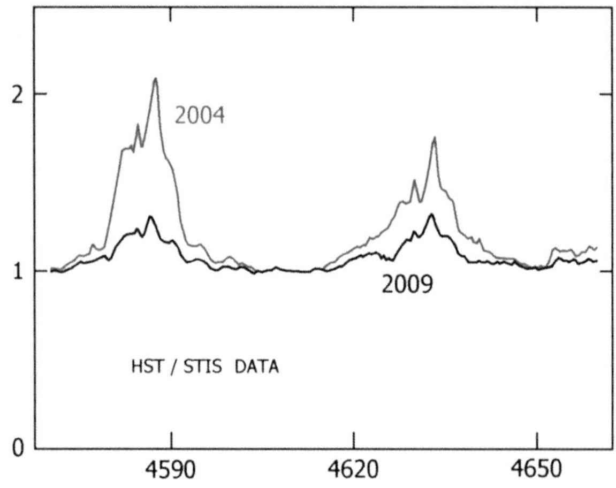

Fig. 3.4 Secular weakening of emission lines in η Car's wind, (Adapted from [58]). These are mostly permitted Fe II lines. Both spectra represent nearly the same phase in the 5.5-year cycle, normalized to have the same continuum level

5. Another episode of rapid change began sometime between 1993 and 1998. In this case the HST made it possible to observe the central stellar wind, whose continuum seemed to brighten by more than a magnitude in only several years (Fig. 3.3, and [52, 53]). Then *the stellar wind emission lines weakened dramatically between 2004 and 2010* – the most remarkable development in this topic in recent years. See Fig. 3.4 and [58,60]. An obvious, but not entirely simple interpretation is that the primary star's mass loss rate has rapidly decreased by a factor of order 3. This implies increased UV radiation, tending to either destroy or inhibit the formation of circumstellar dust grains. According to this interpretation, the apparent brightening is due mainly to decreased circumstellar extinction; the intrinsic continuum has shifted somewhat toward the UV; and the stellar wind emission lines really have weakened substantially. If so, then the wind has at last begun to approach again the state that Halley saw before 1700, Sect. 3.2 above. Indeed, judging from the total amount of change, it may already have reached that condition.

It is a curious fact that η Car has exhibited strange behavior – changes of state, one might say – at intervals of about 54 years: circa 1836, 1890, 1944, and 1998 [32, 35]. This is far too long to be a dynamical wave, but might it be a subtle reverberation timescale in thermal or diffusion phenomena in the outer layers? If so its value, $\sim$50 years and not 25 or 100, may indicate something about the structure.

Even if we ignore the above speculation, the timescale and especially the unsteady character of η Car's "recovery" should provide structural information that is not available in other very massive stars. Yet they have received practically *no* attention among theorists. We do not even know why a huge mass-loss rate – probably too high for a line-driven wind – persisted a century after the Great Eruption.

3.6 Other Puzzles, and Summary

Did η Car suffer earlier explosions? Existing data contain shadowy hints of ejection events before 1800 [64, 90, 96]. It's hard to be sure of these, because one can easily imagine effects that can produce illusory ejection dates, and also because we would need to fit the suspected events into the photometric record [22]. The present-day highly eccentric orbit may also be a hint of considerable mass loss before 1800, as mentioned in Sect. 3.4. Unlike most of the problems featured above, this one needs observational rather than theoretical work.

...And, speaking of "shadowy," here's an archaeological connection. One of the nineteenth century early interpreters of cuneiform suggested that η Car erupted around 5,000 years ago, based on various clues plus a well-known Sumerian myth associated with the city of Eridu [13, 34, 40]. Eighty years later a popular-level author linked the same myth to the Vela SN [62], but the η Car version is far more believable in terms of historical chronology. Altogether this riddle is extraordinary because, although twentieth century archaeology provided no further support, it now seems astrophysically credible in ways that the original author could not foresee! Some sort of miraculous discovery would be delightful, but, alas, the conjecture was more likely wrong than right, and by now it would be very hard either to prove or to disprove.

We *can* do something about the near-future observational record. As noted in Sect. 3.5, η Car's brightness and spectrum have changed dramatically since 2003. This trend may continue or it may already have reached completion. Spectroscopy and UV photometry *of the central wind* are essential, i.e., this is a job for HST and other instruments with spatial resolution better than 0.3 arcsec. As this is written in late 2011, however, well over a year has passed since the last UV photometry and there seems little prospect that more will be done in the forseeable future. This fact is a consequence of decisions made by telescope allocation committees – cf. a parallel instance on p. 224 of [5] – but its effect on the observational record will be permanent. The current behavior cannot be observed again 5, 10, or 40 years from now; and no other object is likely to provide a substitute.

Meanwhile, as emphasized above, the theory of very massive star evolution in general cannot be secure until η Car's mass loss history has begun to be understood. Ultimately it may prove to be merely an abnormal special case; but so far there is no strong evidence to that effect. For all we know, for example, the high masses attributed to Pop III stars might be impossible due to the same eruptive instability. A potential theoretical pitfall this big obviously merits attention.

References

1. M. Akashi, A. Kashi, N. Soker, Accretion of dense clumps in the periastron passage of η Carinae. arXiv:1106.2438
2. E. Artigau, J.C. Martin, R.M. Humphreys et al., Penetrating the Homunculus – Near-infrared adaptive optics images of η Carinae. Astr. J. **141**, 202(2011)

3. J.M. Blondin, I.R. Stevens, T.R. Kallman, Enhanced winds and tidal streams in massive X-ray binaries. Ap. J. **371**, 684–695 (1991)
4. A. Damineli, The 5.52 year cycle of η Carinae. Ap. J. Lett. **460**, L49–L52 (1996)
5. A. Damineli, O. Stahl, B. Wolf, A. Kaufer, F.J. Jablonski, The historical evidence for the 5.52-year cycle. in *Eta Carinae at the Millennium*, ed. by J.A. Morse, R.M. Humphreys, A. Damineli. ASP Conference Series, vol. 179 (ASP, San Francisco, 1999), pp. 221–226
6. A. Damineli, D.J. Hillier, M.F. Corcoran et al., The periodicity of the η Carinae events. Mon. Not. Roy. Astr. Soc. **384**, 1649–1656 (2008)
7. K. Davidson, The relation between apparent temperature and mass-loss rate in hypergiant eruptions. Ap. J. **317**, 760–764 (1987)
8. K. Davidson, Plinian eruptions a la η Carinae. in *Physics of Luminous Blue Variables*, ed. by K. Davidson, A.F.J. Moffat, H.J.G.L.M. Lamers. IAU Colloq. vol. 113 (Kluwer, Dordrecht, 1989), pp. 101–107
9. K. Davidson, Why the binary hypothesis isn't a panacea. in *Eta Carinae at the Millennium*, ed. by J.A. Morse, R.M. Humphreys, A. Damineli. ASP Conference Series, vol. 179 (ASP, San Francisco, 1999), pp. 304–315
10. K. Davidson, Chandra meets η Carinae. in *The High Energy Universe at Sharp Focus*, ed. by E.M. Schlegel, S.D. Vrtilek. ASP Conference Series, vol. 262 (ASP, San Francisco, 2002), pp. 267–274
11. K. Davidson, The physical nature of η Carinae. in *The Fate of the Most Massive Stars*, ed. by R.M. Humphreys, K.Z. Stanek. ASP Conference Series, vol. 332 (ASP, San Francisco, 2005), pp. 101–109
12. K. Davidson, R.M. Humphreys, Eta Carinae and its environment. Ann. Rev. Astron. Astrophys. **35**, 1–32 (1997)
13. K. Davidson, K. Davidson, R.J. Dufour, N.R. Walborn, T.R. Gull, Ultraviolet and visual wavelength spectroscopy of gas around η Carinae. Ap. J. **305**, 867–879 (1986)
14. K. Davidson, D. Ebbets, G. Weigelt et al., HST/FOS spectroscopy of η Carinae: the star itself, and ejecta within 0.3 arcsec. Astr. J. **109**, 1784–1796 (1995)
15. K. Davidson, K. Ishibashi, M.F. Corcoran, The relationship between two periodicities observed in η Carinae. New Astr. **3**, 241–245 (1998)
16. Davidson, K., Smith, N., Gull, T.R., Ishibashi, K., Hillier, DJ.: The shape an orientation of the Homunculus nebula based on spectroscopic velocities. Astr. J. **121**, 1569–1577 (2001)
17. R.J. Dufour, K. Davidson, N.R. Walborn, Kinematics of the η Car condensations. Bull. Am. Astr. Soc. **17**, 884 (1985)
18. R.J. Dufour, T.W. Glover, J.J. Hester, D.G. Currie, D. van Orsow, D.K. Walter, HST-FOS UV-optical spectra of ejecta from η Carinae. in *Eta Carinae at the Millennium*, ed. by J.A. Morse, R.M. Humphreys, A. Damineli. ASP Conference Series, vol. 179 (ASP, San Francisco, 1999), pp. 134–143
19. W.J. Duschl, K.-H. Hofmann, F. Rigaut, G. Weigelt Morphology and kinematics of η Carinae. Rev. Mex. Astr. Astrof. Ser. Conf. **2**, 17–22 (1995)
20. V.V. Dwarkadas, S.P. Owocki, Radiatively driven winds and the shaping of bipolar luminous blue variable nebulae. Ap. J. **581**, 1337–1343 (2002)
21. M. Feast, P. Whitelock, F. Marang, Variability of η Carinae – III. Mon. Not. Roy. Astr. Soc. **322**, 741–748 (2001)
22. D.J. Frew, The historical record of η Carinae, I. The visual light curve, 1595–2000. J. Astr. Data **10**(6), 1–76 (2004)
23. A.W. Fullerton, D.L. Massa, R.K. Prinja, The discordance of mass-loss estimates for Galactic O-type stars. Ap. J. **637**, 1025–1039 (2006)
24. A. Gal-Yam, D.C. Leonard, A massive hypergiant star as the progenitor of SN 2005gl. Nature **458**, 865–867 (2009)
25. W. Glatzel, On the origin of strange modes and the mechanism of related instabilities. Mon. Not. Roy. Astr. Soc. **271**, 66–74 (1994)
26. W. Glatzel, Instabilities in the most massive evolved stars. in *The Fate of the Most Massive Stars*, ed. by R.M. Humphreys, K.Z. Stanek. ASP Conference Series, vol. 332 (ASP, San Francisco , 2005), pp. 22–32

27. W. Glatzel, M. Kiriakidis, Stability of massive stars and the Humphreys/Davidson limit. Mon. Not. Roy. Astr. Soc. **263**, 375–384 (1993)
28. T.W. Glover, R.J. Dufour, J.J. Hester et al., HST FOS spectroscopy of η Carinae's northeast jet. Rev. Mex. Astr. Astrof. Ser. Conf. **7**, 158–162 (1998)
29. J.A. Guzik, A.N. Cox, K.M. Despain, Pulsation-driven outflows in luminous blue variables. in *Eta Carinae at the Millennium*, ed. by J.A. Morse, R.M. Humphreys, A. Damineli. ASP Conference Series, vol. 179 (ASP, San Francisco, 1999) pp. 347–353
30. J.F.W. Herschel, *Results of Astronomical Observations Made in the Years 1834–1838, at the Cape of Good Hope* (Smith, Elder, & Co., London, 1847)
31. D.J. Hillier, K. Davidson, K. Ishibashi, T.R. Gull, On the nature of the central source in η Carinae. Ap. J. **553**, 837–860 (2001)
32. R.M. Humphreys, Eta Carinae – the observational story, 1600 to 2004. in *The Fate of the Most Massive Stars*, ed. by R.M. Humphreys, K.Z. Stanek. ASP Conference Series, vol. 332 (ASP, San Francisco, 2005), pp. 14–21
33. R.M. Humphreys, K. Davidson, Studies of luminous stars in nearby galaxies. III – comments on the evolution of the most massive stars in the milky way and the large magellanic cloud. Ap. J. **232**, 409–420 (1979)
34. R.M. Humphreys, K. Davidson, The luminous blue variables: astrophysical geysers. Publ. Astr. Soc. Pac. **106**, 1025–1051 (1994)
35. R.M. Humphreys, K. Davidson, N. Smith, Eta Carinae's second eruption and the light curves of the η Carinae variables. Publ. Astr. Soc. Pac. **111**, 1124–1131 (1999)
36. R.M. Humphreys, K. Davidson, M. Koppelman, The early spectra of η Carinae 1892 to 1941 and the onset of its high excitation emission spectrum. Astr. J. **135**, 1249–1263 (2008)
37. K. Ishibashi, M.F. Corcoran, K. Davidson, S.A. Drake, J.H. Swank, R. Petre, Resolving X-ray temporal variations in η Carinae. in *Eta Carinae at the Millennium*, ed. by J.A. Morse, R.M. Humphreys, A. Damineli. ASP Conference Series, vol. 179 (ASP, San Francisco, 1999), pp. 266–274
38. K. Ishibashi, M.F. Corcoran, K. Davidson, J.H. Swank, R. Petre, S.A. Drake, A. Damineli, S. White, Recurrent X-ray emission variations of η Carinae and the binary hypothesis. Ap. J. **524**, 983–987 (1999)
39. K. Ishibashi, T.R. Gull, K. Davidson, Discovery of a little homunculus within the homunculus nebula of η Carinae. Astr. J. **125**, 3222–3236 (2003)
40. P.C.A. Jensen, in *Die Kosmologie der Babylonier* (Trübner, Strassburg, 1890), pp. 24–28
41. A. Kashi, N. Soker, Possible implications of mass accretion in η Carinae. New Astr. **14**, 11–24 (2009)
42. A. Kashi, N. Soker, Explaining the early exit of η Carinae from its 2009 X-ray minimum with the accretion model. Ap. J. **701**, L59–L62 (2009)
43. A. Kashi, N. Soker, Using X-ray interactions to explore the binary interaction in η Carinae. Mon. Not. Roy. Astr. Soc. **397**, 1426–1434 (2009)
44. A. Kashi, N. Soker, Periastron passage triggering of the 19th century eruptions of η Carinae. Ap. J. **723**, 602–611 (2010)
45. H.J.G. Lamers, A.W.A. Pauldrach, The formation of outflowing disks around early-type stars by bi-stable radiation driven winds. Astr. Astrophys. **244**, L5–L8 (1991)
46. M. Livio, J.E. Pringle, Can η Carinae be a triple system? Mon. Not. Roy. Astr. Soc. **295**, L59–L60 (1998)
47. A. Maeder, G. Meynet, The evolution of rotating stars. Annu. Rev. Astron. Astrophys. **38**, 143–190 (2000)
48. A. Maeder, G. Meynet, Stellar evolution with rotation. Astr. Astrophys. **361**, 159–166 (2000)
49. A. Maeder, G. Meynet, Mass loss and the evolution of massive stars. in *Mass Loss from Stars and the Evolution of Stellar Clusters*, ed. by A. de Koter, L. Smith, R. Waters. ASP Conference Series, vol. 388 (ASP, San Francisco, 2008), pp. 3–19.
50. A. Maeder, G. Meynet, R. Hirschi, Evolution of the most massive stars. in *The Fate of the Most Massive Stars*, ed. by R.M. Humphreys, K.Z. Stanek. ASP Conference Series, vol. 332 (ASP, San Francisco, 2005), pp. 3–12

51. J.C. Martin, K. Davidson, R.M. Humphreys, D.J. Hillier, K. Ishibashi, On the He II emission in η Carinae and the origin of its spectroscopic events. Ap. J. **640**, 474–490 (2006)
52. J.C. Martin, K. Davidson, M.D. Koppelman, The chrysalis opens? Photometry from the HST η Carinae treasury project, 2002–2006. Astr. J. **132**, 2717–2728 (2006)
53. J.C. Martin, K. Davidson, R.M. Humphreys, A. Mehner, Mid-cycle changes in η Carinae. Astr. J. **139**, 2056–2065 (2010)
54. F. Martins, D. Schaerer, D.J. Hillier, A new calibration of stellar parameters of Galactic O stars. Astr. Astrophys. **436**, 1049–1065 (2005)
55. J. Meaburn, G. Gehring, J.R. Walsh et al., An episodic jet from η Carinae. Astr. Astrophys. **276**, L21–L24 (1993)
56. J. Meaburn, P. Boumis, J.R. Walsh et al., Highly supersonic motions within the outer features of the η Carinae nebulosity. Mon. Not. Roy. Astr. Soc. **282**, 1313–1320 (1996)
57. A. Mehner, K. Davidson, G.J. Ferland, R.M. Humphreys, High-excitation emission lines near η Carinae, and its likely companion star. Ap. J. **710**, 729–742 (2010)
58. A. Mehner, K. Davidson, G.J. Ferland, A sea change in η Carinae. Ap. J. Lett. **717**, L22–L25 (2010)
59. A. Mehner, K. Davidson, G.J. Ferland, N II $\lambda\lambda$5668-5712, a new class of spectral features in η Carinae. Ap. J. **737**, 70–XXX (2011)
60. A. Mehner, K. Davidson, J.C. Martin, R.M. Humphreys, I. Kazunori, G.J. Ferland, Critical differences and clues in η Car's 2009 event. Ap. J. **740**, 2 (2011)
61. G. Meynet, A. Maeder, G. Schaller, D. Schaerer, C. Charbonnel, Grids of massive stars with high mass loss rates. Astr. Astrophys. Suppl. Ser. **103**, 97–105 (1994)
62. G. Michanowsky, *The Once and Future Star: The Mysterious Vela X Supernova and the Origin of Civilizations* (Barnes and Noble, New York, 1979)
63. J.A. Morse, K. Davidson, J. Bally et al., Hubble space telescope wide field planetary camera observations of η Carinae. Astron. J. **116**, 2443–2461 (1998) [See also erratum with better images, Astr. J. **117**, 1949–1955 (1999).]
64. J.A. Morse, J.R. Kellogg, J. Bally et al., Hubble space telescope proper-motion measurements of the η Carinae nebula. Ap. J. **548**, L207–L211 (2001)
65. S. Owocki, K.G. Gayley, The physics of stellar winds near the Eddington Limit. in *Luminous Blue Variables: Massive Stars in Transition*, ed. by A. Nota, H.J.G.L.M. Lamers. ASP Conference Series, vol. 120 (ASP, San Francisco, 1997), pp. 121–127
66. E.R. Parkin, J.M. Pittard, M.F. Corcoran, K. Hamaguchi, I.R. Stevens, 3D modeling of the colliding winds in η Carinae – evidence for radiative inhibition. Mon. Not. Roy. Astr. Soc. **394** 1758–1774 (2009)
67. A. Pastorello, S.J. Smartt, S. Mattila et al., A giant outburst two years before the core collapse of a massive star. Nature **447**, 829–832 (2007)
68. A.W.A. Pauldrach, J. Puls, Radiation-driven winds of hot stars. VIII – The bistable wind of P Cygni. Astr. Astrophys. **237**, 409–424 (1990)
69. J.M. Pittard, M.F. Corcoran, In hot pursuit of the hidden companion of η Carinae. Astr. Astrophys. **383**, 636–647 (2002)
70. J. Puls, J.S. Vink, F. Najarro, Mass loss from hot massive stars. Astr. Astrophys. Rev. **16**, 209–325 (2008)
71. A. Rest et al., Light echoes reveal an unexpectedly cool η Carinae during its 19th-century Great Eruption. Nature, in press (2012)
72. N.J. Shaviv, The porous atmosphere of η Carinae. Ap. J. **532**, L137–K140 (2000)
73. N. Smith, The structure of the homunculus. I. Shape and latitude dependence from H_2 and [Fe II] velocity maps of η Carinae. Ap. J. **644**, 1151–1163 (2006)
74. N. Smith, A blast wave from the 1843 eruption of η Carinae. Nature **455**, 201–203 (2008)
75. N. Smith, G.J. Ferland, The Structure of the homunculus. II. Modeling the physical conditions in η Carinae's molecular shell. Ap.J. **655**, 911–919 (2007)
76. N. Smith, D.J. Frew, A revised light curve of η Carinae and the timing of close periastron encounters. Mon. Not. R. Astron. Soc. **415**, 2009–2019 (2011)

77. N. Smith, S.P. Owocki, On the role of continuum-driven eruptions in the evolution of very massive stars and Population III stars. Ap. J. **645**, L45–L48 (2006)
78. N. Smith, K. Davidson, T.R. Gull, K. Ishibashi, D.J. Hillier, Latitude-dependent effects in the stellar wind of η Carinae. Ap. J. **586**, 432–450 (2003)
79. N. Smith, R.D. Gehrz, P.M. Hinz et al., Mass and kinetic energy of the Homunculus nebulae around η Carinae. Astr. J. **125**, 1458–1466 (2003)
80. N. Soker, The departure of η Carinae from axisymmetry and the binary hypothesis. Mon. Not. Roy. Astr. Soc. **325**, 584–588 (2001)
81. N. Soker, Accretion-indiced collimated fast wind model for η Carinae. Ap. J. **597**, 513–517 (2003)
82. N. Soker, The binarity of η Carinae and its similarity to related astrophysical objects. Ap. J. **619**, 1064–1071 (2005)
83. N. Soker, Accretion onto the companion of η Carinae during the spectroscopic event. IV. The disappearance of highly ionized lines. Ap. J. **661**, 482–489 (2007)
84. N. Soker, E. Behar, Accretion onto the companion of η Carinae during the spectroscopic event. III. The He II λ4686 line. Ap. J. **652**, 1563–1571 (2006)
85. I.R. Stevens, T.R. Kallman, X-ray illuminated stellar winds – Ionization effects in the radiative driving of stellar winds in massive X-ray binary systems. Ap. J. **365**, 321–331 (1990)
86. I.R. Stevens, J.M. Pittard, Eta Carinae: a colliding wind binary? in *Eta Carinae at the Millennium*, ed. by J.A. Morse, R.M. Humphreys, A. Damineli. ASP Conference Series, vol. 179 (ASP, San Francisco, 1999), pp. 295–303
87. I.R. Stevens, J.M. Blondin, A.M.T. Pollock, Colliding winds from early-type stars in binary systems. Ap. J. **386**, 265–287 (1992)
88. N.R. Walborn, The stellar content of the Carina Nebula. Rev. Mex. Astr. Astrof. Ser. Conf. **2**, 51–55 (1995)
89. N.R. Walborn, M.H. Liller, The earliest spectroscopic observations of η Carinae and its interaction with the Carina Nebula. Ap. J. **211**, 181–183 (1977)
90. N.R. Walborn, B.M. Blanco, A.D. Thackeray, Proper motions in the outer shell of η Carinae. Ap. J. **219**, 498–503 (1978)
91. K. Weis, W.J. Duschl, Y.-H. Chu, The nature of strings in the nebula around η Carinae. Astr. Astrophys. **349**, 467–474 (1999)
92. K. Weis, W.J. Duschl, D.J. Bomans, High velocity structures in, and the X-ray emission from the LBV nebula around η Carinae. Astr. Astrophys. **367**, 566–576 (2001)
93. P.A. Whitelock, M.W. Feast, C. Koen, G. Roberts, B.S. Carter, Variability of η Carinae. Mon. Not. Roy. Astr. Soc. **270**, 364–372 (1994)
94. S.E. Woosley, S. Blinnikov, A. Heger, Pulsational pair instability as an explanation for the most luminous supernovae. Nature **450**, 390–392 (2007)
95. R. Zanella, B. Wolf, O. Stahl, Spectroscopy of the shell episode of Eta Car (1981–1983). Astr. Astrophys. **137**, 79–84 (1984)
96. T. Zethson, S. Johansson, K. Davidson, R.M. Humphreys, K. Ishibashi, D. Ebbets, Strange velocities in the equatorial ejecta of η Carinae. Astr. Astrophys. **344**, 211–220 (1999)

Chapter 4
The Winds of Eta Carinae and Other Very Luminous Stars

Francisco (Paco) Najarro and D. John Hillier

Abstract The winds of Eta Carinae, other related very luminous stars, and candidate LBVs are reviewed. While ground-based spectra of η Car are influenced by multiple emission regions, HST observations by the Space Telescope Imaging Spectrograph allow us to investigate the primary star's spectrum in depth; the spectrum that is of its opaque stellar wind. Detailed modeling of the primary star and its dense wind is discussed. We illustrate the influence of the mass-loss rate on the spectrum of the primary star, and show that an accurate estimate for the hydrostatic radius cannot be obtained. Deficiencies in the model are highlighted and we discuss the influence of the binary companion and possible departures of the wind from spherical symmetry. A review of the stellar and wind properties of other very luminous Galactic stars is included to illustrate that η Car is not unique and has some close relatives.

4.1 Introduction

All evolved, massive hot stars have winds, although the wind density varies greatly from object to object. In the hottest stars P Cygni profiles are more prevalent in the UV, but for lower temperature stars with strong winds (e.g., P Cygni [88], HDE 316285 [48]) the optical spectra are dominated by classical P Cygni profiles – line profiles showing both emission and blue-shifted absorption. The line spectrum of η Car, due to its dense wind, is more dominant than in other stars, and differs

F. Najarro (✉)
Departamento de Astrofísica Centro de Astrobiología, CSIC-INTA Ctra.
de Ajalvir a Torrejón km. 4 28850 Torrejón de Ardoz, Madrid Spain
e-mail: najarro@damir.iem.csic.es

D.J. Hillier
University of Pittsburgh, 100 Allen Hall, 3941 O'Hara St., Pittsburgh PA 15260, USA
e-mail: hillier@pitt.edu

K. Davidson and R.M. Humphreys (eds.), *Eta Carinae and the Supernova Impostors*, Astrophysics and Space Science Library 384, DOI 10.1007/978-1-4614-2275-4_4,

from other similar stars in that at most epochs Hα and other prominent lines do not display P Cygni absorption. In this chapter we describe this spectrum and wind. We discuss a model for η Car's wind (Sect. 4.2.1), the hydrostatic radius (Sect. 4.2.2), the He/H abundance (Sect. 4.2.3), non-spherical effects (Sect. 4.2.4), the accuracy of the mass-loss rate determination, and the influence of the mass-loss rate on the observed spectrum (Sect. 4.2.5), the UV spectrum (Sect. 4.2.6), and the potential influence of the binary companion on the observed spectrum (Sect. 4.2.7).

We then review a sample of other, perhaps less complex very luminous stars in our galaxy which may provide further insight into η Car's spectrum. Our starting point will be P Cygni, a "well behaved LBV" (i.e., less complex than η Car) which last had an eruption in the 1600s. We then proceed to complicate the picture with a "half way" object: HDE 316285, and finally, we introduce some relatively new Galactic objects which have been discovered thanks to the IR detector revolution that occurred in the 1990s.

4.2 Eta Carinae

Ground-based spectra of η Car are dominated by broad emission lines of H, He I, and Fe II, and narrow lines of the same elements, together with strong [Fe II] [3, 34, 38, 44, 91]. Many of the latter lines also show a weak broad component. Interpreting the complex line morphology is difficult; more than one emission region contributes to the observed spectra and the lack of spatial resolution means that it is impossible to determine a unique model.

The situation improved when Hillier and Allen [2, 44] undertook spectroscopic mapping of the Homunculus. Their observations confirmed earlier work [66, 92] that the Homunculus was primarily a reflection nebula. Surprisingly, however, the narrow components and [Fe II] lines were much weaker in the Homunculus than they were in the central core. The reflected spectrum resembled that of the LBV candidate HDE 316285, which is dominated by broad P Cygni profiles of H, He I, and Fe II, with only weak [Fe II] [48] (see Sect. 4.3.2 below). Therefore Hillier and Allen proposed that the line of sight to η Car, the central star, has much greater extinction than that for the region emitting narrow lines (See also [19]).

Improved spatial resolution became available with the advent of Hubble Space Telescope (HST). Spectroscopy by Davidson et al. [19, 20] spatially separated the star itself and showed that the narrow emission components arise from nearby slow ejecta including the "Weigelt blobs." Those early results were later confirmed by excellent data from the Space Telescope Imaging Spectrograph (STIS) [21, 37]. The star's spectrum (or rather, that of its opaque wind) is indeed similar to that of HDE 316285, though η Car's emission lines are much stronger, particularly Hα [49, 50]. There is another important difference: the spectrum of η Car varies, showing strong P Cygni absorption for Hα and Fe II only at some epochs [50, 76, 86].

The HST observations show that the broad emission components of the [Fe II] lines arise from the outer wind. They are easily visible in ground-based spectra because of differential extinction by dust – the continuum emitting region suffers a higher extinction than does the [Fe II] emitting region. As a consequence the [Fe II] emission lines appear stronger against the background stellar continuum [49].

With optical HST and UV spectra, it is feasible to model the central star to determine its parameters.

4.2.1 *Modeling the Wind*

Our analysis of the observed spectrum employs the radiative transfer code CMFGEN [45], which assumes spherical geometry, solves for the radiation field in the comoving frame, and allows for "line-blanketing" by tens of thousands of bound-bound transitions. We typically included H I–III, C II–V, N I–V, O I–IV, Na I–II, Mg I–III, Al II–IV, Si II–V, Ca II–III, and Fe II–VI while large models also contained Ti, Cr, Mn, Co, and Ni ions [50]. The modeling reviewed here concerns only the unstable primary star. The influence of the companion star is discussed in Sect. 4.2.7.

We adopted $L = 5.0 \times 10^6 L_\odot$ and $d = 2.3$ kpc. Free parameters were the radius R_* for a given small flow velocity, the mass-loss rate $\dot{M}$, chemical abundances, and the velocity law. We generally assumed a β velocity law, $V(r) = (1 - R_*/r)^\beta$, except that an exponential density distribution was adopted below the sonic point, and some models had an extended acceleration zone in the outer wind. The wind of η Car is extremely optically thick; in LBV jargon, it has a "false photosphere" [53]. This fact, together with uncertainties concerning what is being observed [49], make it extremely difficult to determine stellar parameters for the primary star.

The model adopted by Hillier et al. in 2001 [50] had the following parameters:

$\log L_*/L_\odot = 6.74$,
$R_*(\tau = 155, V = 0.32\,\mathrm{km\,s^{-1}}) = 60.0 R_\odot$ $(\mathrm{T}_{2/3} = 35{,}310\,\mathrm{K})$,
$R_*'(\tau = 10, V = 177\,\mathrm{km\,s^{-1}}) = 99.4 R_\odot$ $(\mathrm{T}_{2/3} = 27{,}433\,\mathrm{K})$,
$R_*''(\tau = 0.67, V = 375\,\mathrm{km\,s^{-1}}) = 881.0 R_\odot$ $(\mathrm{T}_{2/3} = 9{,}210\,\mathrm{K})$,
$\dot{M} = 1.0 \times 10^{-3} M_\odot\,\mathrm{year}^{-1}$,
$v_\infty = 500\,\mathrm{km\,s^{-1}}$,
H/He $\sim$ 5.0 by number ($Y_{\mathrm{He}} \sim 0.45$),
Volume filling factor $f(r) = 0.1 + 0.9\exp(-v(r)/v_{cl})$, where
$v_{cl} = 100\,\mathrm{km\,s^{-1}}$ and $r(v_{cl}) \sim 1.25 R_*$.

In the above, τ is the Rosseland mean optical depth, which is primarily set by electron scattering. The strong wavelength dependence of the absorptive opacity can be seen in Fig. 8 of Hillier et al. [50]. Classically the effective temperature of a star is defined at $\tau = 2/3$, and provides a rough characterization (through the analogy with the temperature of a black-body) of the energy distribution of the star. However, in an extended scattering atmosphere the analogy breaks down, and no single number can necessarily be used to characterize the spectrum. This is the case for η Car.[1]

The volume filling factor, $f(r)$, allows for clumping in the stellar wind; in the present models it reaches a minimum value of 0.1.

Reasonable fits to the observed spectra were obtained, with the H I and Fe II emission line strengths in good agreement with data. To fit the March 1998 HST/STIS spectrum a visual extinction of seven magnitudes was used; this highlights the importance of circumstellar extinction. To characterize the unusually weak wavelength dependence of η Car's extinction, we used the extinction law from Cardelli et al. [6] with R = 5.0, E(B-V) = 1.0, and an additional 2 magnitudes of grey extinction. Two major discrepancies were found. First, the calculated P Cygni absorption components tend to be stronger than observed. Second, the model was not in agreement with the observed UV spectrum shortward of 1,600 Å. This model was fit to a spectrum obtained soon after the 1998.0 spectroscopic event which differs from the "normal" spectrum; the "normal" spectrum has similar emission line strengths but the P Cygni absorption components on Hα, Hβ, and many Fe II lines are very weak, if not absent. The lack of P Cygni absorption components in the "normal" spectrum is very difficult to explain with a spherical model (see Sect. 4.2.5). Interferometric observations support the basic size scale [98].

The predicted theoretical spectrum is not that of a black body (Fig. 4.1), which has important implications when modeling the excitation of the Weigelt blobs (see Chap. 5). The spectrum illuminating the Weigelt blobs must depend strongly on the location and nature of the cavity created by the bow shock arising from interaction between the winds of the primary and secondary stars, since this influences the shape and location of the UV photosphere. Moreover, at some UV wavelengths the wind is opaque out to quite large radii.

4.2.2 The Hydrostatic Radius

The wind is so thick that for most models the emergent spectrum is insensitive to the radius of the star's quasi-hydrostatic interior (i.e., R_S = radius where $V(r) \approx$ sound speed). The emergent spectra for winds of Wolf-Rayet (W-R) stars, by contrast, depend on R_S, though derived values may depend on the adopted β [41].

---

[1] [Editor's note:] $\tau = 2/3$ is standard usage for photospheres and diffuse winds, but it has little mathematical or physical significance in an opaque wind. The parameter T_c, defined by $2.7kT_c = L/$(photon output), is somewhat more representative, because a diffuse spherical configuration allows *much* easier photon escape than a plane-parallel geometry, see [13, 52, 53]. For a gray opacity T_c is much hotter than the temperature at $\tau = 2/3$. Hillier (priv. comm.) remarks that heavy reprocessing of the UV prevents this from being true in his model for η Car.

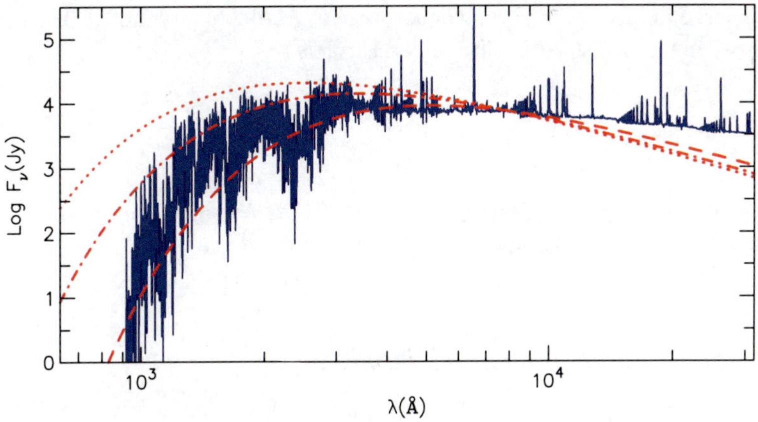

Fig. 4.1 Comparison of theoretical spectrum with black body spectra at 10^4 K, 1.5×10^4 K and 2.0×10^4 K. The black body curves have been shifted vertically to match the model spectrum at 8,000 Å, while the model spectrum was smoothed ($R/\Delta R = 1{,}000$) for clarity

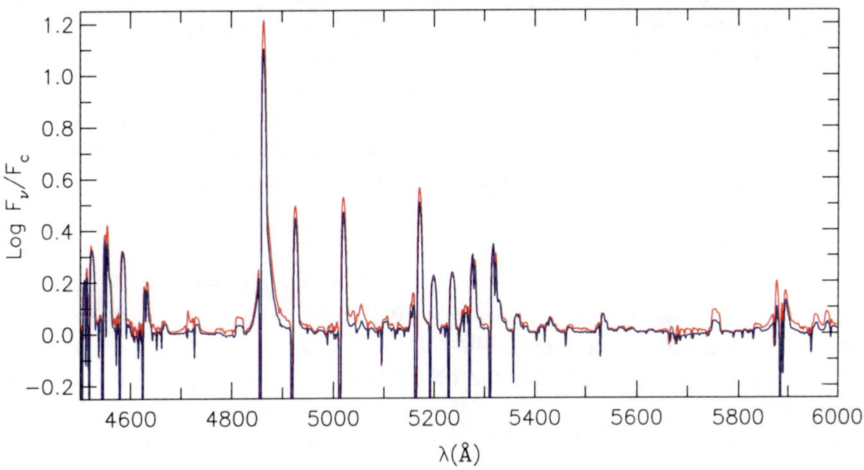

Fig. 4.2 Comparison between model spectra with hydrostatic radii of $60\,R_\odot$ (*grey line*) and $480\,R_\odot$ (*thin black line*). With the exception of the He I lines (e.g., at 5,876 Å) the spectra are very similar. If most of the He I emission is related to the companion star, as now believed, very large radii for the primary star are favored. With the exception of Hβ and the He I/Na D complex at 5,900 Å, most of the features in the spectra are due to Fe II

With the adopted velocity law, sizes smaller then $60\,R_\odot$ tend to produce He II λ4686 emission exceeding the observations [22, 64, 89]. Also, we now know that most of the He I emission does not come from the primary (Sect. 4.2.3). Thus $R_s \sim 240$–$480\,R_\odot$, about 1–2 AU, would be consistent with the observations (Fig. 4.2).

This large radius has important consequences for the binary model. The separation between the two stars' centers at periastron is

$$r_{12} \approx 3.3 \left(\frac{M_1 + M_2}{150 M_\odot}\right)^{1/3} \left(\frac{1-e}{0.2}\right) \mathrm{AU} \approx 700 \left(\frac{M_1 + M_2}{150 M_\odot}\right)^{1/3} \left(\frac{1-e}{0.2}\right) R_\odot,$$

which is not very much larger than the primary R_s (cf. tidal comments in [14]). Almost all authors favor orbital eccentricities e in the range 0.75–0.9.

4.2.3 H/He Ratio and CNO Abundances

Since He is not fully ionized in the wind, there is a degeneracy between the derived mass-loss rate and the N(H)/N(He) abundance ratio. The same uncertainty was found earlier for HDE 316285 [48]. The value of N(H)/N(He) adopted for η Car is consistent with that inferred from outlying ejecta [18], and gives consistent metal abundances at the factor-of-two level. Since the eruption of the 1840s may have ejected more than $10 M_\odot$ of material [87], it is possible that N(H)/N(He) for the primary star should be lower than that inferred from the nebular lines.

It now appears that most of the He I emission is due to the radiation field of a hot companion star, not the primary [16, 23, 51, 54, 76]. In the observed spectrum, the He I lines are much weaker than the main H I features. Throughout most of the 5.5 year cycle, they are blue shifted and often show double peaked profiles [23, 76]. They are most easily interpreted as arising near the shocked wind-wind interface, in a region where helium in the primary wind is photoionized by the companion star [54, 76], although other interpretations have also been proposed [57]. Thus an accurate determination of the H/He ratio for the primary star may need to await 3D modeling of the wind–wind interaction.

The blue shift of the He I emission lines is unlikely to be explained by a more highly ionized equatorial wind [86]. While it is true the emission from the red side will be occulted from the observer by the central star, the corresponding emission from the front side will also not be observed since the line source function will generally be less than that of the continuum. This will be true unless the He I emission arises from a hot region, with an excitation temperature greater than $T_{2/3}$. Blue shifted emission does occur in Type II SN, but in that case the line formation region is less extended, and the optical continuum and Hα line flux form in a similar region [26].

The abundances of C, N, and O are inferred from nebular spectra in the outer ejecta [17, 18]. Those observations, being consistent with material processed in the CNO cycle (i.e., $N > C + O$), were the first proof that η Car is an evolved object.

4.2.4 A Non-spherical Wind?

A major constraint on the system is provided by the absence of P Cygni absorption associated with Hα, Hβ and Fe II emission lines for most of the 5.5 year cycle [16, 76, 86]. The presence of strong and broad wind Fe II emission and the absence of P Cygni absorption components are inconsistent with a spherical wind. Theoretically, the H I and Fe II profiles observed *near the event,* with their strong P Cygni absorption components, are much more typical of those expected from a spherical wind, and more closely resemble those of HDE 316285.

There are two possible scenarios for the lack of P Cygni absorption components.

1. The wind is aspherical, and we happen to look along a direction in which the intrinsic wind absorption is low (i.e., the wind is more highly ionized than on average). The problem with this scenario is that our viewpoint is about 40° from the major axis of the Homunculus [24, 83, 85] – an intermediate latitude that favors neither the equatorial or polar direction. Smith et al. [86] showed that the polar wind is denser than the equatorial flow, but the fractional volume of the wind above 40° is less than 25%. Restricting the iron emitting volume to only these regions requires an extremely dense polar wind.
2. The wind is ionized on our side by the companion star, and/or we look through the companion wind for a large fraction of the orbit. This is the explanation most favored by the authors. This model also explains the presence of P Cygni absorption, at all epochs, on higher members of the Balmer series.

Obviously, both mechanisms may operate at the same time.

Evidence for an aspheric wind comes from HST observations of the spectrum reflected by the Homunculus [86]. They show that Hα has P Cygni absorption at polar directions, even during the middle of the cycle, and also suggest that V_∞ increases towards the pole. Such a scenario is expected if η Car is a fast rotator [62,77]. In a rapidly rotating star the effective surface temperature and gravity varies with latitude, being highest at the pole [62, 96]. Standard radiation driven wind theory then predicts that both the mass loss rate and the terminal velocity should be larger at the pole.[2] A possible defect of this polar-wind interpretation is that the Hα line widths appear to be independent of latitude in the data shown by [86]. In simple 2D model calculations, with a latitude-dependent terminal velocity, there are easily measurable changes in line width as a function of viewing angle [43].

An ambiguity in interpreting η Car's polar reflected lines arises from deciding which part of the P Cygni absorption profile we should associate with the terminal speed V_∞ – does it correspond to the deepest absorption, or the extreme short-wavelength edge of the absorption profile? In O and W-R stars, V_∞ is usually

[2]Possible exception: This tendency might be counteracted by a latitude dependence of the ionization state; in that case the wind-driving acceleration would depend on latitude.

measured at the blue edge where the absorption is still black, while higher-velocity (non-saturated) absorption is attributed to turbulence (e.g., [79, 80]). In η Car's reflected spectrum, the edge velocity shows a much larger variation with co-latitude than does the velocity corresponding to the minimum of the P Cygni profile [86]. A key observation, which would lead to a better understanding of the outflow velocities, would be to examine observations of higher order Balmer lines as a function of latitude, since they show P Cygni absorption components at all times.

Further observational evidence for non-sphericity comes from infrared interferometric images, which show elongation along the polar axis [93].

4.2.5 The Mass-Loss Rate, and Its Influence on Observed Spectra

Various estimates of the current mass-loss rate exist in the literature. These are based on the observed free-free millimeter emission ($2.4 \times 10^{-4}\,M_{\odot}\,\mathrm{year}^{-1}$ [8]); on the strengths and shapes of the Balmer lines ($10^{-3}\,M_{\odot}\,\mathrm{year}^{-1}$ [19,50]); on interferometric angular diameter measurements at 2.2 µm combined with continuum modeling ($1.6 \times 10^{-3}\,M_{\odot}\,\mathrm{year}^{-1}$ [93]); and on the X-ray flux cycle ($2.5 \times 10^{-4}\,M_{\odot}\,\mathrm{year}^{-1}$ [78]). The range in these values is unsurprising, since a number of factors contribute to possible errors [50]. The worst ambiguities involve the uncertain $N(\mathrm{H})/N(\mathrm{He})$ abundance ratio and the assumption of spherical geometry, which affects both the continuum and line fluxes. The filling-factor derivation may, in effect, partially compensate for departures from spherical geometry.

Strong emission lines in W-R spectra show evidence for Thomson scattering by free electrons. In LBVs, where V_{∞} is comparable to the thermal velocity of an electron ($\sim$500 km s^{-1}), the line wings extend to $\sim$2000 km s^{-1} and are fairly symmetrical about line center. In W-R winds, where V_{∞} is much larger than the electron thermal velocities, the wings are weak on the short-wavelength side [4,40].

In homogeneous models the theoretical Thomson scattering wings are much stronger than observed. Because the line emission usually scales as density squared, while the electron scattering optical depth scales as the density, it is possible to reconcile the relative strength of the scattering wings by using an inhomogeneous wind [40]. For modeling simplicity, a filling factor approach is adopted, in which we assume that the clumps are much smaller than the photon mean free path [46]. For η Car a volume-filling factor $f \sim 0.1$ was found [50].

Other sources of uncertainty in the mass-loss rate are the two assumptions that the primary star dominates the luminosity of the system, and that the mid-IR flux of the Homunculus does truly give, with only a minor correction, the total luminosity. For different choices of the primary luminosity, we can transform the mass-loss rate found in [50] according to[3]

[3]This $\dot{M} \sim L^{0.75}$ relation agrees with an old semi-analytic criterion for different opaque winds to have the same characteristic temperature [13].

$$\dot{M} = 1.0 \times 10^{-3} \left(\frac{L}{5 \times 10^6 L_\odot} \right)^{0.75} M_\odot\,\mathrm{year}^{-1}. \tag{4.1}$$

Because the mass-loss rate of η Car is very large, it has a profound influence on the observed spectrum. Let us assume a fixed hydrostatic radius ($60\,R_\odot$), and keep all parameters constant except the mass-loss rate.

- For low $\dot{M}$ (e.g., $\dot{M} \approx 2.5 \times 10^{-4}\,M_\odot\,\mathrm{year}^{-1}$), hydrogen is fully ionized in the wind. Fe II emission is weak, He I emission is prominent, and He II 4686 and N III 4640 emission lines are visible. Hα and other Balmer lines show only weak P Cygni absorption.
- As we increase $\dot{M}$, the He I emission weakens and Balmer lines strengthen.
- For a sufficiently high mass-loss rate, the H^0/H^+ ratio rises to about 10^{-3} instead of the 10^{-6} that is typical in lower-mass-loss models. When this occurs, the very efficient charge exchange reaction $Fe^{2+} + H^0 \leftrightarrow Fe^+ + H^+$ causes Fe II emission to strengthen dramatically.

Most of the blue-to-red Fe II emission lines are formed via continuum fluorescence – that is, the UV Fe II lines absorb continuum radiation and redistribute it to longer wavelengths via other bound-bound transitions. This process is also efficient for other species, in η Car, and in other hot stars [48]. We expect the recombination contribution to the observed Fe II lines to be weak, simply because Fe is far less abundant than H.

[Editors' note: Recent observations with HST/STIS show that the wind emission lines have decreased in strength by factors of more than 2 in the past decade, most likely due to a decrease in the wind density and thus the mass loss rate [69]. See Fig. 1.1 in the Chap. 1.]

4.2.6 *The UV Spectrum*

In our model described in Sect. 4.2.1 above [50], most of the observational constraints were supplied by the violet-to-red spectrum, with UV data used only as a consistency check. We found reasonable agreement down to 1600 Å, but substantial disagreements at shorter wavelengths. Because the reddening law is unique and unknown, absolute flux comparisons are not possible – rather, only spectral features can be compared. Originally two explanations were invoked to explain the discrepancies – an aspherical wind and/or the influence of the companion star on the wind. Later work revealed a third explanation: some UV features originate at such large radii that one needs to allow for the limited width of the HST/STIS spatial coverage [51]. It is also possible that circumstellar dust blocks some of the source regions more than others. Surprisingly, we were able to model the UV spectral features using only the spectrum originating outside $r \approx 0.3''$, though

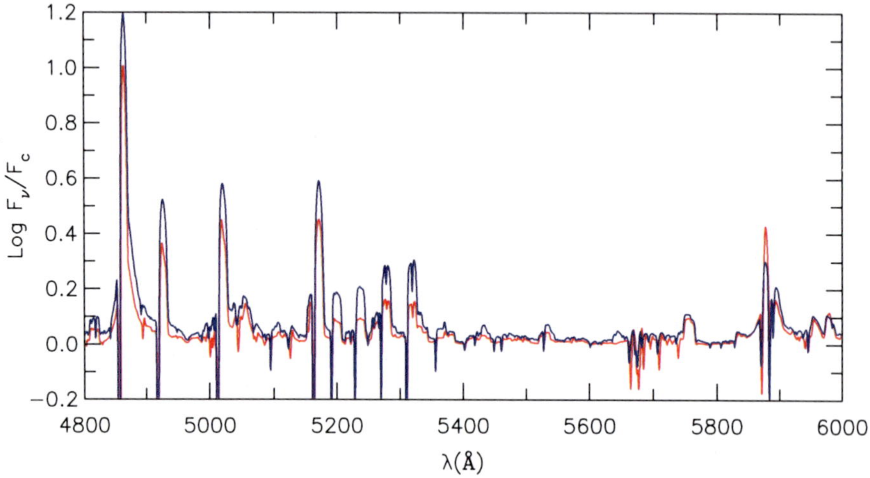

Fig. 4.3 Comparison between two models with $\dot{M} = 1.0 \times 10^{-3}\, M_{\odot}\,\text{year}^{-1}$ (*blue line*) and $\dot{M} = 5.0 \times 10^{-4}\, M_{\odot}\,\text{year}^{-1}$ (*red line*). In the lower-mass-loss-rate model Fe II lines are becoming weaker (e.g., the complex near 5200Å) and He I lines (e.g., λ5876) stronger. When the mass-loss rate is decreased by another factor of 2, the Fe II lines disappear, He I lines increase further in strength, and He II 4686 is in emission

with substantially less circumstellar extinction than was used for the violet-to-red data. This simple model is unlikely to be right; more likely all four effects influence the observed spectrum.

Dust is known to have a strong effect on η Car's spectrum. Since the effective extinction varies on a scale of $0.1''$, based on comparisons between the Weigelt blobs and the central star [19, 97], it may affect comparisons of UV vs. blue-to-red data.

Another potentially important influence is the companion star, whose fast wind and ionizing UV flux must carve out a concave region on its side of the dense primary wind. Since this zone is inside the UV-emitting radii, it may significantly influence the emergent UV spectrum. Continuity between optical and UV wavelengths indicates that the primary star probably dominates the observable UV spectrum. Detailed modeling confirms this conclusion [51], provided that circumstellar extinction is similar for both stars. As the secondary is inside the UV source region, its spectrum will be absorbed by the primary wind, at least for some directions. During the 2003 spectroscopic event, FUSE observations showed that flux in the 920–1100 Å spectral region virtually disappeared; Ipping et al. [55] argue that this was due to an eclipse of the secondary by the primary. An alternative explanation is that the far-UV flux declined because of changes in the ionization of the primary wind, as signified by changes in the hydrogen P Cygni absorption profiles. Modeling by Hillier et al. [51] shows that the FUSE spectral interval is the most promising for a direct detection of the companion star.

4.2.7 Consequences of Binarity

There is now overwhelming evidence that η Car is a binary system with a period of 5.5 years [1,7,9–12,56,76,78,99], although some relevant anomalies in the historical record have not been explained [28, 54] (see Chap. 1), and we do not quantitatively understand the radial velocity variations [14,23,76]. There is also evidence for non-periodic behavior of line profiles, suggesting the possibility of a shell ejection [25]. The strongest evidence for binarity is the X-ray light curve. Eta Car is a variable X-ray source [7, 56, 78], most easily interpreted as a consequence of a wind-wind collision in a binary system [10]. Most of the observable X-rays ($E > 1$ keV) are produced on the secondary star's side of the shocked wind-wind interface. The 2–10 keV X-ray spectrum implies a wind terminal velocity close to $3000\,\mathrm{km\,s^{-1}}$ for the secondary star, and the X-ray flux implies a secondary mass-loss rate on the order of $10^{-5}\,M_\odot\,\mathrm{year}^{-1}$ [78]. (See Corcoran and Ishibashi, this volume, for these constraints.) The luminosity ratio of the two stars is unknown; the primary is generally assumed to be far more luminous (e.g., [10, 15, 76]). The relative masses and radii of the two stars are also unknown. The absence of obvious spectral features belonging to the secondary star does not yield strong constraints, since most of its luminosity may occur at unobservable wavelengths below 900 Å.

The secondary star's ionizing UV radiation probably accounts for the high-excitation nebular lines seen in the Weigelt blobs. Assuming that this is the case, Mehner et al. [68] used photoionization models to constrain its stellar characteristics. The center of the allowed region in parameter space is near $T_{\mathrm{eff}} \approx 40{,}000$ K and the suitable luminosity range is $10^{5.2}$–$10^{6.0}\,L_\odot$, based on this reasoning. A value close to $10^6\,L_\odot$ with a cooler temperature was proposed in an earlier analysis [95], but those parameters together with the wind speed and mass-loss rate cited above, do not correspond to a normal O-type star.

A star like HD32195A, with $L/L_\odot \approx 1.5 \times 10^6$, $V_\infty = 3200\,\mathrm{km\,s^{-1}}$, and $\dot{M} \approx 2.6 \times 10^{-5}\,M_\odot\,\mathrm{year}^{-1}$ [81], would be a suitable companion object. Note, however, that such a star may be quite massive. If the X-ray constraints on $\dot{M}$ and V_∞ can be lowered (e.g., to $\dot{M} =$ few $\times\,10^{-6}\ M_\odot\,\mathrm{year}^{-1}$, and $V_\infty \sim 2500\,\mathrm{km\,s^{-1}}$) then there are many known stars in Carina that might be similar to the binary companion (see [81, 84]). It is also possible, given the extensive mass-loss history of η Car, that, due to mass-transfer in the system, the secondary star is not normal.

4.2.8 Unanswered Questions

The most fundamental questions that need to be answered are the mass, spectral types, and evolutionary status of the two component stars. Without these we cannot fully address issues such as the cause of the major outburst in the 1840s that generated the Homunculus, the cause of the more minor outburst that occurred in the 1890s, and the future evolution of the system. Of course numerous other questions also require answers. For example, what is the latitude dependence of the mass-loss in the primary star? What causes the circumstellar extinction toward the primary star, why is it so localized, and why is the circumstellar extinction declining with time [63,65]? What is the origin of the broad He II λ4686 emission that occurs near the "low excitation" event [22, 64, 89]?

4.3 Eta Carinae-Like Objects

Further insight into η Car may be gained by looking at other, less complex, but luminous unstable stars. In this section we approach the problem "stepwise", starting with P Cygni, which experienced an eruption in the 1600s but since then has been relatively quiescent. Its spectrum is thus less confusing and less contaminated by nearby ejecta. We then proceed to complicate the picture with an object that shares characteristics with P Cygni and η Car, already mentioned in Sect. 4.2, HDE 316285. Finally, we introduce some relatively new members of the club.

4.3.1 P Cygni

Historically, P Cygni has been in a secondary position relative to η Car. It has been classed as an LBV because of its two major outbursts in 1600 and 1660 which were followed by three decades of irregular variations. After 1700 no major eruption occurred, but the star has gradually brightened at a rate of 0.15 ± 0.02 mag per century (6% decrease in $T_{\rm eff}$ per 100 years [60]). It would not be considered an unstable star or an LBV without the observational records of the seventeenth century [42]. Polarization data suggest that while variable asymmetric structures exist in the wind, there is no preferred axis of symmetry [90]; spherical geometry may be a suitable first approximation. Like other LBVs and candidate LBVs, P Cygni is surrounded by faint nebular shells [67, 82] which neither contaminate much, nor reprocess, the stellar spectrum. The stellar spectrum can be easily observed from the UV to the radio region (see discussions by Vink and by Weis, this volume).

P Cygni can thus be considered an ideal candidate for quantitative spectroscopic studies of the wind of an LBV in quiescence. It was the first LBV to be fully analyzed [72] by means of the new generation of blanketed models [45]. Najarro et al. [72] included H, He, C, N, O, Mg, Al, Si, S, Cr, Mn, Ti, Co, Ni and Fe in their calculations and used the full mean averaged spectra covering the range between 1100 Å and 45 μm as observational constraints. A summary of their model fits are shown in Fig. 4.4, and the derived parameters are in Table 4.1 together with those obtained from previous optical [73] and IR studies [61]. All three analyses agree quite well. P Cygni can thus be used as a guideline to the analysis of other LBVs.

Figure 4.4 demonstrates the tremendous potential of the blanketing method to perform detailed quantitative spectroscopic investigations in LBVs. Overall, there is excellent agreement between the model and the observations. Notice that the IUE UV and ISO fluxes displayed in Fig. 4.4 are the true stellar fluxes corrected for reddening ($E(B-V) = 0.534$, $R = 3.2$) and assuming a distance of 1.8 kpc. The crucial result is the consistency between wavelengths from UV to IR. Thus, for LBVs and other hot stars where only IR spectra are available due to the high extinction (e.g. LBVs in obscured clusters in our Galaxy) we may be confident of the results.

One of our most important results, obtained with allowance for the effects of clumping, was the derivation of an accurate mass-loss rate. Using the strong electron scattering wings of the Balmer lines, Najarro [72] was able to constrain the wind clumping and derived a volume filling factor of $f = 0.5$. This value is a factor of 5 larger than the one typically found for W-R stars (e.g. [39, 46]) The sensitivity and leverage of the clumping determination is displayed in Fig. 4.5. The resulting clumping implied a final mass-loss rate of $\dot{M}_C = 2.35 \times 10^{-5}\, M_\odot\, \mathrm{year}^{-1}$.

Table 4.1 Derived stellar parameters for P Cygni

Model	R_* ($R_\odot$)	L_* ($L_\odot$)	T_* (10^4 K)	n_{He}/n_H	$\dot{M}/f^{1/2}$ ($M_\odot$ year^{-1})	V_∞ (km s^{-1})	β
Optical	76	7.0×10^5	1.92	0.29	3.2×10^{-5}	185	4.5
ISO-SWS	76	5.6×10^5	1.81	0.30	3.0×10^{-5}	185	2.5
Blanketed	76	6.1×10^5	1.87	0.29	3.3×10^{-5}	185	2.5

β is the shape parameter of the velocity field

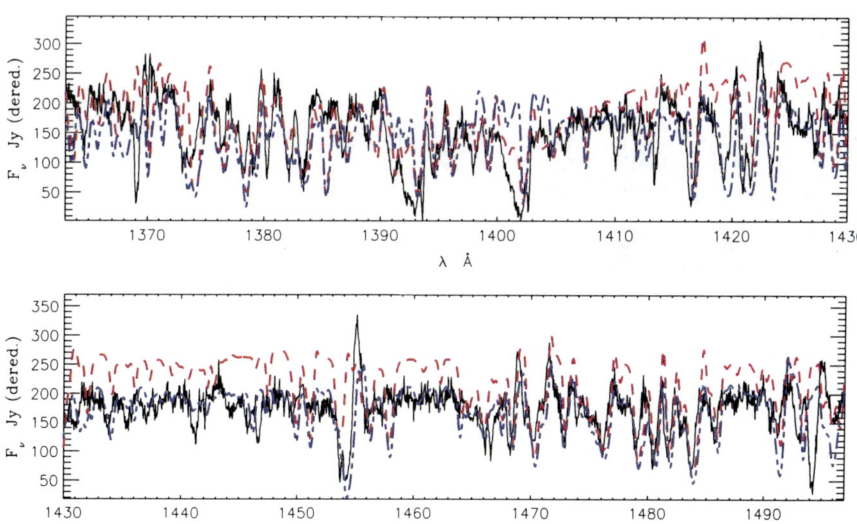

Fig. 4.4 Comparison of the blanketed model (*dashed*) and the averaged observed (*solid*) UV to IR spectrum of P Cygni. A model including Nickel and Cobalt (*dashed-dotted*) is also displayed in the IUE-SWP region (See discussion in text)

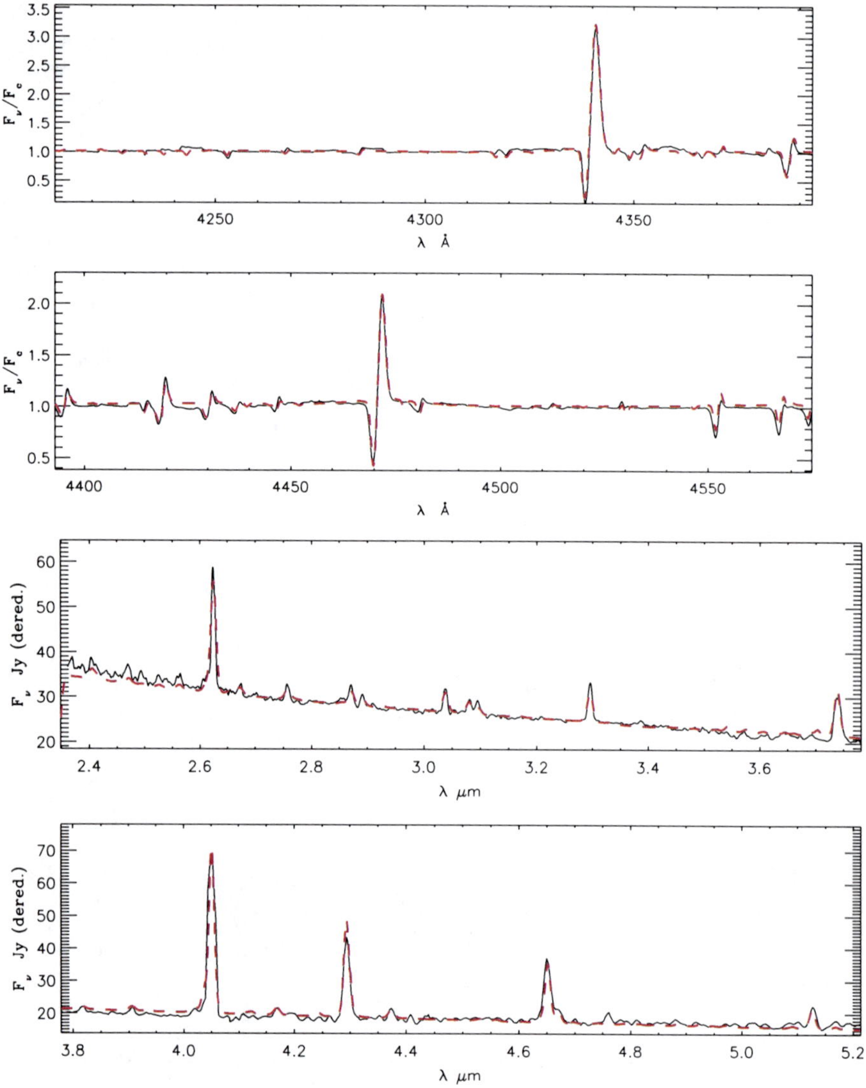

Fig. 4.4 (continued) – OPT and IR region

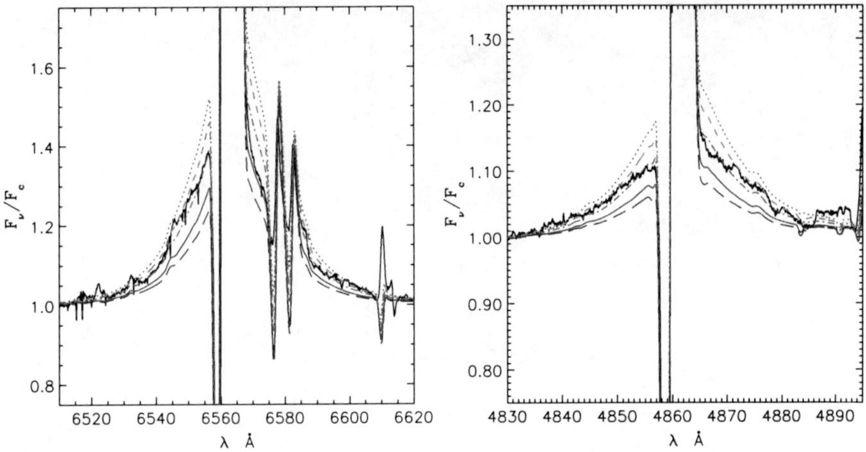

Fig. 4.5 Determination of wind clumping in P Cygni. Observations of Hα and Hβ vs models with the following volume-filling factors: $f = 1.0$ (*dotted*), $f = 0.75$ (*dashed*), $f = 0.50$ (*dashed-dotted*), $f = 0.25$ (*dashed-dotted-dotted*) and $f = 0.15$ (*long-dashed*). Models with $f = 0.50$ reproduce best the observations

4.3.2 A Close Twin: HDE 316285

As mentioned in Sect. 4.2 the spectrum of HDE 316285 has been compared with that of η Car [50]. This early-type Galactic supergiant is also considered to be an LBV candidate. Despite its close spectroscopic similarity to P Cygni, its properties remained poorly determined for a long time, largely because of the high extinction associated with its location in the Galactic plane [47]. The first detailed study of the object was made by Hillier et al. [48]. Their preferred model for HDE 316285 yielded the following parameters:

$$T_* = 15{,}000\ \mathrm{K}, \qquad L_*/L_\odot = 2.8\times10^5, \qquad \dot{M} = 2.4\times10^{-4} M_\odot\,\mathrm{year}^{-1},$$
$$v_\infty = 410\ \mathrm{km\,s^{-1}}, \qquad E_{\mathrm{B-V}} = 1.81\ \mathrm{mag}, \qquad \mathrm{H/He} \sim 1.5 \text{ by number.}$$

They found a so-called H/He degeneracy, or coupling, between the derived mass-loss rate and H/He abundance, caused by the low degree of He ionization in the wind. Thus, models with H/He = 10–0.5 were equally capable of explaining the spectrum provided the mass-loss rate was scaled according to the formula $\dot{M} \approx 9.1 + 26.3(\mathrm{He/H} - 0.1) \times 10^{-5} M_\odot\,\mathrm{year}^{-1}$. These results, with a slight downward revision of the temperature, were confirmed with blanketed models [74]. Figure 4.6 shows the excellent quality of the model fits to the optical and NIR spectra of the star, reproducing the H I, He I, Fe II, N II, and Mg II features.

Except for the stellar luminosity, the parameters [48,74] are intermediate between those of P Cygni and η Car. Indeed, from the point of view of performance number, wind density and location of $\tau = 2/3$ at a radius where V = a significant fraction

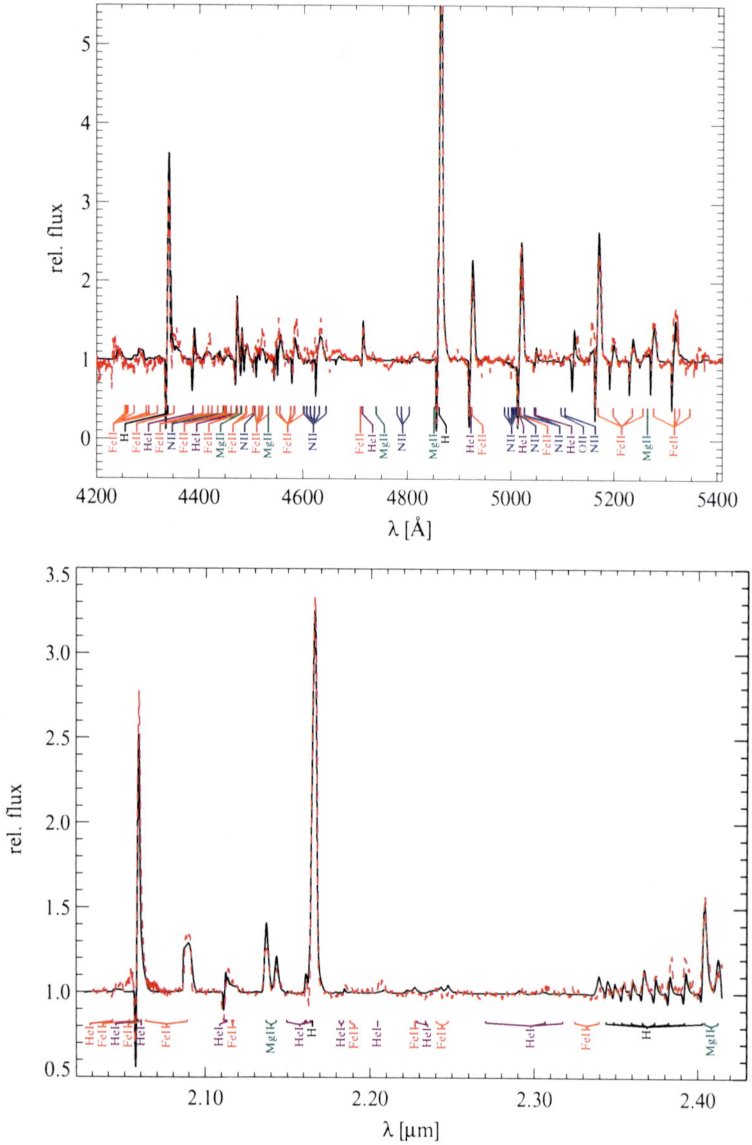

Fig. 4.6 Model fits (*dashed lines*) to the observed optical and IR diagnostic lines (*solid lines*) of HDE 316285 (Adapted from [74])

of V_∞, HDE 316285 is closer to η Car than to P Cygni. Another property shared with η Car and not found in P Cygni is the presence of a dense circumstellar nebula. Further, the lack of specific dust bands that are the signatures of crystalline material, which normally condenses in the thick slowly-expanding envelopes of

RSGs, suggests that the material forming the present-day nebula was rapidly ejected and could not crystallize under conditions of rapid cooling and low densities [71]. Such lack of crystalline silicates has also been confirmed in the explosive outflows from η Car.

Thus, the obvious question arising from the comparison of these three objects is why HDE 316285 is less luminous than P Cygni when all of its other spectral and morphological features suggest a luminosity closer to that of η Car? The answer is related to its assumed distance. Hillier et al. [48] adopted 2 kpc as a rough average of the literature values that lie between 1 and 3.4 kpc, recognizing that there was no real constraint on the distance.

Recently, Morris [71] pointed out that HDE 316285 lies in close (apparent) proximity to the infrared SFR Sgr D. Assuming that HDE 316285 is equidistant with Sgr D places it in the same dynamic ecosystem as the SFR and neighboring YSOs and maser sources, as well as a supernova remnant SNR 1.4–0.1. Its 8 kpc estimated distance is based on H_2CO and CS line kinematics measured at the VLA [70] . At this distance, HDE 316285 would have a luminosity of 6.6 log $L_\odot$ making it one of the most luminous stars known. However, Morris also showed that the interstellar extinction to HDE 316285 is substantially lower than the extinction to Sgr D. So, unless the star is located in an extinction hole it is likely in the foreground. Further, the velocity of the Na I D lines also supports a closer distance. In summary, the distance and luminosity of HDE 316285 remain uncertain.

4.3.3 *Relatives at the Galactic Center*

The lack of a good distance estimate to a star may result in an uncertainty of more than one dex in its luminosity. In our galaxy, one of the places where we think we know the distance is the Galactic Center (8 kpc). Unfortunately, due to the high extinction ($A_V \sim 30$) we had to wait for the IR detector revolution in the 1990s to obtain spectra of the less reddened ($A_K \sim 3$) spectra of sources in this region. If we want to look for LBVs at the Galactic Center we have to look for massive clusters at least 3 Myr old. Two clusters fulfilling this condition have been found in the inner 30 pc of our Galaxy; the Central Parsec Cluster and the Quintuplet Cluster. In this section, we will concentrate on the two candidate LBVs, the Pistol Star [29] and FMM362 [35], which belong to the Quintuplet Cluster ([30, 31, 36]). This cluster contains a variety of massive stars, including WN, WC, WN9/Ofpe, and less evolved blue-supergiants. Their presence constrains the cluster age to about 4 Myr old, assuming coeval star formation.

The Pistol Star was first studied by Figer et al. [29] who used non-blanketed models to fit the main H I and He I lines of the spectrum. Due to the lack of good temperature diagnostics, they found two solutions which could equally well reproduce the H I and He I spectra of the star: a hot, high luminosity solution with $\log L_*/L_\odot = 7.2$ and a cooler, lower luminosity one with $\log L_*/L_\odot = 6.6$, which makes the object, together with η Car, one the most luminous stars known.

Table 4.2 Derived stellar parameters

Parameter	Pistol	FMM362
$L_*(10^6 L_\odot)$	1.60	1.77
$R_{2/3}$ $(R_\odot)$	306	350
$T_{2/3}(10^4\,K)$	1.18	1.13
H/He	1.5	2.8
$Fe/Fe_\odot$	1.1	1.1
$Mg/Mg_\odot$	2.2	1.5
$Si/Si_\odot$	1.8	2.1
$\dot{M}(10^{-5} M_\odot\, year^{-1})$	2.1	1.2
β	3.0	1.3
$V_\infty(km\, s^{-1})$	105	170
$D_{mom}=\log(\dot{M}V_\infty(R/R_\odot)^{1/2})$	29.39	29.38
M_{Edd} $(M_\odot)$	22.5	30.5
CL_1	0.08	0.08
CL_2	2.5	2.00

β is the exponent describing the velocity field, D_{mom} is the modified wind momentum [58], CL_1 and CL_2 are clumping parameters, M_{Edd} is the Eddington mass, H/He is the ratio by number, and other abundances are relative to solar.

The Pistol Star and FMM362 have recently been reanalyzed by Najarro et al. [75] with blanketed models. The new blanketed models provide a significant improvement in our knowledge of the physical properties of these two stars. The degeneracy of the "high" and "low" luminosity ($T_{2/3}$) solutions for the Pistol Star presented by [29] is broken by the Si II, Mg II and Fe II lines, which are clearly more consistent with the "low" solution (see the excellent fits in Fig. 4.7). Najarro et al. derived a luminosity of $\sim$1.6 $\times$ $10^6\, L_\odot$, an effective temperature of $\sim$11,800 K, and an initial mass of 100 $M_\odot$ (see Table 4.2). The stellar luminosity is reduced by a factor of 2 compared with the previous estimate, illustrating the importance of the new generation of line-blanketed models. For star FMM362 ([35]), a close twin of the Pistol Star, Najarro et al. [75] also obtained a luminosity around $1.7 \times 10^6\, L_\odot$ and an effective temperature of $T_{2/3}$ $\sim$10,500 K. From Table 4.2 it is clear that the Pistol Star and FMM362 have very similar properties, with the exceptions of the Pistol Star's significantly higher wind density (evidenced by its stronger spectral lines) and its higher He content which may denote a slightly more advanced evolutionary stage.

In our IR analysis of the Pistol Star and FMM362 we developed the necessary techniques for future quantitative spectroscopic and metallicity studies of luminous, hot stars in heavily obscured regions. For example, we were able to break the H/He degeneracy mentioned in previous sections and obtain robust estimates of the He content. Due to the high degree of clumping found in both objects, the $\tau = 2/3$ radius (defined using the Rosseland mean opacity), where $T_{2/3}$ is defined, is reached at considerably lower velocities than for classical LBVs. Thus, wind speeds roughly between half and one-third of the sound speed are found in the Pistol Star and FMM362 while classical LBVs have wind speeds well above the

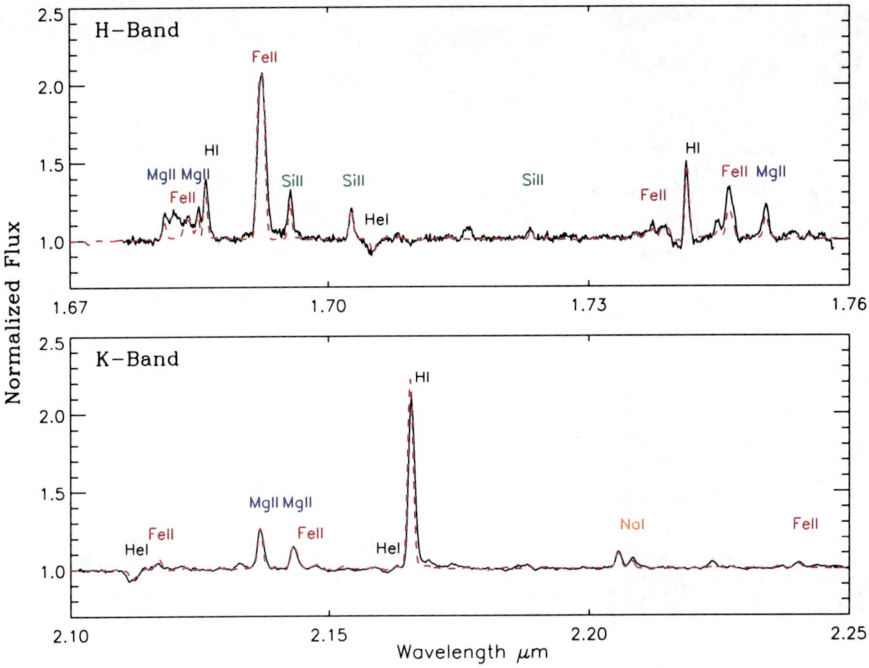

Fig. 4.7 Model fits (*dashed lines*) to the observed infrared diagnostic lines (*solid lines*) of the Pistol Star (See [75])

speed of sound [48, 72]. This enables quasi-photospheric absorption lines to form. The He I 2.112 μm line is the key in breaking the degeneracy. Najarro et al. [75] showed that while identical H I and He I (He I 1.700 μm) line profiles (also for the metal lines) are obtained for all H/He ratios considered, the absorption *depths* of the He I 2.112 μm and He I 2.15 μm lines react sensitively to the He abundance.

One of the main characteristics of LBVs is the change in their spectral type and apparent temperature through their excursions to the red in the HR-diagram. Figure 4.8 displays H- and K-Band spectra of the Pistol Star and star FMM362 obtained in 1998 (CGS4-data) and 2003 (NIRSPEC data). The spectra have been degraded to the same resolution. From Fig. 4.8 we see that while the Pistol Star has barely changed its spectroscopic appearance (minor changes only in Brγ), FMM362 shows a large change in its spectral type. The He I lines have vanished, the Mg II lines have weakened drastically, and the H-Bracket lines have gone into absorption. The 2003 data show that the star has started an excursion to the red in the HR-diagram, with a lower apparent temperature and an expanded envelope significantly reducing the stellar wind density. This result is consistent with photometric data which show an increase in the K magnitude of the star. The 2003 data presents the possibility to obtain an estimate of the gravity of an LBV, for the first time, using the Brackett absorption lines.

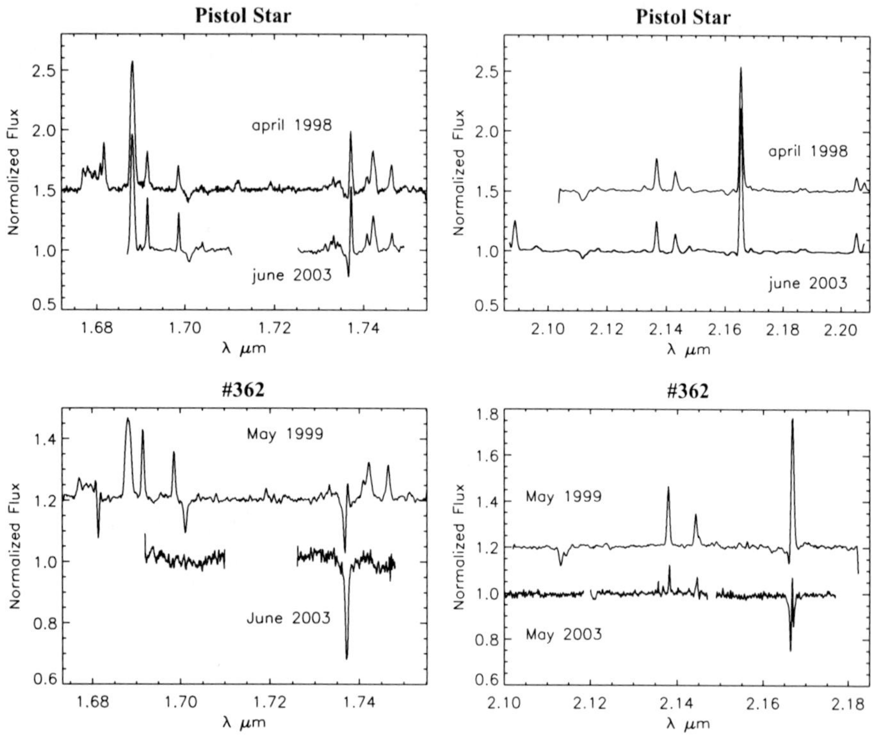

Fig. 4.8 Variability of the Pistol Star and FMM362

4.3.4 Other Monsters: LBV 1806-20

LBV 1806-20 was first identified as a potential counterpart to the soft gamma-ray repeater SGR 1806-20 in 1995 [59], with a high near-infrared brightness (K = 8.4 mag) despite significant absorption from interstellar dust in the Galactic Plane. Moderate resolution infrared spectroscopy [94] revealed it to be an LBV candidate and one of the most luminous stars in the Galaxy, with an estimated luminosity of $L_* \sim 10^6 L_\odot$.

Subsequent observations of the star and surrounding cluster [27, 32, 33] were obtained about a decade later. Eikenberry et al. [27] estimated a luminosity for LBV 1806-20 of greater than $5 \times 10^6 L_\odot$ which was supported by the kinematic distance (15.1 kpc) and the inferred spectral types and absolute magnitudes of the cluster members. Using the spectral types of the clusters stars they found $d \geq 9.5$ kpc ($L \geq 2 \times 10^6 L_\odot$), while the presence of an ammonia absorption feature requires $d \geq 5.7$ kpc ($L \geq 7 \times 10^5 L_\odot$). Eikenberry et al. [27] also obtained speckle images with high angular resolution that demonstrated that LBV 1806-20 was not a cluster, but rather a single or binary system with a mass exceeding $190 M_\odot$.

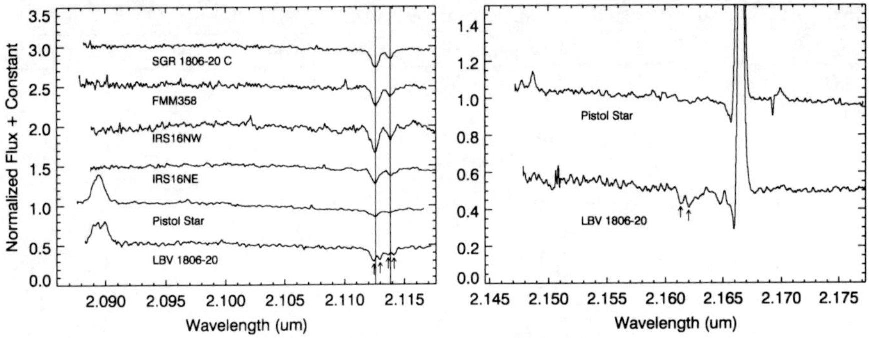

Fig. 4.9 *Left* Spectra of massive stars near the 2.112/2.113 μm He I doublet. The Pistol Star, IRS16NE, and LBV 1806-20 are LBVs or LBV-candidates, and the remaining stars are B-type supergiants. All the spectra show a single He I doublet near 2.1125 μm, except for LBV 1806-20, in which it is obviously double (marked by *arrows*). *Right* Spectra of the Pistol Star and LBV 1806-20 near the Brackett$-\gamma$ line. The two absorption lines near 2.160 μm in the spectrum for 1806-20, (marked by *arrows*) are due to two separate transitions of He I (Adapted from [33])

To test whether LBV 1806-20 could indeed be the "most massive star known", Figer et al. [33] obtained high-resolution near infrared spectroscopy that showed it could be a massive binary system with components separated in velocity by $\sim$70 km s^{-1}(see Fig. 4.9). They [33] also argued that a binary, or merged, system is more consistent with the ages of nearby stars in the cluster [32]. In addition, their derived VLSR = 36 km s^{-1} on the Galactic rotation curve, lowered the distance to 11.8 kpc, the luminosity to $10^{6.3} L_\odot$, and the system mass to $\sim 130 M_\odot$. Recent NIR spectroscopy and subsequent absolute K-band calibration [5] of a sample of W-R and B-supergiants in the same cluster further revised the distance downwards to 8.7 kpc which, scaling the luminosity estimate from [33], would result in a final luminosity of $\log L_*/L_\odot = 6.0$ for LBV 1806-20.

As we have seen for HDE 316285 and LBV 1806-20, not only the distance but also the possible binary nature of the object can have a major effect on the derived luminosity and other parameters.

4.4 Summary

Eta Carinae, while perhaps the most extreme example, appears to be related to other very luminous stars (Table 4.3) that are suffering extensive mass loss. These stars are luminous, show strong optical H and He I emission lines that are generally associated with P Cygni absorption, and may be variable. Unfortunately the precise evolutionary state, and the cause of the variability (when present) is unknown. Several difficulties contribute to this uncertainty. First, some of the stars are binaries. This contributes to uncertainties in both the mass and luminosity of the primary

Table 4.3 Comparison of stellar parameters for the stars discussed in the text

Parameter	Eta Car	P Cygni	HDE 316285	Pistol	FMM362	LBV 1806-20
$\log(L_*/L_\odot)$	6.74	5.79	5.60	6.20	6.25	6.0
d(kpc)	2.3	1.8	2.0	8.0	8.0	8 .7
$R_{2/3}$ $(R_\odot)$	880	105	128	306	350	213
$T_{2/3}(10^4\,K)$	[0.92+]	1.57	1.29	1.18	1.13	1.25
T_c $(10^4$ K)	0.98+	1.80	1.38	1.29	1.31	1.42
H/He	5.0	3.4	2.5	1.5	2.8	1.5
$\dot{M}(10^{-5}M_\odot\,\text{year}^{-1})$	100	2.4	8.0	2.1	1.2	1.8
f_{min}	0.1	0.5	0.1	0.08	0.08	0.1
$V_\infty(\text{km s}^{-1})$	500	185	370	105	170	160

The effective temperature and $R_{2/3}$ are both defined at $\tau(\text{Rosseland}) = 2/3$. The characteristic temperature T_c is defined by $2.7kT_c$ = (escaping luminosity)/(escaping photons)[a]. The HDE 316285 stellar properties have been updated to the ones presented by Hillier et al. [48] according to our current best blanketed models

[a]See footnote 1, Sect. 4.2.1.

component, and can also influence spectroscopic diagnostics, and potentially may have altered (or will alter) the evolution of the stars. Second, many of the stars have uncertain distances. While the uncertain distance does not strongly influence the modeling, since derived parameters can be simply scaled to a new distance, it does have a major influence on comparisons with evolutionary calculations. Third, because of solution degeneracies, great care must be taken in deriving the n(H)/n(He) abundance ratio. Fourth, many of the stars are heavily reddened, limiting observations to the IR. This necessarily reduces the confidence of the solutions, and necessarily limits the metal abundances which can be determined. Finally, the stars can suffer large eruptions. The timescale and multiplicity of these eruptions are not known.

Many of these difficulties can, and are being, overcome. Studies of the LBV environment provide clues to the evolutionary state. Distances can be constrained by cluster studies, by studying the dynamics of the ejected nebula, and by future astrometric observations. Careful observations, over as wide a passband as possible, can be used with line-blanketed radiative transfer codes to obtain accurate parameters. Long-term monitoring, particularly radial velocity studies, is required to detect binarity. Imaging studies can be used to detect earlier shell ejections, and thus provide some constraints on the size and frequency of eruptions. Importantly, we need to study other stars with similar luminosities and enhanced helium abundances in order to gain an understanding of the evolution of these stars.

Acknowledgements F. Najarro acknowledges support from AYA2008-06166-C03-02 grant. Hillier gratefully acknowledges support from the STScI which has supported the studies of Eta Carinae (GO-08327.03; GO-9420 and GO-9973 programs, HST-AR-10957).

References

1. Z. Abraham, D. Falceta-Gonçalves, T.P. Dominici, L.Å. Nyman, P. Durouchoux, F. McAuliffe, A. Caproni, V. Jatenco-Pereira, Millimeter-wave emission during the 2003 low excitation phase of η Carinae. A&A. **437**, 977–981 (2005)
2. D.A. Allen, D.J. Hillier, The shape of the homunculus nebula around Eta-Carinae. Proc. Astron. Soc. Aust. **10**, 338–341 (1993)
3. L.H. Aller, T. Dunham, The spectrum of Eta Carinae in 1961. ApJ. **146**, 126–141 (1966)
4. L.H. Auer, D. van Blerkom, Electron scattering in spherically expanding envelopes. ApJ. **178**, 175–182 (1972)
5. J.L. Bibby, P.A. Crowther, J.P. Furness, J.S. Clark, A downward revision to the distance of the 1806-20 cluster and associated magnetar from Gemini near-infrared spectroscopy. MNRAS. **386**, L23–L27 (2008)
6. J.A. Cardelli, G.C. Clayton, J.S. Mathis, The relationship between infrared, optical, and ultraviolet extinction. ApJ. **345**, 245–256 (1989)
7. M.F. Corcoran, K. Ishibashi, J.H. Swank, R. Petre, The X-ray light curve of η Carinae: refinement of the orbit and evidence for phase-dependent mass loss. ApJ. **547**, 1034–1039 (2001)
8. P. Cox, P.G. Mezger, A. Sievers, F. Najarro, L. Bronfman, E. Kreysa, G. Haslam, Mil-limeter emission of Eta Carinae and its surroundings. A&A. **297**, 168–174 (1995)
9. A. Damineli, The 5.52 year cycle of Eta Carinae. ApJL. **460**, L49–L52 (1996)
10. A. Damineli, P.S. Conti, D.F. Lopes, Eta Carinae: a long period binary? New Astr. **2**, 107–117 (1997)
11. A. Damineli et al., A multispectral view of the periodic events in η Carinae. MNRAS. **386**, 2330–2344 (2008)
12. A. Damineli et al., The periodicity of the η Carinae events. MNRAS. **384**, 1649–1656 (2008)
13. K. Davidson, The relation between apparent temperature and mass-loss rate in hypergiant eruptions. ApJ. **317**, 760–764 (1987)
14. K. Davidson, Is Eta Carinae a long-period binary? New Astr. **2**, 387–396 (1997)
15. K. Davidson, Why the binary hypothesis isn't a panacea, in *Eta Carinae at the Millennium*, eds. by J.A. Morse, R.M. Humphreys, A. Damineli. Astronomical Society of the Pacific Conference Series, vol. 179 (Astronomical Society of the Pacific, San Francisco, 1999), pp. 304–315
16. K. Davidson, Unique spectroscopic problems related to Eta Carinae, in *Eta Carinae and Other Mysterious Stars*, eds. by T.R. Gull, S. Johansson, K. Davidson. Astronomical Society of the Pacific Conference Series, vol. 242 (Astronomical Society of the Pacific, San Francisco, 2001), pp. 3–13
17. K. Davidson, N.R. Walborn, T.R. Gull, The remarkable spectrum of some material ejected by Eta Carinae. ApJL. **254**, L47–L51 (1982)
18. K. Davidson, R.J. Dufour, N.R. Walborn, T.R. Gull, Ultraviolet and visual wavelength spectroscopy of gas around ETA Carinae. ApJ. **305**, 867–879 (1986)
19. K. Davidson, D. Ebbets, G. Weigelt, R.M. Humphreys, A.R. Hajian, N.R. Walborn, M. Rosa, HST/FOS spectroscopy of ETA Carinae: the star itself, and ejecta within 0.3 arcsec. AJ. **109**, 1784–1796 (1995)
20. K. Davidson, D. Ebbets, S. Johansson, J.A. Morse, HST/GHRS observations of the compact slow ejecta of Eta Carinae. AJ. **113**, 335–345 (1997)
21. K. Davidson, K. Ishibashi, T.R. Gull, R.M. Humphreys, HST/STIS observations of the star during its recent event, in *Eta Carinae at the Millennium*, eds. by J.A. Morse, R.M. Humphreys, A. Damineli. Astronomical Society of the Pacific Conference Series, vol. 179 (Astronomical Society of the Pacific, San Francisco, 1999), pp. 227–235
22. K. Davidson, T.R. Gull, R.M. Humphreys, K. Ishibashi, P. Whitelock, L. Berdnikov, P.J. Mc-Gregor, T.S. Metcalfe, E. Polomski, M. Hamuy, An unusual brightening of Eta Carinae. AJ. **118**, 1777–1783 (1999)

23. K. Davidson, T.R. Gull, K. Ishibashi, HST/STIS spectroscopy of η car, in *P Cygni 2000: 400 Years of Progress*, eds. by M. de Groot, C. Sterken. Astronomical Society of the Pacific Conference Series, vol. 233 (Astronomical Society of the Pacific, San Francisco, 2001), pp. 173–183.
24. K. Davidson, N. Smith, T.R. Gull, K. Ishibashi, D.J. Hillier, The shape and orientation of the homunculus nebula based on spectroscopic velocities. AJ. **121**, 1569–1577 (2001)
25. K. Davidson, J. Martin, R.M. Humphreys, K. Ishibashi, T.R. Gull, O. Stahl, K. Weis, D.J. Hillier, A. Damineli, M. Corcoran, F. Hamann, A change in the physical state of η Carinae? AJ. **129**, 900–906 (2005)
26. L. Dessart, D.J. Hillier, Quantitative spectroscopy of photospheric-phase type II supernovae. A&A. **437**, 667–685 (2005)
27. S.S. Eikenberry et al., Infrared observations of the candidate LBV 1806-20 and nearby cluster stars. ApJ. **616**, 506–518 (2004)
28. M. Feast, P. Whitelock, F. Marang, Variability of η Carinae – III. MNRAS. **322**, 741–748 (2001)
29. D.F. Figer, F. Najarro, M. Morris, I.S. McLean, T.R. Geballe, A.M. Ghez, N. Langer, The pistol star. ApJ. **506**, 384–404 (1998)
30. D.F. Figer, S.S. Kim, M. Morris, E. Serabyn, R.M. Rich, I.S. McLean, Hubble space telescope/NICMOS observations of massive stellar clusters near the Galactic center. ApJ. **525**, 750–758 (1999)
31. D.F. Figer, I.S. McLean, M. Morris, Massive stars in the quintuplet cluster. ApJ. **514**, 202–220 (1999)
32. D.F. Figer, F. Najarro, T.R. Geballe, R.D. Blum, R.P. Kudritzki, Massive stars in the SGR 1806-20 cluster. ApJL. **622**, L49–L52 (2005)
33. D.F. Figer, F. Najarro, R.P. Kudritzki, The double-lined spectrum of LBV 1806-20. ApJL. **610**, L109–L112 (2004)
34. E. Gaviola, Eta Carinae. II. The spectrum. ApJ. **118**, 234–251 (1953)
35. T.R. Geballe, F. Najarro, D.F. Figer, A second luminous blue variable in the quintuplet cluster. ApJL. **530**, L97–L101 (2000)
36. I.S. Glass, R.M. Catchpole, P.A. Whitelock, J, H and K maps of the Galactic centre region. II – Qualitative aspects of the interstellar absorption. MNRAS. **227**, 373–379 (1987)
37. T.R. Gull, K. Ishibashi, K. Davidson, First observations of Eta Carinae with the space telescope imaging spectrograph, in *Eta Carinae at the Millennium*, eds. by J.A. Morse, R.M. Humphreys, A. Damineli. Astronomical Society of the Pacific Conference Series, vol. 179 (Astronomical Society of the Pacific, San Francisco, 1999), pp. 144–154
38. F. Hamann, D.L. Depoy, S. Johansson, J. Elias, High-resolution 6450-24500 a spectra of Eta Carinae. ApJ. **422**, 626–641 (1994)
39. J.E. Herald, D.J. Hillier, R.E. Schulte-Ladbeck, Tailored analyses of the WN 8 Stars WR 40 and WR 16. ApJ. **548**, 932–952 (2001)
40. D.J. Hillier, The effects of electron scattering and wind clumping for early emission line stars. A&A. **247**, 455–468 (1991)
41. D.J. Hillier, Theory of wolf-rayet atmospheres (review), in *Wolf-Rayet Stars and Interrelations with Other Massive Stars in Galaxies*, eds. by K.A. van der Hucht, B. Hidayat. IAU Symposium, vol. 143 (Kluwer, Dordrecht, 1991), pp. 59–73
42. D.J. Hillier, Luminous blue variables (review), in *The Atmospheres of Early-Type Stars*, eds. by U. Heber, C.S. Jeffery. Lecture Notes in Physics, vol. 401 (Springer, Berlin, 1992), pp. 105–120
43. D.J. Hillier, The envelopes of B[e] stars, in *Stars with the B[e] Phenomenon*, eds. by M. Kraus, A.S. Miroshnichenko. Astronomical Society of the Pacific Conference Series, vol. 355 (Astronomical Society of the Pacific, San Francisco, 2006), pp. 39–50
44. D.J. Hillier, D.A. Allen, A spectroscopic investigation of Eta Carinae and the Homunculus Nebula. I – Overview of the spectra. A&A. **262**, 153–170 (1992)
45. D.J. Hillier, D.L. Miller, The treatment of non-LTE line blanketing in spherically expanding outflows. ApJ. **496**, 407–427 (1998)

46. D.J. Hillier, D.L. Miller, Constraints on the evolution of massive stars through spectral analysis. I. The WC5 Star HD 165763. ApJ. **519**, 354–371 (1999)
47. D.J. Hillier, P.J. McGregor, A.R. Hyland, HD 316285 – an extreme P Cygni star, in *Mass Outflows from Stars and Galactic Nuclei*, eds. by L. Bianchi, R. Gilmozzi. Astrophysics and Space Science Library, vol. 142 (Kluwer, Dordrecht, 1988), pp. 215–218
48. D.J. Hillier, P.A. Crowther, F. Najarro, A.W. Fullerton, An optical and near-IR spectroscopic study of the extreme P Cygni-type supergiant HDE 316285. A&A. **340**, 483–496 (1998)
49. D.J. Hillier, K. Davidson, K. Ishibashi, T. Gull, Eta Carinae: the central star, in *Eta Carinae and Other Mysterious Stars: The Hidden Opportunities of Emission Spectroscopy* eds. by T.R. Gull, S. Johannson, K. Davidson. Astronomical Society of the Pacific Conference Series, vol. 242 (Astronomical Society of the Pacific, San Francisco, 2001), pp. 15–27
50. D.J. Hillier, K. Davidson, K. Ishibashi, T. Gull, On the nature of the central source in η Carinae. ApJ. **553**, 837–860 (2001).
51. D.J. Hillier et al., The UV scattering halo of the central source associated with η Carinae. ApJ. **642**, 1098–1116 (2006)
52. D.G. Hummer, G.B. Rybicki, Radiative transfer in spherically symmetric systems. The conservative grey case. MNRAS. 152, **152**, 1–20 (1971)
53. R.M. Humphreys, K. Davidson, The luminous blue variables: astrophysical geysers. PASP. **106**, 1025–1051 (1994)
54. R.M. Humphreys, K. Davidson, M. Koppelman, The early spectra of Eta Carinae 1892 to 1941 and the onset of its high excitation emission spectrum. AJ. **135**, 1249–1263 (2008)
55. R.C. Iping, G. Sonneborn, T.R. Gull, D.L. Massa, D.J. Hillier, Detection of a hot binary companion of η Carinae. ApJL. 633, **633**, L37–L40 (2005)
56. K. Ishibashi, M.F. Corcoran, K. Davidson, J.H. Swank, R. Petre, S.A. Drake, A. Damineli, S. White, Recurrent x-ray emission variations of η Carinae and the binary hypothesis. ApJ. **524**, 983–987 (1999)
57. A. Kashi, N. Soker, The source of the helium visible lines in η Carinae. New Astr. **12**, 590–596 (2007)
58. R.P. Kudritzki, J. Puls, Winds from hot stars. ARA&A. **38**, 613–666 (2000)
59. S.R. Kulkarni, K. Matthews, G. Neugebauer, I.N. Reid, M.H. van Kerkwijk, G. Vasisht, Optical and infrared observations of SGR 1806-20. ApJL. **440**, L61–L64 (1995)
60. H.J.G.L.M. Lamers, M.J.H. de Groot, Observed evolutionary changes in the visual magnitude of the luminous blue variable P Cygni. A&A. **257**, 153–162 (1992)
61. H.J.G.L.M. Lamers, F. Najarro, R.P. Kudritzki, P.W. Morris, R.H.M. Voors, J.I. van Gent, L.B.F.M. Waters, T. de Graauw, D. Beintema, E.A. Valentijn, D.J. Hillier, The ISO-SWS spectrum of P Cygni. A&A. **315**, L229–L232 (1996)
62. A. Maeder, G. Meynet, Stellar evolution with rotation. VI. The Eddington and Omega -limits, the rotational mass loss for OB and LBV stars. A&A. **361**, 159–166 (2000)
63. J.C. Martin, M.D. Koppelman, η Carinae's brightness variations since 1998: Hubble space telescope observations of the central star. AJ. **127**, 2352–2361 (2004)
64. J.C. Martin, K. Davidson, R.M. Humphreys, D.J. Hillier, K. Ishibashi, On the He II Emis-sion in η Carinae and the origin of its spectroscopic events. ApJ. **640**, 474–490 (2006)
65. J.C. Martin, K. Davidson, M.D. Koppelman, The chrysalis opens? photometry from the η Carinae Hubble space telescope treasury project, 2002–2006. AJ. **132**, 2717–2728 (2006)
66. J. Meaburn, R.D. Wolstencroft, J.R. Walsh, Echelle and spectropolarimetric observations of the Eta Carinae nebulosity. A&A. **181**, 333–342 (1987)
67. J. Meaburn, J.A. Lopez, M.J. Barlow, J.E. Drew, The expansion of the outer circumstellar shell of P Cygni. MNRAS. **283**, L69–L71 (1996)
68. A. Mehner, K. Davidson, G. Ferland, R.M. Humphreys, High excitation emission lines in the close environment of Eta Carinae. ApJ. **710**, 729–742 (2010)
69. A. Mehner, K. Davidson, R.M. Humphreys, J.C. Martin, K. Ishibashi, G.J. Ferland, N.R. Walborn, A sea change in Eta Carinae. ApJ. Lett. **717** L22–L25 (2010)
70. D.M. Mehringer, W.M. Goss, D.C. Lis, P. Palmer, K.M. Menten, VLA observations of the sagittarius D star-forming region. ApJ. **493**, 274–290 (1998)

71. P. Morris, The *Spitzer* WRRINGS Team, Infrared tracers of mass-loss histories and Wind-ISM interactions in hot star nebulae, eds. by F. Bresolin, P.A. Crowther, J. Puls., *IAU Symposium*, vol. 250 (International Astronomical Union, Paris, 2008), pp. 361–366
72. F. Najarro, Spectroscopy of P Cygni, in *P Cygni 2000: 400 Years of Progress*, eds. by M. de Groot, C. Sterken. Astronomical Society of the Pacific Conference Series, vol. 233 (Astronomical Society of the Pacific, San Francisco, 2001), pp. 133–146
73. F. Najarro, D.J. Hillier, O. Stahl, A spectroscopic investigation of P Cygni. I. H and HeI lines. A&A. **326**, 1117–1134 (1997)
74. F. Najarro, D.J. Hillier, D.F. Figer, T.R. Geballe, Theoretical modelling of hot stars, in *The Central Parsecs of the Galaxy*, eds. by H. Falcke, A. Cotera, W.J. Duschl, F. Melia, M.J. Rieke. Astronomical Society of the Pacific Conference Series, vol. 186 (Astronomical Society of the Pacific, San Francisco, 1999), pp. 340–+
75. F. Najarro, D.F. Figer, D.J. Hillier, T.R. Geballe, R.P. Kudritzki, Metallicity in the Galactic center: the quintuplet cluster. ApJ. **691**, 1816–1827 (2009)
76. K.E. Nielsen, M.F. Corcoran, T.R. Gull, D.J. Hillier, K. Hamaguchi, S. Ivarsson, D.J. Lindler, η Carinae across the 2003.5 minimum: spectroscopic evidence for massive binary interactions. ApJ. **660**, 669–686 (2007)
77. S.P. Owocki, K.G. Gayley, S.R. Cranmer, Effects of gravity darkening on radiatively driven mass loss from rapidly rotating stars, in *Properties of Hot Luminous Stars*, eds. by I. Howarth. Astronomical Society of the Pacific Conference Series, vol. 131 (Astronomical Society of the Pacific, San Francisco, 1998), pp. 237–244
78. J.M. Pittard, M.F. Corcoran, In hot pursuit of the hidden companion of Eta Carinae: an X-ray determination of the wind parameters. A&A. **383**, 636–647 (2002)
79. R.K. Prinja, M.J. Barlow, I.D. Howarth, Terminal velocities for a large sample of O stars, B supergiants, and Wolf-Rayet stars. ApJ. **361**, 607–620 (1990)
80. R.K. Prinja, M.J. Barlow, I.D. Howarth, Terminal velocities for a large sample of O stars, B supergiants, and Wolf-Rayet stars: Erratum. ApJ. **383**, 466 (1991)
81. T. Repolust, J. Puls, A. Herrero, Stellar and wind parameters of Galactic O-stars. The influence of line-blocking/blanketing. A&A. **415**, 349–376 (2004)
82. C.J. Skinner, R.H. Becker, R.L. White, K.M. Exter, M.J. Barlow, R.J. Davis, A large radio nebula around P Cygni. MNRAS. **296**, 669–688 (1998)
83. N. Smith, Dissecting the Homunculus nebula around Eta Carinae with spatially resolved near-infrared spectroscopy. MNRAS. **337**, 1252–1268 (2002)
84. N. Smith, A census of the Carina Nebula – I. Cumulative energy input from massive stars. MNRAS. **367**, 763–772 (2006)
85. N. Smith, The structure of the homunculus. I. Shape and latitude dependence from H2 and [Fe II] velocity maps of η Carinae. ApJ. **644**, 1151–1163 (2006)
86. N. Smith, K. Davidson, T.R. Gull, K. Ishibashi, D.J. Hillier, Latitude-dependent effects in the stellar wind of η Carinae. ApJ. **586**, 432–450 (2003)
87. N. Smith, R.D. Gehrz, P.M. Hinz, W.F. Hoffmann, J.L. Hora, E.E. Mamajek, M.R. Meyer, Mass and kinetic energy of the homunculus nebula around η Carinae. AJ. **125**, 1458–1466 (2003)
88. O. Stahl, H. Mandel, B. Wolf, T. Gaeng, A. Kaufer, R. Kneer, T. Szeifert, F. Zhao, Long-term spectroscopic monitoring of P Cygni-type stars. I - Spectral atlas of P Cygni. A&AS. **99**, 167–177 (1993)
89. J.E. Steiner, A. Damineli, Detection of he II λ 4686 in η Carinae. ApJL. **612**, L133–L136 (2004)
90. M. Taylor, K.H. Nordsieck, R.E. Schulte-Ladbeck, K.S. Bjorkman, A study of the asymmetric wind of P Cygni. AJ. **102**, 1197–1206 (1991)
91. A.D. Thackeray, Identifications in the spectra of Eta Carinae and RR telescopii. MNRAS. **113**, 211–236 (1953)
92. A.D. Thackeray, Spectra of the polarized halo around Eta Carinae. Observatory. **81**, 99–102 (1961)

93. R. van Boekel, P. Kervella, M. Schller, T. Herbst, W. Brandner, A. de Koter, L.B.F.M. Waters, D.J. Hillier, F. Paresce, R. Lenzen, A.M. Lagrange, Direct measurement of the size and shape of the present-day stellar wind of Eta Carinae. A&A. **410**, L37–L40 (2003)
94. M.H. van Kerkwijk, S.R. Kulkarni, K. Matthews, G. Neugebauer, A luminous companion to SGR 1806-20. ApJL. **444**, L33–L35 (1995)
95. E. Verner, F. Bruhweiler, T. Gull, The Binarity of η Carinae revealed from photoionization modeling of the spectral variability of the weigelt blobs B and D. ApJ. **624**, 973–982 (2005)
96. H. von Zeipel, The radiative equilibrium of a rotating system of gaseous masses. MNRAS. **84**, 665–683 (1924)
97. G. Weigelt, J. Ebersberger, Eta Carinae resolved by speckle interferometry. A&A. **163**, L5–L6 (1986)
98. G. Weigelt et al., Near-infrared interferometry of η Carinae with spectral resolutions of 1500 and 12000 using AMBER/VLTI. A&A. **464**, 87–106 (2007)
99. P.A. Whitelock, M.W. Feast, F. Marang, E. Breedt, The 2003 shell event in η Carinae. MNRAS. **352**, 447–456 (2004)

Chapter 5
Physics of the Inner Ejecta

Fred Hamann

Abstract Eta Carinae's inner ejecta are dominated observationally by the bright Weigelt blobs and their famously rich spectra of nebular emission and absorption lines. They are dense ($n_e \sim 10^7$–$10^8\,\mathrm{cm}^{-3}$), warm ($T_e \sim 6{,}000$–$7{,}000\,\mathrm{K}$) and slow moving ($\sim 40\,\mathrm{km\,s}^{-1}$) condensations of mostly neutral (H^0) gas. Located within 1,000 AU of the central star, they contain heavily CNO-processed material that was ejected from the star about a century ago. Outside the blobs, the inner ejecta include absorption-line clouds with similar conditions, plus emission-line gas that has generally lower densities and a wider range of speeds (reaching a few hundred $\mathrm{km\,s}^{-1}$) compared to the blobs. The blobs appear to contain a negligible amount of dust and have a nearly dust-free view of the central source, but our view across the inner ejecta is severely affected by uncertain amounts of dust having a patchy distribution in the foreground. Emission lines from the inner ejecta are powered by photoionization and fluorescent processes. The variable nature of this emission, occurring in a 5.54 years "event" cycle, requires specific changes to the incident flux that hold important clues to the nature of the central object.

5.1 Introduction

The "inner ejecta" of η Car reside in the bright core of the Homunculus Nebula, spatially unresolved in seeing-limited ground-based images. Spectra show a complex amalgam of features, including broad emission lines from the stellar wind and a vast number of narrower lines from the ejecta [1, 14, 36, 46, 52, 95]. High-resolution images using speckle techniques [56, 103], and later the Hubble Space

F. Hamann (✉)
Department of Astronomy, University of Florida, 211 Bryant Space Science Center, Gainesville, FL 32611-2055, USA
e-mail: hamann@astro.ufl.edu

K. Davidson and R.M. Humphreys (eds.), *Eta Carinae and the Supernova Impostors*, Astrophysics and Space Science Library 384, DOI 10.1007/978-1-4614-2275-4_5,

Telescope *(HST)* [31,78,91,104], revealed several bright objects less than 0.3″ apart, customarily labeled A, B, C, D. The first *HST* spectra showed that A, the brightest object, is the central star, while the others – the "Weigelt knots" or "Weigelt blobs" – are slow-moving nebular ejecta that produce strong narrow emission lines while also reflecting the star's light [24,25]. Their origin has not been explained, and this article is concerned mainly with their present-day nature.

The blobs appear to be located near the Homunculus' midplane, which is usually assumed to lie close to the star's equatorial plane and the orbital plane of the binary [21]. They are on the near side, moving away from the star at speeds of 30–50 $\mathrm{km\,s^{-1}}$ – less than a tenth as fast as the Homunculus lobes [25, 80, 109] Ejection dates based on proper motions have ranged from 1890 to 1940, well after Great Eruption in the 1840s [25, 29, 91, 104].[1] The Weigelt Knots are most often linked with the "Little Homunculus" ejected during the second eruption in the 1890s ([59, 61]; see Chaps. 6 and 7).

Objects B, C and D are located 0.1–0.3″ northwest of the star, corresponding to 300–1000 AU in deprojected distance or a light travel time of several days.[2] Their apparent sizes are somewhat less than 0.1″ or ∼200 AU, but these are just the brightest peaks in a complex pattern of emission and reflection (affected by extinction) that extends out to 0.4″ or more from the central star. Little is known about the fainter associated emission/reflection regions, but altogether we call this ensemble of nebular material the "inner ejecta."[3]

Spectroscopic studies of the inner ejecta have pursued three main goals. The most basic is to estimate physical properties: density, temperature, ionization, kinematics, composition, and mass. These parameters may be clues to the nature and history of the central star(s). Another goal is to characterize the spectrum of the central source, e.g., as a binary system, by considering the nebular gas as a light reprocessing machine. The excitation, photoionization and specific emissions from the gas depend on illumination by the central source in the UV and unobservable far-UV. These are critical wavelengths for testing models of the central star or stars. Finally, a third goal is to use the exceptionally bright and rich (and sometimes very unusual) line emission to study basic atomic physics and line formation processes.

In this review we focus mostly on spectroscopy of the brightest Weigelt knots B, C and D, including a new analysis of D based on *HST* Treasury spectra obtained in 2002–2003. We also briefly discuss two other phenomena, namely, narrow nebular absorption lines that appear throughout the inner ejecta [26, 38–40, 80] and a remarkable emission line region known as the "Strontium filament" [38, 110].

[1] In principle, long-term acceleration might affect this question [25, 91].

[2] $D = 2{,}300$ pc for η Car, see Chaps. 1 and 2.

[3] Additional fainter knots are noted in some papers. One must be wary, however, because the HST's optical point spread function has a "ring of beads" which is not entirely removed by deconvolution using the standard STScI software.

Strictly speaking these lie outside the inner ejecta as defined above, but they also provide insights into the nature of the central star and inner ejecta.

One important uncertainty is the nature of localized dust extinction. We know that the dust around η Car is patchy on small scales, in order to explain the blobs' high apparent brightness relative to the star. Our line of sight to the latter has several magnitudes more extinction than B, C, and D which are less than $0.3''$ away [23, 24, 52, 55] – or at least this was true a few years ago [74]. Moreover, the visual-wavelength Weigelt blobs appear almost inversely correlated with the spatial distribution of mid-IR (warm) emission by dust. These factors lead to a fundamental ambiguity about whether the observed blobs are distinct gas condensations or simply minima in the intervening dust. Perhaps they are a combination of both.

Another issue to keep in mind is the overall transience of the inner ejecta. The Weigelt blobs were ejected from the star less than ~120 years ago. If the blobs are not confined by surrounding pressures, they should expand and dissipate in roughly a sound-crossing time, on the order of 75 years. Their spatial and spectral appearance has changed in recent years, see Chap. 1. High-ionization emission lines such as He I, which are now trademarks of the knot spectra, did not appear until the 1940s [34, 57]. Continuing changes will occur as the material expands and moves farther from the star. Meanwhile there are cyclical changes with a 5.54-year period, usually attributed to the binarity of the central object (Sect. 5.2 below). Any discussion of the inner ejecta must be framed with reference to the epoch of the observations.

5.2 Spectroscopic Overview of the Weigelt Blobs

The Weigelt blobs produce H I and He I recombination lines and more than 2,000 other identified emission lines spanning a range of ionizations from Ca^{+} and Ti^{+} up to S^{+2}, Ar^{+2}, and Ne^{+2}. Most of the UV, visual, and near-IR features belong to singly-ionized iron group species, notably Fe II and [Fe II]. Extensive line lists are available [14,46,52,98,100,102,111]. A particularly interesting aspect is the variety of strong fluorescent lines, whose upper energy states are vastly overpopulated by photoexcitation because of accidental wavelength coincidences with the H I Lyman series or other strong lines. Fluorescent Fe II $\lambda\lambda 2507, 2509$ are the strongest emission features in near-UV spectra; their enhancements compared to other Fe II lines are larger in η Car than in any other known object [24,66,67,100]. Altogether, the varieties of lines and excitation processes provide a broad array of diagnostics with which to study both the blobs and the central object.

An obstacle to these studies has been the seeing-limited angular resolution of ground-based spectroscopy, typically $\sim 1''$. To some extent the line profiles distinguish between blobs and the stellar wind; velocity dispersions are 40–70 km s^{-1} vs. several hundred km s^{-1} respectively. However, the rich narrow-line spectrum has complex blends that resemble broad features, *HST* spectra of the inner ejecta show differences at sub-arcsec scales, and reflection and projection effects occur at all scales. Moreover, some high ionization forbidden lines originate in high velocity gas, distinct from the blobs and not directly part of the stellar wind (Sect. 5.4.1). Spatial resolution better than 0.2″ is therefore essential for detailed studies.

The spectra of the stellar wind and the blobs vary with a 5.54 years period [13, 16, 17, 72, 75, 106, 108]. This cycle is punctuated by "events" defined by the disappearance of high-ionization emission from the blobs and inner ejecta, notably [Ne III], [Fe III], [Ar III] and He I. These features vanish on time scales of 1–6 weeks and then recover more slowly afterward. Meanwhile other phenomena occur, including an abrupt drop in the 2–10 keV X-ray emission from the colliding winds [10, 11, 51, 60].[4] A major goal of η Car studies since 1997 has been to understand the spectroscopic events, which must be linked to some basic aspect of the central source. The disappearance of high-ionization [Ne III], [Fe III], [Ar III] and He I lines is almost certainly caused by an abrupt drop in the far-UV flux incident on the blobs and inner ejecta as originally proposed by [108] (see Sects. 5.3 and 5.5 below). What physical effects in the central object can change its spectral energy output?

The regularity of the event cycle is usually interpreted as evidence that the central object is a 5.54 year binary [13, 15, 16], but no specific model has emerged that explains the full range of phenomena (see [16, 57, 77] for recent discussions). The proposed companion star is less luminous but hotter than the very massive primary [75]. It contributes most of the helium-ionizing far UV flux and thus controls the high-ionization emission lines. The orbit is highly eccentric, so close interactions occur only for a brief time near periastron. Between spectroscopic events, the stars are widely separated and both contribute to the ionization and excitation of the inner ejecta. During an event, near periastron, the hot companion plunges deep inside the dense primary wind, so its contribution to the emergent far-UV emission is briefly obscured.[5] The inner ejecta then become less ionized because they receive light from only the cooler primary, whose wind structure and spectral energy output might also be altered by the binary encounter [16, 19, 28, 89]. A model of this type can account for the X-ray variations [10, 60, 83]. But there are many uncertainties, including the nature of the hot companion star, the orbit parameters, and properties of both winds. Any model of the binary system must explain the spectroscopic properties of the inner ejecta and, specifically, the changes that occur throughout the 5.54 years cycle. See comments and references in Chap. 3.

[4] See Chap. 9.

[5] (Editors' comment:) Strictly speaking, the disappearance of far-UV near periastron may be caused by mass accretion onto the secondary star as proposed by Soker et al. See [77], Chap. 3, and references therein.

5.3 Blob D and the 2003.5 Spectroscopic Event

The spectroscopic event that occurred in mid-2003 was studied at wavelengths ranging from radio through X-rays.[6] Here we summarize key results for the inner ejecta, especially the almost-resolved ($\sim 0.1''$) spectra of blob D that were obtained as part of the *HST* Treasury program on η Car.[7] This publicly available dataset provides the most complete and reliable existing information of this type. The 2003.5 HST observations covered roughly 1630–10100 Å at resolution $\sim 40\,\mathrm{km\,s^{-1}}$and $\sim 0.1''$; no better data on an event are expected in the forseeable future.[8]

Figures 5.1–5.8 illustrate spectral properties of blob D measured on six occasions across the 2003.5 event. The line with highest ionization, [Ne III] $\lambda 3868$, showed

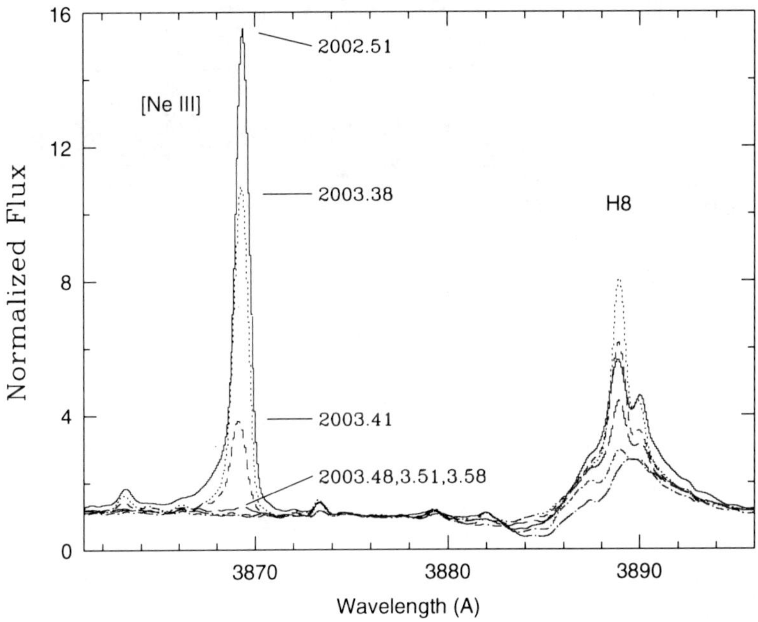

Fig. 5.1 [Ne III] $\lambda 3868$ and H8 emission lines in blob D on the dates 2002.51 (*solid line*), 2003.38 (*dotted*), 2003.41 (*short dash*), 2003.48 (*long dash*), 2003.51 (*short dash-dot*), 2003.58 (*long dash-dot*). The [Ne III] peak heights are marked on the different dates. The weak emission bump at 3,863.2 Å is Si II 3,863.69 ÅThe broad P Cygni profile in H8 is reflected starlight (From [43])

[6]See other chapters in this volume as well as [16, 28, 73] and many refs. therein.

[7]http://etacar.umn.edu/.

[8]The only instrument with adequate spatial resolution, HST/STIS, was inoperative during the subsequent event in 2009. Moreover, the star's rising brightness progressively makes the Weigelt blobs harder to observe [74]. Thus it is very conceivable that no one will ever obtain new "event" spectra of these objects as good as the 2003.5 STIS data.

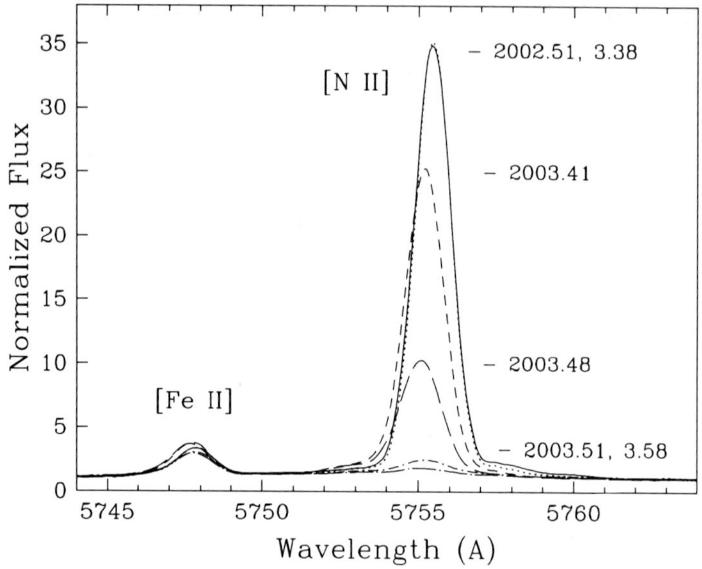

Fig. 5.2 [N II] λ5755 and [Fe II] λ5747 measured in blob D during the 2003.5 spectroscopic event. The labels and line styles for the different dates match Fig. 5.1

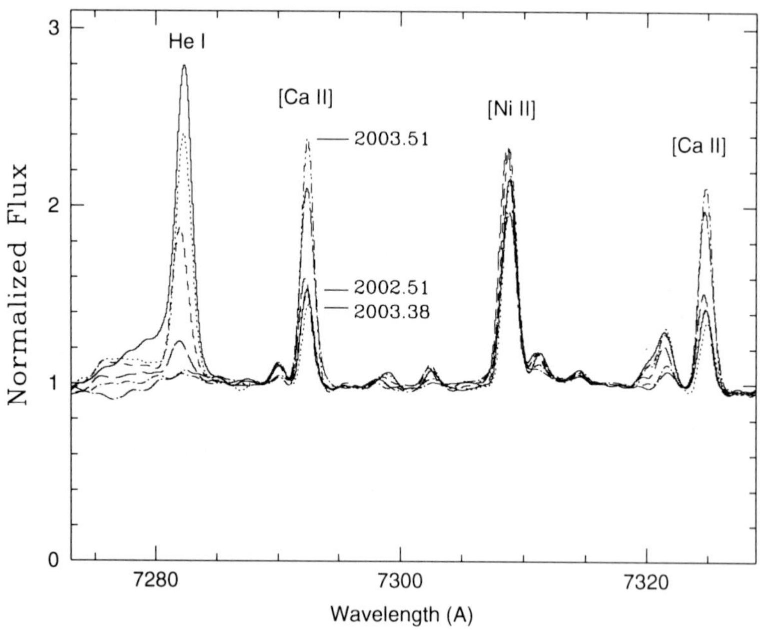

Fig. 5.3 [Ca II] λ7291,7323, [Ni II] λ7307, and He I λ7281 in blob D during the 2003.5 event. The [Ca II] lines tripled in strength, the [Ni II] strengthened slightly, and the narrow He I line became >30 times weaker during the 2003.5 event. Labels and line styles match Fig. 5.1 (From [43])

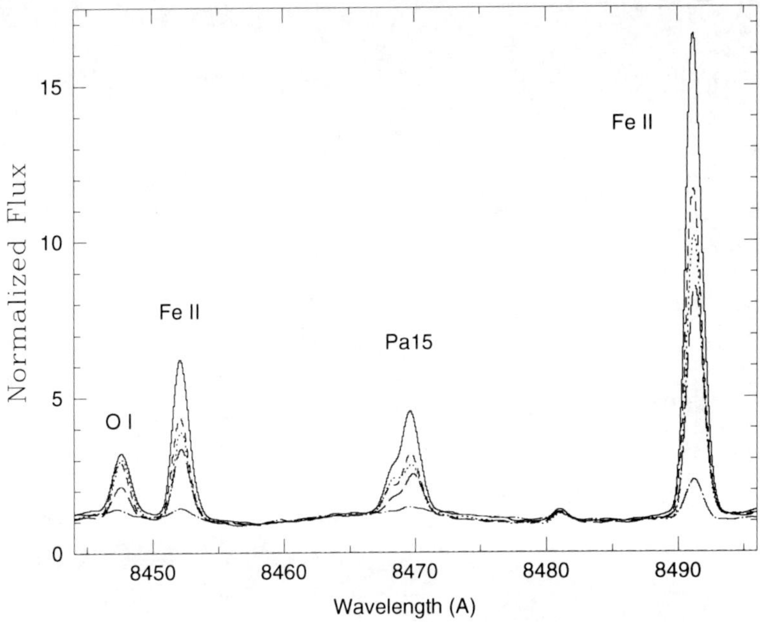

Fig. 5.4 H I Pa15 and the fluorescent lines O I λ8446 pumped by Lyβ and Fe II λ8451 and λ8490 pumped by Lyα in blob D. The different line styles represent different observation dates as in Fig. 5.1. The fluorescent lines decreased much more dramatically during the event than other collisionally-excited lines of Fe II or [Fe II] (From [43])

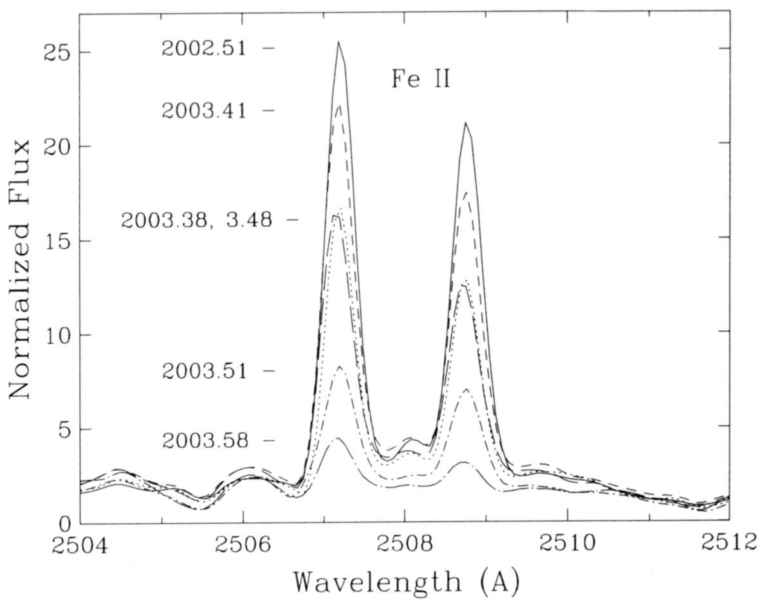

Fig. 5.5 The fluorescent lines Fe II λ2508 and λ2509 pumped by Lyα in blob D. The line labels and styles match Fig. 5.1

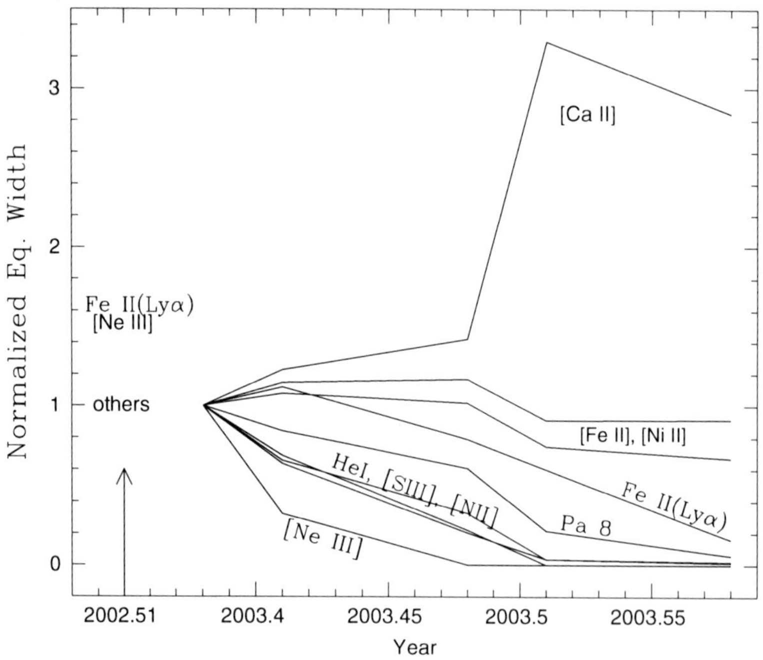

Fig. 5.6 Measured equivalent widths (Å) of representative lines in blob D, normalized to their value in 2003.38. Fe II(Lyα) represents an average of several fluorescent Fe II lines pumped by Lyα. The time scale is distorted for the 2002.51 data, Vertical positions of labels for 2002.51 indicate equivalent widths relative to 2003.38, where "others" refers to all lines plotted here except [Ne III] and Fe II(Lyα). Higher ionization lines weakened sooner and more completely during the event, while the lowest ionization lines of [Ca II] became stronger (From [43])

the most rapid decline and complete disappearance. Figure 5.1 shows it along with the more complex Balmer H8. The narrow in situ H I emission disappeared while the reflected broad stellar wind component weakened in emission but strengthened in its blueshifted P Cyg absorption (see also [28, 55]). Narrow lines with slightly lesser ionization, e.g., [Ar III] λ7135 and [S III] λ9532, disappeared about 2–3 weeks later than [Ne III] (see Fig. 2 in [43]). At lower ionizations, features like [N II] λ5755 lagged even farther, cf. Figs. 5.1 and 5.2. At $t = 2003.38$ the event was significantly underway in [Ne III] but not in [N II], and at 2003.48 the [Ne III] line had disappeared while [N II] emission was still present.

Continuing toward lower ionization, the behavior of Fe II, [Fe II], and [Ni II] ranged from modest weakening to modest *strengthening* during the event. The lowest ionization lines measured in blob D, [Ca II], Ti II and V II, all became stronger (see also [14]). Figure 5.3 shows that [Ca II] λ7291 and λ7323 lines nearly tripled in strength, while [Ni II] λ7303 mildly strengthened and He I λ7281 disappeared. Some Ti II lines (not shown) approximately doubled in strength.

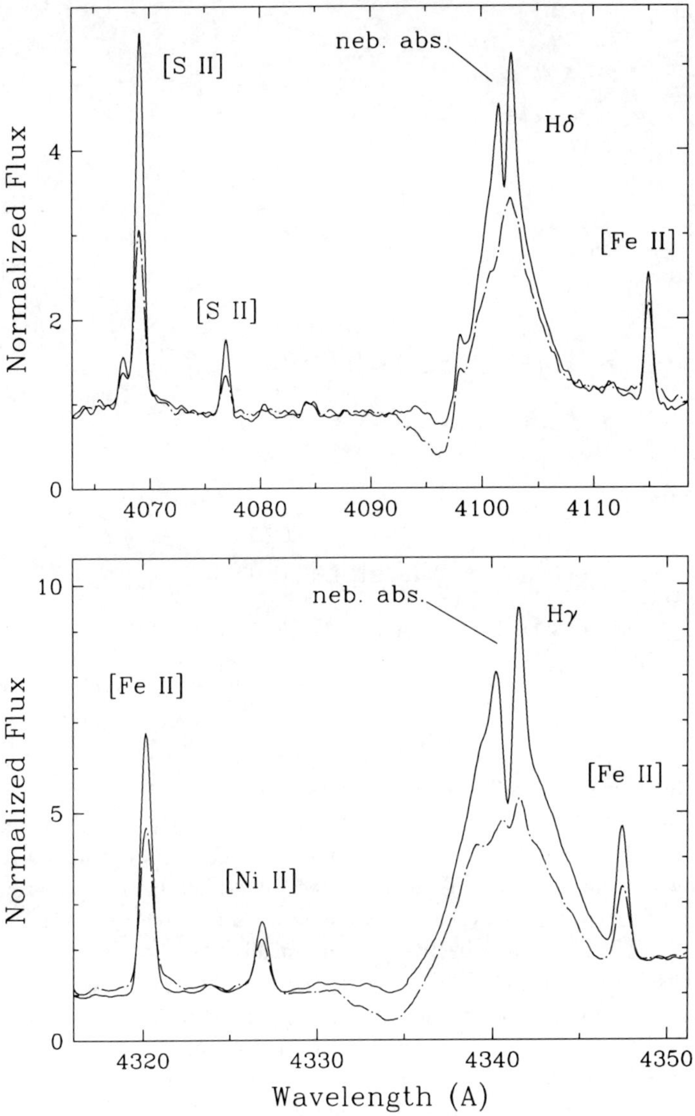

Fig. 5.7 The Hδ (*upper panel*) and Hγ (*lower panel*) lines are plotted along with various lines of [Fe II], [S II] and [Ni II] measured in blob D on the dates 2002.51 (*solid line*) and 2003.58 (*long dash-dot*). The P Cygni shaped Balmer lines are seen in reflected light from the star. Their broad emission weakened and broad absorption strengthened during the event (2003.58). Narrow nebular absorption (*neb. abs.*) in both Hδ and Hγ, at roughly $-50\,\mathrm{km\,s^{-1}}$ heliocentric, disappeared during the event

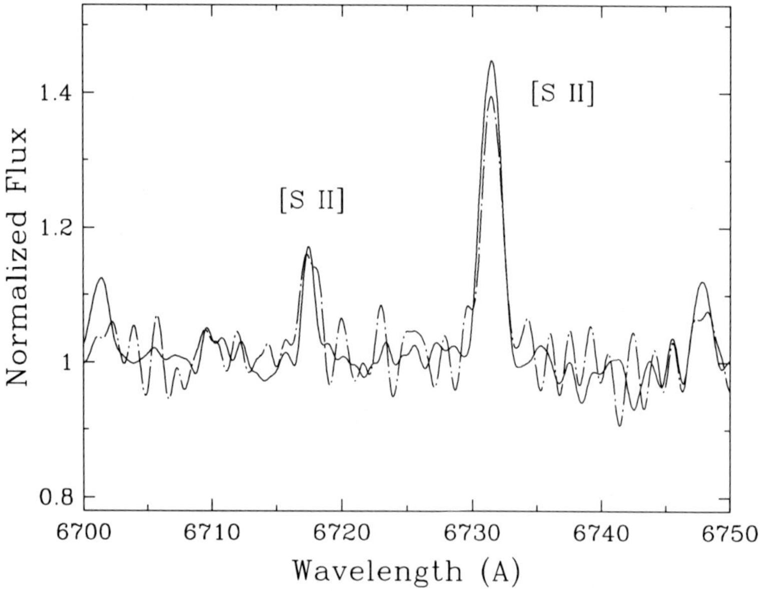

Fig. 5.8 [S II] lines measured in blob D in 2002.51 (*solid line*) and 2003.58 (*log dash-dot*) are plotted for comparison to the higher energy [S II] doublet shown in Fig. 5.7. The different flux changes between these lines suggest that the temperature dropped by roughly 15% during the event

These phenomena were correlated with the ionization energy needed to create each ion. Ne^{+2}, Ar^{+}, S^{+2} and N^{+} require 41.0, 27.6, 23.3 and 14.5 eV, respectively, while Fe^{+} and Ni^{+} require 7.9 and 7.6 eV, and Ti^{+} and Ca^{+} need just 6.8 and 6.1 eV. The narrow in situ (not reflected) emission lines of H I and He I are consistent with this pattern if they form, as expected, by recombination in regions of H^{+} (13.6 eV) and He^{+} (24.6 eV). He I lines behaved approximately like [Ar III] and [S III], while the H I lines declined like [N II] (see also [43]).

The only obvious exceptions to this simple ionization trend were fluorescent features excited by H I Lyman lines. Figures 5.4 and 5.5 show the dramatic weakening of fluorescent O I λ8446 pumped by Lyβ, and Fe II lines excited by Lyα. Their behavior differed from the other Fe II and [Fe II] lines and seems to contradict the ionization trend. However, their emission depends not only on the amounts of Fe^{+} and O^{0}, but also on H^{+} which produces Lyα and Lyβ by recombination (see Sect. 5.4.6). In fact, the fluorescent lines behaved much like the narrow Balmer and Paschen lines in blob D, consistent with the ionization trend. This can be seen in Fig. 5.4, which shows the disappearance of narrow H I Pa15 along with the fluorescent Fe II and O I features. All these lines disappeared when the gas became mostly H^{0}.

Figure 5.6 summarizes the ionization trend for narrow lines in blob D during the 2003.5 event. Damineli et al. [16, 17] produced similar plots based on ground-based observations of a spatially unresolved mixture of regions, but with better temporal sampling and a longer temporal baseline. They show that the high-ionization lines disappeared abruptly, in just 5–10 days. The highest ionization line, [Ne III] λ3868, was extinguished first, followed by He I λ6678 about 5 days later, then [S III] λ6312 6.5 days later, and finally [N II] λ5755 8.5 days after that. The highest ionization lines were also the last to recover after the event, roughly in reverse order of their disappearance; the recovery times were more gradual (months) than the disappearance times (days to weeks).

The shortest disappearance times might be limited by recombination rates, which depend on gas density (see Sect. 5.4.2). But this is unlikely to dominate the other temporal behaviors, since the recovery times (when overall ionization is increasing) are much longer. Therefore, the emission line changes probably trace the central source's spectral changes as viewed by the inner ejecta. In particular, the far-UV flux that regulates [Ne III] emission must have been extinguished faster, earlier and then recovered more slowly, than the lower energy spectrum that controls the lower ions. The incident photon energy distribution must have progressively "softened" until mid-event when all of the far-UV was gone. Then the spectrum hardened again as the far-UV recovered over a period of months (see also [17, 75, 77] and Sect. 5.5 below).

Another spectral change tied to the ionization was the weakening of narrow Balmer *absorption* during the 2003.5 event. Unlike the stellar wind's P Cyg absorption, these lines are formed in regions far outside the wind and even outside the Weigelt blobs (Sect. 5.4.6 below). They require a large column density of dense, partially ionized gas with a significant population of H^0 in the $n = 2$ level. Figure 5.7 shows the disappearance of Hγ and Hδ absorption in blob D during the 2003.5 event. The corresponding Hβ feature (not shown) weakened by a factor of $\sim$2 but did not disappear. This narrow absorption line behavior did not exactly follow the weakening of the narrow H I emission lines or the Fe II emission lines pumped by Lyα, but a decline during the 2003.5 event surely did occur in blob D. Weaker Balmer absorption strength indicates a drop in the H^0 $n = 2$ population, related to a lower degree of ionization.[9] We conclude that the absorbing gas participated in a spectroscopic/ionization event similar to the emission line regions in the inner ejecta.

Closer inspection of these data suggests that the temperature in blob D also fell by a moderate amount during the 2003.5 event. For each ion species, emission lines arising from higher energy states generally faded more dramatically.

[9]Johansson et al. [71] report that corresponding narrow absorption in Hα *strengthened* during the 2003.5 event, and the same is true of ground-based spectra of the star plus ejecta in the 2009.0 event [84]. We cannot make direct comparisons to these results, because the spatial coverage was different and instrumental saturation in the *HST* data may have degraded spectral extractions near the peak of Hα. We can only speculate that the ground-based Hα absorption results were affected by blending with the narrow emission from the blobs, which is time variable and stronger in Hα than in the other Balmer lines. Hα might also be less sensitive to changes in the $n = 2$ population if its larger oscillator strength leads to greater line saturation compared to Hγ and Hδ.

Figures 5.7 and 5.8 show, for example, that [S II] λ4069 and λ4076 declined by a factor of 2 during the event while [S II] λ6716 and λ6731 decreased by only ~10%. These lines are collisionally excited, and the upper states of $\lambda\lambda$4069, 4076 and $\lambda\lambda$6716, 6731 have energies of 3.0 and 1.8 eV repectively. There is also a density dependence [42, 81], but if we assume that the density did not change much during the event, then the observed change in the line ratio indicates a drop in temperature. We cannot derive a specific temperature without knowledge of the density, but if the temperature in the S^+ gas was ~7000 K before the event (2002.51), then during the event (2003.58) it declined by roughly 900 K (or ~700 K if the initial temperature was 6000 K). Similar effects can be seen in [Fe II] and [Ni II], see Sect. 5.4.2 below.

Finally, we note that the Balmer P Cygni absorption lines, formed in the stellar wind and reflected by dust in blob D, varied contemporaneously with the narrow in situ emission lines discussed above (Figs. 5.1 and 5.7). Detailed comparisons have shown that these changes tracked each other to within a month [17, 28, 55, 94]. This relationship indicates again that the blobs were responding to changes in the radiative output from the central source, presumably the opaque primary stellar wind plus the hot secondary star. (Variations in the kinetic energy of the primary wind would affect the blobs only after a travel-time delay of a year or more.) This result implies that ionization and excitation in the blobs (and probably all of the inner ejecta) are dominated by the radiative flux from the central star, rather than shocks or other mechanical processes (see Sect. 5.5).

5.4 Analysis: Physical Properties and Peculiarities

Here we review basic physical properties that can be derived from the spectral lines in the inner ejecta.[10] We discuss blobs B, C and D interchangeably because their spectra are broadly alike.

5.4.1 Kinematics and Location of the Highly Ionized Gas

Kinematic data are essential for defining locations and origins of various types of regions in the inner ejecta. We mentioned some results in Sects. 5.1 and 5.2, and more information can be found elsewhere in this volume, in Chaps. 6 and 7. The Weigelt knots have Doppler velocities around $-40\,\mathrm{km\ s^{-1}}$ in both low and high excitation emission lines. High ionization features such as [Ne III] have Doppler widths of 65–70 $\mathrm{km\,s^{-1}}$ (FWHM), while [Fe II] and non-fluorescent Fe II have FWHM ~55 $\mathrm{km\,s^{-1}}$ (Sect. 5.3, and [91]). We expect features with disparate ionization energies to form in different locations (Sect. 5.5, [77, 99]).

[10] Some of the text in Sect. 5.4.1 was contributed by K. Davidson and A. Mehner.

In one plausible but unproven geometry, the blobs are mostly neutral (H^0) gas with ionized layers facing the central binary system. Since the hot secondary star is the chief source of relevant ionizing photons, the ionized zone varies during the 5.54 years orbit, leading to alternate appearance and disappearance of high-ionization lines. At spatial resolution $\sim 0.05''$, *HST* spectroscopy shows that locations of maximum [Ne III] and [Fe III] brightness approximately match those of the low-excitation lines [75].[11] Higher spatial resolution will be needed to show ionization stratification. The differing line widths suggest that high-ionization zones are only loosely related to the low-ionization material, and possibly ablating from the blobs.

Ground-based spectra show blue wings of [Ne III], [Fe III], [Ar III], and [S III], extending to local peaks near $-380\,\mathrm{km\,s^{-1}}$ [1, 14, 42, 52, 95]. These are unrelated to the Weigelt knots; *HST* spectroscopy shows that the blue-shifted [Fe III] originates in a slightly elongated region with radius $\sim 0.1''$, centered near the star (Figs. 8 and 9 in [75]). The simplest guess is that these features come from our side of a mildly oblate region in the outer wind, with densities low enough to emit forbidden lines [75]. Larger-scale locations in the Homunculus are also possible, however. Helium emission is not useful in this regard, because the complex He I line profiles combine several distinct regions, including absorption in the wind.

Apart from the Weigelt knots, narrow high-excitation forbidden lines appear in *HST* spectra of the star itself [75]. Based on their small widths, low velocities, and de-excitation densities, they represent line-of-sight gas comparable to the Weigelt knots, not the stellar wind. Given *HST's* high spatial resolution, these features' response to the central UV output must be closely correlated with our direct spectroscopic view of the star itself.[12] In fact the line-of-sight [Ne III] and [Fe III] intensities do vary systematically and non-trivially through the 5.54 years cycle [16, 75]. Their growth, broad mid-cycle maximum, and gradual decline seem reasonable in terms of photoionization by the secondary star as it moves along its orbit, but no quantitative model has been developed. The line-of-sight data [75] showed a conspicuous brief secondary maximum in [Ne III] and [Fe III] several months before the 2003.5 event (Fig. 5.10). This may have been the time when the orbiting secondary star was optimally located for photoionizing our line of sight [77].

Narrow Balmer absorption lines also trace the ionized nebular gas (Sect. 5.4.6). These features appear in spectra across the central $1''$ to $2''$, encompassing the star and inner ejecta [27]. In *HST* spectra of the Weigelt blobs, they have heliocentric velocities of roughly -46 to $-50\,\mathrm{km\,s^{-1}}$ (Fig. 5.7). These values are similar to those

---

[11] Some earlier authors assumed that high-ionization lines originate diffusely between the star and the Weigelt knots [99]. To some extent this may be true, but the brightness peaks are located as stated above. *HST* had no imaging filters suitable for isolating the pertinent spectral lines.

[12] Strictly speaking, our view of the dense primary wind plus the hot companion star. The Weigelt knots may differ because they "see" the star from other directions. The primary wind is not spherically symmetric, and local circumstellar extinction may be both patchy and variable.

for the narrow high-ionization emission lines [91]. During the 2003.5 spectroscopic event, narrow Hβ absorption weakened while corresponding Hγ and Hδ absorption disappeared, in a manner similar to the narrow H I emission lines. It therefore seems likely that these absorption lines form in an ionized layer that is loosely related to the blobs. In spectra taken along our direct line of sight to the star, there is additional narrow Balmer absorption at $-146\,\mathrm{km\,s^{-1}}$ [39]. This component of ionized absorbing gas seems to have no relationship to either the stellar wind or the blobs. It probably resides elsewhere in the inner ejecta, see Sect. 5.4.6.

5.4.2 Reddening, Extinction and Temperature

Extinction and reddening by dust can dramatically affect the observations. In principle we can estimate the reddening, E_{B-V}, by comparing emission lines that share the same upper level. If optically thin, then their intrinsic flux ratios are

$$\frac{F_1}{F_2} = \frac{A_1\lambda_2}{A_2\lambda_1} \tag{5.1}$$

where F_1, λ_1, and A_1 are the flux, wavelength and decay rate for line 1, etc. Ideally we would measure enough lines at different wavelengths to characterize the reddening curve. In practice, however, there are not enough well-measured lines of this type. One could simply adopt a standard reddening curve for the interstellar medium (e.g. [7]), but both the cirumstellar and interstellar extinction for η Car are known to be anomalously gray with $A_V/E_{B-V} > 4$ rather than a normal value around 3.1 [21]. Thus A_V was probably about 7 mag for the central star in 1998, but $E_{B-V} \sim 1$ [54]. Both these values appear to have declined since that time [74, 76]. The blobs are thought to have much less extinction, in order to explain their surprisingly large brightnesses compared to the central object [24].

Hamann et al. [47] measured suitable line pairs in HST/STIS spectra of blobs B + D. They found $E_{B-V} \sim 0.6$, 0.7, and 0.8 mag, respectively, based on [Fe II] λ3175/λ5551, [Fe II] λ3533/λ6355, and [Ni II] λ4326/λ7256. Investigations with non-[Fe II] lines, however, have given $E_{B-V} \lesssim 0.2$ [75,98]. It is difficult to reconcile all these results, except to note that they span different wavelengths, and that a STIS instrumental effect tends to counteract reddening for $\lambda \gtrsim 4500$ Å [75].

Another way to estimate both reddening and temperature is to compare the entire rich spectrum of [Fe II] lines to theoretical predictions. The observed [Fe II] lines arise from metastable states at energies $\lesssim$4 eV, and densities in the Weigelt blobs (Sect. 5.4.3 below) exceed the collisional de-excitation values for most of them. Thus we can reasonably assume that the level populations are close to local thermodynamic equilibrium. In that case the relative [Fe II] line strengths are given by (5.1) multiplied by a Boltzmann factor for the upper states' populations. With enough lines spanning a range of wavelengths and excitation energies, we can solve

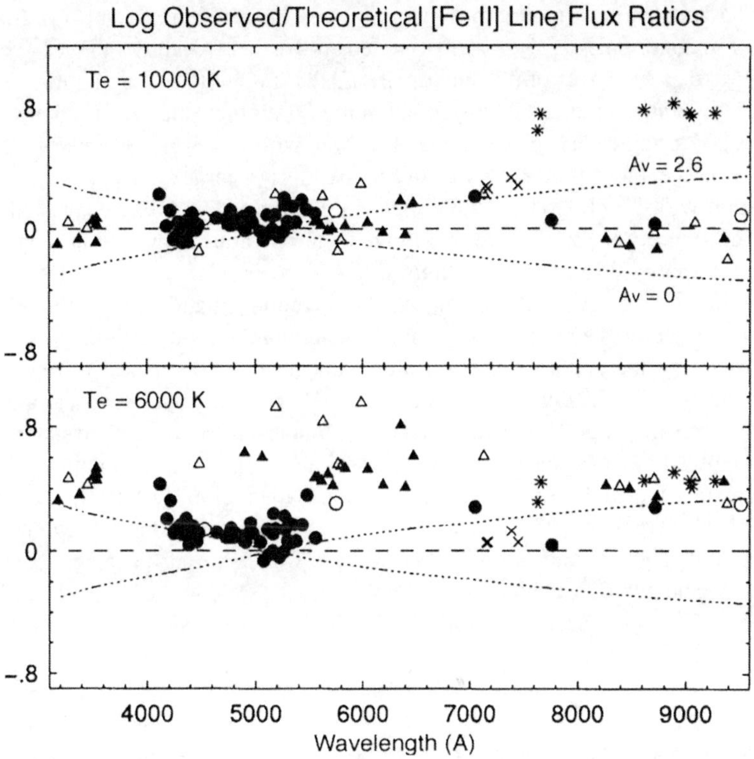

Fig. 5.9 Observed [Fe II] line fluxes compared to LTE predictions at assumed temperatures 10,000 K (*top panel*) and 6,000 K (*bottom*), if $E(B-V) = 0.42$ ($A_V = 1.3$ with a normal reddening law). The vertical scale is $C + \log_{10} F_{\rm obs}/F_{\rm theor}$, normalized to be zero for [Fe II] λ5159. Lines show expected values for $E(B-V) = 0, 0.42$, and 0.84 mag, with a standard reddening curve. Each point represents a different [Fe II] line in the spectrum of Weigelt blobs B + C. Symbols indicate approximate upper-state energies: asterisks for $E_{up} < 1.7$ eV, crosses for 1.7–2.5 eV, circles for 2.5–3.7 eV, and triangles for $E_{up} > 3.7$ eV

for both reddening and temperature. Resulting temperatures are useful because the usual nebular diagnostics [81] don't work for the Weigelt blobs. For example, [O III] λ4363 and λ5007 are too weak because of η Car's low oxygen abundance, and [N II] $\lambda5755/\lambda6583$ is too sensitive to density in this environment.

Using a few [Fe II] lines measured in blobs B + C + D with the pre-1997 HST/FOS instrument, and assuming that $T_e \sim 8000$ K, Davidson et al. estimated $E_{B-V} \sim 0.55$ [24]. Figure 5.9 shows results of a more detailed analysis by Hamann et al. [47] using every reliably measured [Fe II] line in the visual and red spectrum of blobs B + C obtained in March 1998 with HST/STIS. Here each [Fe II] line's observed flux F(observed) is plotted relative to a theoretical value F(theoretical) that assumes LTE conditions at temperatures 6000 and 10000 K, $E_{B-V} = 0.42$, and a standard reddening curve with $A_V/E_{B-V} = 3.1$ (atomic data from [62]). Since η Car has an abnormal A_V/E_{B-V} ratio as noted above, the A_V values in Fig. 5.9 should be regarded as estimates of the quantity $3.1E_{B-V}$, smaller than the true A_V.

No temperature-and-reddening combination fits all the data. Levels above 2.5 eV (circles and triangles in Fig. 5.9) agree fairly well with $T \sim 10000$ K and $E_{B-V} \sim 0.4$, but this choice under-predicts lower-excitation lines (asterisks). The latter are more consistent with $T \sim 6000$ K and $E_{B-V} \sim 0.8$. Qualitatively a discrepancy like this can occur merely because there is a range of temperatures, so the high-excitation lines preferentially represent the highest T – especially if dust within each blob causes $E_{B-V} \sim 0.1$ or 0.2 mag of *local* reddening. Quantitatively, though, this explanation requires a temperature range broader than 6000–10000 K.

Most likely the actual gas temperatures are near 6000 K based on the lowest energy levels (Fig. 5.9), while levels above 2 eV are over-populated by continuum pumping. Permitted Fe II transitions absorb UV star light and populate states above 4 eV, followed by a cascade through the lower states. This "continuum fluorescence" mechanism affects Fe II emission in AGN broad-line regions [79, 107], LBV winds [53, 54], H II regions [97], and probably the "strontium filament" region in the Homunculus ([4, 98] and Sect. 5.5 below). Eta Car has long been recognized as a good locale for fluorescence in general (e.g., [18, 24, 46, 101]), and the likely role of Fe II continuum fluorescence in the Weigelt blobs became clear in the late 1990s [47].

[Ni II] lines provide additional contraints. For example, the strong $\lambda\lambda 7378$, 7412 lines have an upper level about 1.8 eV above the ground state, while $\lambda\lambda 7255$, 7308 share an upper state near 2.9 eV. Their ratios are nearly immune to reddening but sensitive to temperature. In the high density limit with LTE populations, the theoretical $\lambda 7378/\lambda 7308$ ratio is about 21 for $T \approx 5000$ K and 5 for $T \approx 10000$ K [42]. The observed value in blob D at 2002.5 was about 6, broadly consistent with the [Fe II] results if there is some non-LTE over-population of the Ni^+ upper states.

During the 2003.5 event, these line ratios signaled a drop in either the temperature or the amount of fluorescence excitation. In blob D, for example (Sect. 5.3), [Fe II] and [Ni II] lines with higher energy states tended to weaken more than the lower-excitation lines. While this fact might have something to do with declining amounts of continuum photo-excitation, the [S II] changes described in Sect. 5.3 are more readily interpreted as a temperature decrease of $\Delta T \sim -700$ to -900 K.

Thus the observed changes in the line ratios suggest a temperature decline over 2–3 months. This is much longer than the radiative cooling time. For the nominal composition and physical conditions we derive for the blobs (see below), a representative cooling time is 1–10 days based on calculations in [12] and G.J. Ferland's unpublished *Hazy* ionization-code manual. In addition, the blobs are located just several light-days from the star. Thus it appears that the temperature responded to changes in the spectral energy output from the central source, with only small delays due to cooling and light travel times (see also Sect. 5.5).

5.4.3 Densities

The rich emission line spectra provide various density indicators. The most reliable involve ratios of lines with similar upper-state energies, i.e., members of a multiplet, to minimize the temperature sensitivity. Such ratios can be sensitive to electron densities n_e if they are within an order of magnitude or so of the critical densities for collisional de-excitation [81]. Most estimates for the Weigelt blobs provide only lower limits to n_e because each observed line ratio is near the high-density limit. Hamann et al. [47] found these results: $n_e > 10^4\,\mathrm{cm}^{-3}$ based on [S II] $\lambda 6716/\lambda 6731$, $n_e > 10^6\,\mathrm{cm}^{-3}$ from [S II] $\lambda 4069/\lambda 6731$, $n_e > 10^7\,\mathrm{cm}^{-3}$ from [Fe II] $\lambda 7155/\lambda 8617$, and $n_e \sim 10^8$–$10^9\,\mathrm{cm}^{-3}$ from [Ni II] $\lambda 3439/\lambda 3993$ and $\lambda 7412/\lambda 7387$. Wallerstein et al. [102] estimated $n_e \geq 10^7\,\mathrm{cm}^{-3}$ based on [S II] $\lambda 4068/\lambda 4076$. The classic ratio [O II] $\lambda 3729/\lambda 3726$ is not detected due to the low oxygen abundance, and the analogous [N I] $\lambda 5201/\lambda 5198$ lines are severely blended with Fe II and [Fe II]. Later work [98] produced similar results using photoionization models.

The spatial and spectral complexity of the inner ejecta suggest that there is a range of densities. The estimates quoted above apply to low-ionization gas and they are skewed toward high densities which give the largest emissivities. There are no simple line-ratio density indicators for the more highly ionized gas, but we obtain some constraints from the observed time scales for H I and He I emission changes during spectroscopic events. Such changes cannot be much faster than the recombination time $t_{rec} \sim 1/\alpha n_e$, where α is a recombination rate coefficient. At the beginning of the 2003.5 event, the H I and He I line fluxes dropped substantially in less than 10 days [16, 17, 50]. This fact implies densities $n_e > 10^6\,\mathrm{cm}^{-3}$ and $n_e \geq 10^7\,\mathrm{cm}^{-3}$ in the He I and H I emitting regions, respectively.

We conclude that $n_e \sim 10^7$–$10^8\,\mathrm{cm}^{-3}$ is a reasonable estimate for the main emitting regions in the blobs, but other density regimes may also be present.

5.4.4 Composition

The composition of the ejecta is relevant to dust formation in the stellar wind and to nucleosynthesis and mixing processes in the stellar interior. The amount of CNO processing is of particular interest for η Car. Hydrogen burning via the CNO cycle produces no net change in the total number of $\mathrm{C+N+O}$ nuclei, but the reaction rates in equilibrium lead to a net conversion of C and O into N. Spectra of the "outer ejecta" (just outside the nominal outer boundary of the bipolar Homunculus) show that nitrogen exceeds $\mathrm{C+O}$ there [23]. Inside the Homunculus the logarithmic N/O abundance relative to solar is very large, $2.0 \leq [\mathrm{N/O}] \leq 2.5$ [30], while the ejecta far outside it have nearly solar N/O [88, 93]. The outer ejecta thus appear to contain the first CNO processed gas to be expelled, possibly just before the ~1843 eruption (see also [105]). The stellar wind today is also CNO-processed [54].

If the blobs and inner ejecta were expelled after the Homunculus (Sect. 5.1), then they too should contain heavily CNO processed gas. The most reliable abundance estimates rely on lines that form in the same physical conditions. Hamann et al. [47] used the N III] λ1750 and O III] λ1664 inter-combination lines to estimate [N/O] > 1.8 (i.e., $n(\mathrm{N})/n(\mathrm{O}) > 60$) in blobs B + D. They also estimate [Fe/O] ~ 2.0–2.3, based on [Fe II] λ8617/[O I] λ6300 and [Fe II] λ7155/[O I] λ6300, where the factor-of-2 uncertainty comes mainly from the uncertain density. These results are consistent with an estimate that [C/Ne] and [O/Ne] are both roughly −1.7 to −2.0 based on photoionization models [99]. We conclude that the blobs and associated high-ionization gas are heavily CNO-processed.

The dust content of the blobs is more uncertain. The fact that we see a reflected spectrum of the star in the blobs indicates that dust is present. Mid-IR observations show warm dust in the inner ejecta [9, 90]. The strutures seen in the mid-IR images closely correspond with the knots seen in the visible although they are not spatially coincident. The visible structures, dominated by scattering, trace the *walls* of the dense clumps of dust, while the infrared structures are identified with the emission from hot dust, probably the external layers of the clumps. This observational evidence is consistent with the theoretical dust temperatures, which indicate that dust can survive closer to the star than the nearest Weigelt blob B [21,90]. The infrared flux appears to be decreasing from 2002 to 2005 [2], but is not correlated with the 2003.5 event or with orbital phase. It may be due to enhanced dust destruction in response to the increased stellar flux.

To avoid projection effects and determine how much dust actually resides *within* the line-emitting blobs, we can examine the gas phase depletions of refractory elements like Fe, Cr, Ni, Ti and Ca compared to non-refractory elements like C, N, O, Ar and S. In cool interstellar gas clouds in our Galaxy, refractory elements are typically depleted by factors of 10 to >100 because they are locked up in dust grains [86]. Iron, in particular, is depleted by a factor of ~200 in cool Galactic clouds. In the Weigelt blobs, however, one study of [Fe III] lines found the Fe/H abundance to be roughly half solar [99]. If we combine this with an estimate by Hillier et al. [54, 55] for solar Fe/H in the star, we conclude that iron is not strongly depleted in the vicinity of the blobs. This result is consistent with a cursory inspection of the low-ionization blob spectra, wherein emission lines of Fe, Ca and Ti are well represented compared to the lines of non-refractory elements like S and even the grossly overabundant N. Thus the depletions of refractory elements are much smaller in the blobs than in cool Galactic clouds.

There are two other hints that the dust-to-gas ratio is small in the blobs. First, the strong fluorescent emission lines of Fe II and O I pumped by Lyα and Lyβ, respectively, require many scattering events in the Lyman lines (Sect. 5.4.5 below). A "normal" Galactic amount of dust would destroy the Lyman line photons before they are absorbed substantially Fe II or O I. Second, energy budget considerations indicate that the blob material (like the Homunculus on much larger scales) has a relatively unobscured view of the central object [24]. If much dust is present, it must have a patchy distribution so that starlight can largely avoid it by scattering [55,74].

5.4.5 Fe II and Fluorescent Line Emission

Spectra of the blobs and inner ejecta are strongly affected by resonant fluorescence. Fe II has by far the richest spectrum of known fluorescent lines, including a spectacular pair at 2507.6 and 2509.1 Å (Fig. 5.5, [24,25,46,64,100]). There are also fluorescent lines of O I, Cr II, Fe III, Ni II and possibly Mn II [68, 70, 111]. Their upper energy states are vastly over-populated because of accidental wavelength coincidences with H I Lyman lines. An interesting exception is Mn II, which apparently absorbs a strong UV line of Si II [68]. Fe II and O I fluorescence has been discussed thoroughly [33, 37, 81, 87] for a variety of stellar environments, including symbiotic stars and of other early-type stars with dense circumstellar envelopes [8, 44, 45, 48, 63, 65, 82, 85]. In general this phenomenon can help us diagnose physical and radiative conditions in the emitting regions.

The fluorescence in Fe^+ is "pumped" by Lyα and yields a unique pattern of emission lines. The primary cascade lines appear in the UV between ~1800 and 3000 Å and also in the far-red between ~8000 and ~10000 Å ([46] and refs. therein). Figures 5.4 and 5.5 above show examples of Fe II fluorescent lines in the blob D spectrum. These lines would not be detectable without this form of excitation.

The strongest fluorescent lines in η Car are the Fe II $\lambda\lambda$2508,2509 shown in Fig. 5.5. Their particular excitation has been discussed extensively by Johansson et al. [64, 69]. They constitute an interesting puzzle because they appear far too strong compared to other lines arising from the same upper states; transitions to particular lower states are anomalously favored. Johansson et al. [69] and Johansson and Letokhov [66] proposed that stimulated emission is responsible – a natural UV laser! This hypothesis is controversial because it seems incompatible with simple estimates of the photon densities [21, 109]; more work is needed. In any case, Lyα fluorescence clearly does control the excitation of $\lambda\lambda$2508,2509.

Figure 5.4 also shows fluorescent O I λ8446, which represents a secondary cascade from an energy state pumped by H I Lyβ. The importance of fluorescent excitation in this case can be deduced from the relative strengths of the primary cascade lines in the near-IR (e.g., λ11287/λ13165) and from the strength of λ8446 compared to non-fluorescent O I λ7773 [46]. The O I resonance wavelength differs from Lyβ by only 0.04 Å or 12 km s^{-1}. Given the high Lyβ opacity expected in these regions, Lyβ photons incident from the outside would have little effect on the O^0 excitation. Therefore O I fluorescence must be driven by Lyβ photons created within the blobs; we will return to this point below.

Simple considerations of the fluorescence processes lead to useful constraints on the physical conditions. For example, the Fe^+ transitions that absorb Lyα arise from metastable lower states which must be significantly populated. These populations can be maintained by collisions if the gas densities are above the critical values for those levels, $n_e \geq 10^6$ cm^{-3}, consistent with our estimates in Sect. 5.4.3 above.

Another constraint involves ionization. Strong O I emission requires a significant amount of neutral gas, since $n(O^0)/n(O^+)$ is closely coupled to $n(H^0)/n(H^+)$ by charge exchange reactions. The Fe II emission regions are also expected to have appreciable amounts of H^0 (Sect. 5.5 below and [98]). However, Lyα and Lyβ photons must be abundant in order to drive the fluorescence. Any Lyman line that is incident from the outside will be blocked by the extremely large line opacities of H^0. Therefore, the fluorescence observed in Fe II, O I, etc. is caused by Lyman line photons that are created locally inside the emitting regions (see also below and [46]). This requires a particular ionization balance with enough neutrals to maintain sufficient O^0, Fe^+, etc., but also enough H^+ to produce Lyman line emission. In a photoionized gas, this balance is achieved in zones of partial ionization just behind (i.e., on the more-neutral side of) the H^+–H^0 recombination front [3, 98].

One can think of Fe II fluorescence as an escape route for Lyα photons that are otherwise trapped. The low Fe/H $\sim Fe^+/H^0$ abundance ratio means that such photons will scatter many times from hydrogen atoms before being absorbed by Fe^+. Some of this absorption occurs in Fe II lines that have relatively poor wavelength coincidences with Lyα (1215.67 Å). For instance, the strong Fe II λ2508 line is pumped by a UV transition offset by 630 km s^{-1}. Therefore, if the Lyα line profile in this gas is symmetric, it cannot be much less than 1260 km s^{-1} wide. For another strong fluorescent line, Fe II λ9123, the corresponding value is 1340 km s^{-1}. Since fluorescent Fe II lines with poorer wavelength coincidences are absent, 1300 km s^{-1} is a fair estimate of the full width of the exciting Lyα line within the gas [46,50]. Similar results have been derived from the Cr II fluorescence features [111].

Locally-emitted Lyα photons can indeed be distributed across this wide range in apparent velocities because of natural broadening. If the line-center optical depth is large, $\tau_0 \geq 10^4$, then the Lyα absorption profile is dominated by damping wings and FWHM $\sim 0.18\, v_D\, \tau_0^{1/3}$, where v_D is a characteristic doppler width [32]. Our estimate FWHM(Lyα) $\sim$ 1300 km s^{-1} implies that Lyα has $\tau_0 \sim 3 \times 10^8$, which corresponds to an H^0 column density of $N(H^0) \sim 3 \times 10^{21}$ cm^{-2} if the velocity distribution is thermal and $T_e \sim 7000$ K. This estimate agrees well with theoretical predictions for dense Fe II emitting regions [3, 33]. If the Fe II emitting region in η Car is ~20% ionized [98], then the total hydrogen column would be roughly $N(H) \sim 4 \times 10^{21}$ cm^{-2}. This is much smaller than the probable column density through an entire Weigelt blob (Sect. 5.1), which is consistent with the fluorescent lines forming in boundary layers between the neutral and ionized gas.

Another important constraint comes from the O I fluorescence. In low-density H II regions where the optical depth in Hα is not large, Lyβ is converted, after only about 10 scattering events, into Hα plus Lyα photons which then escape: "Case B recombination." This situation cannot produce O I fluorescence because a typical Lyβ photon is scattered by H^0 about 5×10^4 as often as by O^0, leaving little opportunity for O^0 to absorb Lyβ [22, 37]. O I fluorescence therefore requires large optical depths in Hα in order to trap the Hα photons and inhibit the conversion of Lyβ into Hα + Lyα. This in turn requires a large population in the $n = 2$ level of H^0. If we could view this gas against the background of a bright continuum source, we should see strong narrow absorption in the Balmer lines (see Sect. 5.4.6 below)!

Finally, it has been suggested that the Lyα photons needed for the Fe II fluorescence come from the stellar wind rather than in situ emission in or near the Weigelt blobs [20, 50, 66, 98]. This might seem plausible because (1) the Lyα intensity from the star should be much stronger than the adjacent stellar continuum [54], and (2) the width of the Lyα profile can be FWHM $\sim$ 1340 km s^{-1} because the wind speeds can exceed 500 km s^{-1} [54, 89]. However, it is not clear that the wind's Lyα profile really is this broad; the observed stellar Balmer lines are narrower (Fig. 5.7, [28, 55]). Several other stars with strong nebular Fe II fluorescence have even narrower H I lines than η Car, though their fluorescent spectra also require Lyα with FWHM $\sim$ 1300 km s^{-1} [44, 45]. Therefore an external source of broad H I emission lines does not seem to be important for the fluorescent Fe$^+$ excitation. A more serious concern is that the large Lyman line opacities would prevent external photons from penetrating the blobs to drive the fluorescence. We noted above that Lyβ photons from the outside cannot play any role in the O I pumping. The large Lyα opacities in the blobs imply that external Lyα photons could pump Fe II only in transitions that are far removed from the Lyα line center; the other Fe II lines need to be pumped by locally created Lyα. Thus there would need to be two fluorescent processes operating at the same time and varying in unison during the spectroscopic events. Moreover, it is not obvious why the star's Lyα intensity would vary as needed because, for example, the stellar Balmer line fluxes change by only a factor of $\sim$2. Overall, the observed Fe II fluorescent lines are more easily understood if they are linked to the ionization and the local creation of Lyα photons.

5.4.6 Narrow Nebular Absorption Lines

Narrow absorption lines of H I and some low-ionization metals appear across the 1–2$''$ core of the Homunculus, including the central star and the Weigelt blobs [19, 26,38,40,47,71,80]. Figure 5.7 shows, for example, narrow absorption in Hδ and Hγ in the spectrum of blob D. These features are clearly not related to the broad P Cygni wind profiles. At least some of them form in the inner ejecta (see below), while others apparently arise farther out in the "Little Homunculus" or in the outer shell surrounding the Homunculus itself [38, 40, 80]. This situation is highly unusual; one does not generally see Balmer absorption in the ISM or even in denser-than-average nebulae because very few H I atoms there are in the $n = 2$ level.

The metal absorption lines have low ionizations typified by Fe II and Ti II. In spectra towards the star they show at least 30 distinct velocity components, with the strongest features near -146 and -513 km s^{-1}. In the blob spectra the absorption components are less distinct and their velocities are different [80]. A detailed analysis of the stellar spectrum [40, 80] shows that many of the Fe II and Ti II lines arise from metastable excited states, so densities in the absorbing gas must be near or above those states' critical values. Gull et al.[40] estimate that the strong system at -146 km s^{-1} has $n_e \sim 10^7$–10^8 cm^{-3}and $T \sim$ 5700–7300 K, located roughly 1300 AU from the star. The relevant column density in Fe$^+$ is

$\sim 5 \times 10^{15}\,\mathrm{cm}^{-2}$. If we assume that Fe/H is solar and all of the iron is singly ionized, then the corresponding hydrogen column is $N_H \sim 2 \times 10^{20}\,\mathrm{cm}^{-2}$. They also note that metastable Fe^+ levels below $\sim$3.2 eV are approximately in LTE, while those at higher energies are overpopulated compared to LTE – similar in this respect to the Weigelt blobs as discussed above. The estimated location places the absorbing gas within the inner ejecta, at $r < 1''$, but it is clearly distinct from the blobs since it has a different velocity and lies along our line of sight to the central star.

Balmer absorption lines in a nebular environment are surprising because they require significant populations in the $n = 2$ level of H^0. A line-center optical depth of $\tau_0 \geq 1$ in Hγ, for example, implies a column density $N(n = 2) \geq 3 \times 10^{13}\,\mathrm{cm}^{-2}$ if the doppler velocities are thermal with $T \approx 7000$ K. This requires a dense gas that is neutral enough to have H^0 but also ionized enough to populate the $n = 2$ level. This situation is believed to occur in the broad emission line regions of quasars [22, 35], and Balmer line absorption has been observed directly in quasar outflows where the densities and ionizations might be similar to the inner ejecta of η Car [41, 58].

One way to populate the H I $n = 2$ level is by collisions in a warm gas where this level is thermalized, such that the downward rate due to electron collisions exceeds the net rate for radiative decays. At first sight this would require an absurdly large electron density – except that almost every radiative decay is nullified, macroscopically speaking, when the fresh Lyα photon is immediately absorbed in an excitation event. Therefore our *net* radiative decay rate includes only the few Lyα photons that escape from the vicinity. In this case the minimum density for strong collisional de-excitation is roughly given by

$$\frac{n_e q_{21}}{A_{21}\beta} \approx \frac{n_e \tau_0}{n_{cr}} \geq 1\,, \tag{5.2}$$

where q_{21} is the downward collision rate coefficient, τ_0 is the line-center optical depth in Lyα, $\beta \sim 1/\tau_0$ is the escape probability for Lyα photons, and $n_{cr} \approx A_{21}/q_{21} \sim 10^{17}\,\mathrm{cm}^{-3}$ is the critical density for the $n = 2$ level at $T_e \sim 7000$ K in the absence of Lyα entrapment. For $n_e \leq 10^9\,\mathrm{cm}^{-3}$, thermalization requires $\tau_0 \geq 10^8$ and thus a total H^0 column density $N(n = 1) \sim N(\mathrm{H}^0) \geq 10^{21}\,\mathrm{cm}^{-2}$, assuming a thermal velocity dispersion. This corresponds to $N(n = 2) \geq 2 \times 10^{14}\,\mathrm{cm}^{-2}$ in LTE at 7,000 K, easily sufficient to produce $\tau_0 > 1$ for the Balmer absorption lines.

These physical conditions are reasonable for the partially ionized gas associated with the Weigelt blobs. In fact, they resemble what we inferred from the fluorescent emission lines (Sect. 5.4.5). However, thermalization is an extreme requirement. Balmer absorption lines can occur at values of $N(\mathrm{H}^o)$ and $\tau_o(\mathrm{Ly}\alpha)$, below the thermalization limit if recombination is also important for creating Lyα photons that are subsequently trapped. The inability of these Lyα photons to escape can lead to $n = 2$ populations that are significantly enhanced relative to LTE [41].

This general scheme for observable Balmer absorption is supported by measurements of damped Lyα and Lyβ absorption lines in spectra of the central star [55]. The origin of the damped lines is uncertain because they are too broad to measure their kinematics. However, they clearly form in the nebular environment of η Car and a likely location is in the Balmer line absorber discussed above. The neutral hydrogen column density derived for the damped absorber, $N(\mathrm{H}^o) \sim 3\times10^{22}\,\mathrm{cm}^{-2}$, is also consistent with the conditions needed for Balmer line absorption.

Another factor that might play a role is the metastable nature of the H^0 $2s$ state [71]. Radiative decays from $2s$ occur primarily by 2-photon emission with transition probability $\sim 8\,\mathrm{s}^{-1}$, compared to $\sim 6\times10^8\,\mathrm{s}^{-1}$ for Lyα decay from $2p$. Therefore the $2s$ level thermalizes at much lower density than $2p$. On the other hand, collisional mixing between the l states (e.g. $2s \leftrightarrow 2p$) may help to depopulate $2s$. Detailed calculations are needed to examine the various processes controlling H^0 ionization and $n = 2$ population in environments consistent with the absorbers in η Car.[13]

In any case, the narrow absorption lines clearly represent important gas components in the Homunculus. Some of the absorbers reside in the inner ejecta with large column densities and physical conditions similar to the Weigelt blobs. They appear to be blob-like material that happens to be viewed against a bright background of direct or reflected starlight. The narrow Balmer absorption lines seen toward the blobs might actually form in an outer ionized layer of the blobs themselves (Sect. 5.4.1). However, the distributed appearance of the Balmer and other narrow absorption lines across the inner ejecta, with a range of velocities, shows that these absorbing regions are much more extended than the individual Weigelt blobs. They might also contain a significant amount of mass (Sect. 5.6 below).

5.4.7 Mass of the Weigelt Blobs and Inner Ejecta

The mass of ejected material has direct implications for the nature, evolution, and instabilities of the central star. We can estimate this mass from the strengths of forbidden emission lines. This is fairly straightforward because the relevant level populations are close to LTE (see above); the observed flux is therefore proportional to mass rather than mass times density. Davidson et al. [24] used the [Fe II] $\lambda5376$ flux measured in *HST* observations of a $0.3''$ region including Weigelt blobs B + C + D, and found $\sim 0.002\,\mathrm{M}_\odot$ assuming all of the iron is Fe^+, solar Fe/H abundances, 3 mag of extinction, and LTE at $T_e = 8000\,\mathrm{K}$. This estimate is surely a lower limit because it ignores possible depletion of iron into dust grains and it applies only to the emitting regions of [Fe II] $\lambda5376$. (It excludes the more highly ionized zones, see Sect. 5.5). Nonetheless, if we adopt a diameter $0.1'' \approx 230\,\mathrm{AU}$

[13] Johansson et al. [71] argued that the $2s$ population is regulated by absorption of Lyβ photons from the central object. However, that scheme ignores the large Lyman line opacities in the absorbing nebula (see also Sect. 5.4.5). We also note that most of the environments favored by their calculations for Balmer line absorption would be optically thick to Thomson scattering at all wavelengths.

for each blob (Sect. 5.1), we find that this mass corresponds to an average density $n_H \sim 3 \times 10^7\,cm^{-3}$within them, reasonably consistent with our estimates above (Sect. 5.4.3).

We can estimate the mass outside the blobs by using emission lines measured through a larger 1″aperture. Starting with the measured flux in [Fe II] λ7155 [46], we reduce this flux by 20% to represent only the narrow (non-stellar) emission component, correct for 2 mag of red extinction, assume solar Fe/H, and LTE populations with $T_e = 7000$ K. The result, ∼0.006 $M_\odot$, is three times larger than the Davidson et al. estimate for the inner 0.3″. A similar calculation applied to [Ar III] λ7136 in the same ground-based data (but including its entire flux because it appears to be entirely nebular, Sect. 5.4.1), yields ∼0.002 $M_\odot$ for higher ionization gas.

The narrow absorption lines might provide a rough estimate of the nebular mass independent of extinction. For instance, if the Balmer line absorber covers an area of $1'' \times 1'' \approx 2300^2$ AU^2 and its average column density is $N(H^0) \sim 5 \times 10^{21}\,cm^{-2}$ (well below the value measured toward the star but above the minimum needed for thermalization, Sect. 5.4.5), then the total mass in this absorber is $\sim 0.005/f_0\,M_\odot$, where $f_0 = H^0/(H^0 + H^+) < 1$ is the neutral fraction.

To some degree, we can simply add these mass estimates together because they represent different gas components. Doing this we find a minimum total mass of ∼0.013 $M_\odot$ within the central $1'' \times 1''$. However, the values based on emission line fluxes are only lower limits because (1) they probe just the optimal emitting regions for particular lines, and (2) the extinction corrections may be larger if, as expected, the obscuration is substantially gray (Sect. 5.4.2) or the dust distribution is patchy (Fainter regions may be those with more extinction rather than less emission, Sect. 5.1). Based on these considerations, the total mass in the *inner* ejecta is most likely in the range ∼0.02–0.05 $M_\odot$ (see also [25]).

5.4.8 The Strontium Filament

The "strontium filament" is a patch of nebulosity located several arcsec northwest of the star. It is remarkable for its Sr II and [Sr II] emission lines as well as some extraordinarily low-ionization features [4–6, 49, 110, 111]. Not really a filament, this structure is much larger than the Weigelt blobs and has more complicated kinematics. Studies of the Sr filament may be helpful for understanding the inner ejecta.

Its spectrum is essentially a lower-ionization version of that emitted by the Weigelt blobs. It is dominated by emission from species such as C^0, Mg^0, Ca^0, Ca^+, Sc^+, Ti^+, V^+ and Mn^+, in addition to the signature Sr^+ lines [49]. Its iron spectrum has more Fe I and [Fe I] than Fe II and [Fe II], and there are none of the fluorescent lines of Fe II, O I, etc., that require partially ionized gas (Sect. 5.4.5). There are also no H I or He I emission lines; the hydrogen must be essentially neutral.

Spectroscopically, the Weigelt blobs most resemble the Sr filament during a spectroscopic event. At such a time the blobs' H I and He I lines become extremely weak, fluorescent lines disappear, and the lowest ionization features like [Ca II] and Ti II and [Ti II] strengthen (Sect. 5.3). The blobs never become as neutral as the Sr filament, but overall the conditions then appear to be similar. Calculations of multi-level Sr^+ and Ti^+ atoms in the Sr region [4, 5] suggest that the free electron densities there are of order $n_e \sim 10^7\,\mathrm{cm}^{-3}$ at temperatures of $T_e \sim 6{,}000\,\mathrm{K}$ (cf. Sects. 5.4.2 and 5.4.3). Considering that the gas is mostly neutral, the total hydrogen density may be substantially higher. The energy source for this region's line emission is believed to be incident stellar radiation. However, the lower ionization in the filament cannot be explained merely by its distance from the star. The stellar spectrum seen by the filament appears to cut off sharply above $\sim$8 eV, limiting the ionization state to neutrals and some singly-ionized species in the iron group [49]. The Weigelt blobs see a harder stellar spectrum, even during a spectroscopic event.

The strong metal line emissions from both the blobs and the Sr filament are believed to be excited by a combination of collisions and photo-absorption of the stellar visible and near-UV flux, i.e., by continuum pumping (see below).

5.5 Line Formation Physics

The rich emission-line spectra discussed above present many diagnostic opportunities but also a basic problem: How are they produced? The importance of radiation from the star is evident in the spectroscopic events. For example, the stellar wind P Cygni features reflected by dust in the blobs change at roughly the same time as the in situ narrow emission lines (Sect. 5.3 and Fig. 5.7). Changes in the star and in the blobs track each other within a month or less (see also [16, 28, 55, 94]). This fact probably indicates that the blob spectra are responding to changes in radiation from the star (or rather the two stars). Evidently this is the energy source for ionization and excitation inside the blobs. The other possibility, kinetic energy in the stellar wind, is less powerful and is ruled out by the event timings (also Sect. 5.3).

Verner et al. [98, 99] and Mehner et al. [75] used photoionization codes to show that most properties of the blob spectra can, indeed, be matched by dense clouds irradiated by sources like those expected for the primary star and its putative hot companion. In these models the high-ionization lines of [Ne III], [Ar III], He I, etc., form in (H^+) layers directly exposed to the central source, while Fe II, [Fe II], Ni II, [Ni II], [Ca II], Ti II, [Ti II], etc., form in a warm, partially ionized environment behind the H^0–H^+ recombination front. The predicted temperature there is roughly 5,000–7,000 K and the H^+/H fraction ranges from $\sim$50% in the warmest regions to $\sim$15% farther behind the front [98]. Collisions and UV continuum pumping together produce strong emission from Fe II, [Fe II] and similar ions.

Circumstances like these have often been discussed for the Fe II emission regions of active galactic nuclei [3, 35, 79, 96, 107]. Continuum pumping dominates the excitation of Fe^+ and similar ions for energy levels above a few eV [4, 5, 53, 98, 99]. Thermal collisional excitation populates metastable lower states, which serve as launching pads for continuum pumping to the higher states [3, 98].

The well-studied line absorber at $-146\,\mathrm{km\,s^{-1}}$ appears to be blob-like material seen against the background of the stellar continuum (Sect. 5.4.6). Specific calculations for that environment indicate, again, that low states of Fe^+ are populated by collisions at $T_e \sim 6400\,\mathrm{K}$ and $n_e \sim 10^7$–$10^8\,\mathrm{cm^{-3}}$. UV absorption lines directly measure continuum pumping out of these states [39]. Narrow Balmer absorption lines at this same velocity in the stellar spectrum indicate that the low-ionization absorber is accompanied by enough hydrogen ionizations to populate the $n = 2$ level of H^0.

These results take us a long way toward understanding the spectra of the Sr filament and inner ejecta, but there are complications. The most basic is that the local ionization waxes and wanes with the 5.54 years event cycle. To some degree we can think of this as a spatial movement of the H^0–H^+ recombination front. Between spectroscopic events, such a front must exist somewhere in the region of interest, because we see both low and high ionization lines.[14] During each spectroscopic event, however, the spectral signatures of the (H^+) zone dramatically weaken throughout the inner ejecta. The recombination front must then be much closer the central star.

A key to understanding these ionization changes is in the detailed timing of the transition from spectroscopic high to low states. First one sees a decline in the [Ne III], [Ar III] and He I lines, followed in order by [Fe III], Si III], N II, [N II] and H I, according to their ionization energies (Sect. 5.3, also [16] and refs. therein). Before the event there must be a significant flux with $h\nu > 41\,\mathrm{eV}$ to maintain the Ne^{++}. But then, over about 2 weeks, the source cutoff energy slides down to the Lyman limit at 13.6 eV. This behavior reverses during the recovery phase, but over a longer time. Meanwhile, most low-ionization features, notably [Fe II] and the non-fluorescent Fe II lines, remain fairly steady. Their ionization and excitation must be less affected by changes in the flux above the Lyman limit. The lowest-ionization lines, [Ca II] and [Ti II], strengthen during an event – most likely due to a general shift in ionization from Ca^{+2} and Ti^{+2} toward Ca^+ and Ti^+.

Mehner et al. [75] reported the [Ne III] and [Fe III] behavior shown in Fig. 5.10, including a strong peak several months before the 2003.5 event (see Sect. 5.4.1 above). These *HST/STIS* data refer specifically to gas along our line of sight to the star, with a velocity of $-40\,\mathrm{km\ s^{-1}}$ like the Weigelt blobs. The blobs vary in a similar way, but the peak just mentioned was not reported in ground-based data [16].

The Sr filament, on the other hand, has a lower ionization spectrum and remains that way throughout the event cycle [49]. Evidently the H^0–H^+ ionization front never reaches that region, although the stellar flux would normally be strong enough to do so. Some intervening gas must block the stellar UV continuum and impose a cutoff above $\sim 8\,\mathrm{eV}$ at that location of the filament [49]. The shield must be

[14] H^0 and H^+ can be diffusely mixed in comparable amounts, with no well-defined ionization front, only if the "photoionization parameter" U_{H} is very small [22]. Straightforward models do not allow such small values in η Car's inner ejecta [75].

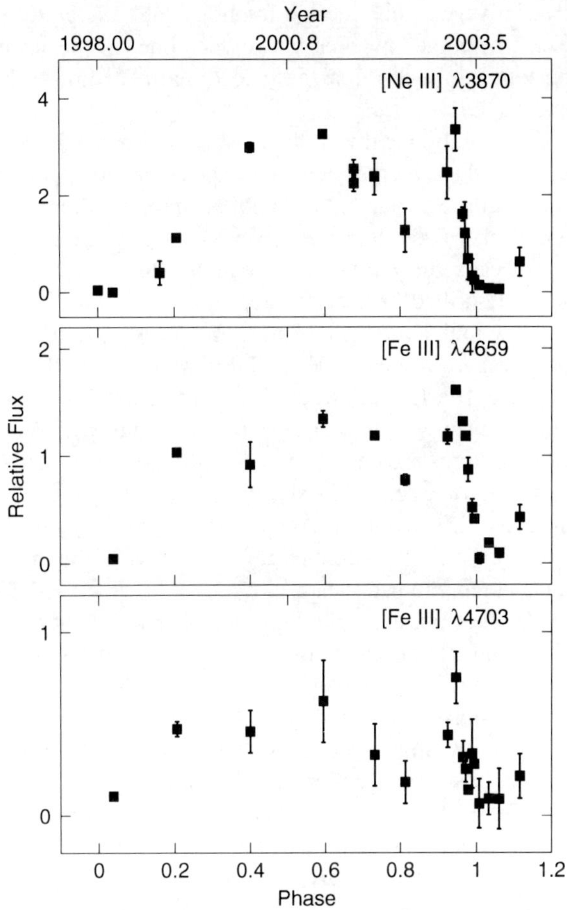

Fig. 5.10 The strengths of the narrow [Ne III] and [Fe III] emission lines measured in *HST/STIS* spectra throughout its spectroscopic cycle. The lines in the blobs vary in a similar way (From [75])

substantially non-ionized, with large bound-free opacities not just in H I but also in complex atoms with ionization thresholds near 8 eV. The leading candidates are Mg^0, Si^0, and Fe^0 with ionization energies 7.6, 8.1, and 7.9 eV. The nature of this UV shielding medium is almost as intriguing as the Sr filament itself, but we should keep in mind that UV extinction may occur throughout a range of locations.

The most puzzling aspect of line emission from the Sr Filament and the Weigelt blobs during an event is that neither region appears to contain a significant amount of H^+. Hydrogen recombination lines are very weak or absent (Sects. 5.3 and 5.4.7). The photoionization calculations cited above generically predict that the low-ionization emitting zones should be partially ionized. Some amount of ionization is necessary to provide photoelectric heating and free electrons. If hydrogen and helium are entirely neutral, then the burden for these tasks falls entirely upon the heavy elements and possibly dust grains. This seems problematic because it would lead to very small electron fractions and temperatures insufficient to collisionally excite even the low-energy forbidden lines in species like Fe^+ or Fe^0. There should be some amount of ionization leading to H I line emission during all phases of the event cycle.

The solution to this problem is not obvious. One possibility is that significant H^+ exists in these regions but has escaped detection. Quantitative upper limits to Balmer lines in the Sr Filament have not been reported; Hartman et al. [49] imply that they are weaker than other lines. Weak nebular H I lines are difficult to measure near η Car because they are blended with the dust-reflected broad stellar wind lines. Blob D as described in Sect. 5.3 appears to have narrow Hα and Hβ emission during the low-ionization states, but these are weaker than before the event and Hγ and Hδ are hard to detect (Fig. 5.7). If we compare the weak-state narrow Hβ flux to a typical [Fe II] line, we find that the ionized fraction H^+/H in blob D during the 2003.5 event was less than 1/10 of the values 15–50% predicted by models of the Fe^+ emitting region [98].[15] The disappearance of Fe II fluorescence pumped by in situ Lyα (Sect. 5.4.5) also suggests that the Fe II zone is effectively H^0 during an event. Meanwhile the temperature drops by only about 10% and collisionally excited [Fe II] lines remain nearly steady, as though they have nothing to do with the changing ionization.

Another possibility is that shocks or turbulence in the outflow provide just the right amount of heat and free electrons without significantly ionizing hydrogen. This process might be supplemented by absorption of the stellar flux by heavy elements and possibly dust grains. It is not at all clear whether this scenario is viable, but it is worth investigating because these processes are unrelated to the incident far-UV flux and the hydrogen ionization. They might provide a natural explanation for the steadiness of some low-ionization emission throughout the 5.54 years event cycle.

A third possibility is that the gas is too cool for collisional excitation but some other process drives the low-ionization emission. Continuum pumping is expected to play an important role, which may be enhanced if non-thermal motions (e.g., turbulence) broaden the lines and thus enhance the photo-excitation rates [3, 79, 98]. Continuum pumping ties low-ionization lines directly to the relatively stable near-UV and visible spectrum of the central object. However, collisional excitation at some reasonable temperature is still needed to populate the low-energy metastable states, facilitating continuum pumping to higher states. As noted earlier, low metastable states of Fe^+, Ti^+, etc., in the blobs and Sr Filament appear to have LTE-like populations with $T \sim 6000$–7000 K (Sect. 5.4.2, and [4]). Somehow this occurs without much hydrogen ionization. This presents a problem because at 6000–7000 K with 10^7 electrons per cm^3, the Saha eqn predicts more H^+ than H^0!

[15] Here we assume Case B recombination for Hβ and LTE for [Fe II], with $n_H \sim 10^7$ cm^{-3} and $T \approx 7{,}000$ K.

5.6 Summary: The Nature and Origin of the Inner Ejecta

The inner ejecta are dominated observationally by the Weigelt blobs, which appear to be concentrations of warm, relatively dense gas that is heated, photo-excited and usually (apart from the spectroscopic events) photo-ionized by the central continuum source. Considerable amounts of gas also exist outside the blobs, including prominent absorption line regions and some faster and less dense gas that emits [Ne III], [Ar III], etc. The blobs, at least, represent heavily CNO processed gas emitted from the primary star roughly a century ago. There appears to be little dust within the main emitting and absorbing condensations of the inner ejecta, but our knowledge of the amount and spatial distribution of this material is limited by uncertainties in the patchiness and amount of the (rather gray) foreground extinction.

There are at least two leading unsolved puzzles in the emission line physics. The most fundamental concerns the heating and weak ionization of the gas that produces strong emission from ions like Fe^+, Ni^+, Ca^+, Ti^+ and Sr^+. Existing photoionization and photo-excitation models imply that this gas should be partially ionized, with significant amounts of H^+; but the data indicate that hydrogen is practically non-ionized in the blobs during the spectroscopic events (and in the Sr filament at all times). Another puzzle involves the bizarre line ratios emitted from Lyα-pumped levels of Fe^+ that produce the strong $\lambda 2508$ and $\lambda 2509$ lines.

We conclude with a reminder about the broader goal of studies of the inner ejecta – to understand the nature and evolution of the central object. Here let us mention one particular topic that deserves more study. During the transitional phases at the beginning and end of a spectroscopic event, the UV flux from the central star(s) is extinguished by varying amounts/properties of a shielding gas. Moreover, the event timings observed in various ions (Sect. 5.3) imply that the cutoff energy slides from the far-UV to the near UV and back again. Calculations are needed to see what might cause this behavior. It cannot occur merely by variable column densities in a neutral medium. A more realistic scenario would involve column densities in partially ionized gas, possibly combined with varying degrees of ionization.

In the binary model of the central object, radiative shielding occurs when the hot companion star plunges deep inside the dense wind of the primary (Sect. 5.2). The shielding medium is the dense wind, perturbed or enhanced by its interaction with the hot binary. Quantitative spectral studies of the inner ejecta, especially during the transition phases, should be very helpful for constraining basic properties of the companion star, its wind, and the binary orbit/orientation. For a few days before and after periastron passage, some regions of the inner ejecta should be lit up by far more UV radiation than others. New observations with enough spatial and temporal resolution might allow us to see this pattern of illumination move across the inner ejecta, as has been suggested already for some nebulosity farther out in the Homunculus [92]. The best tracer for these effects is probably the [Ne III] $\lambda 3868$ line because its emission is tied directly to the far-UV output from the hot companion and its profile is not blended with reflected features from the stellar wind (see [75]).

Acknowledgements FH grateful to the HST–Eta Carinae Treasury Team, especially Kris Davidson and Bish Ishibashi, for their help and guidance with the HST spectra. Brian Cherinka also helped with some of the data processing. FH had valuable discussions about nebular physics with Gary Ferland and Pat Hall. Andrea Mehner contributed recent information, especially for Sect. 5.4.1. Finally, I thank the editors Roberta Humphreys and Kris Davidson for useful comments.

References

1. L.H. Aller, T. Dunham, The spectrum of Eta Carinae in 1961. ApJ. **146**, 126–141 (1966)
2. É. Artigau, J.C. Martin, R.M. Humphreys, K. Davidson, O. Chesneau, N. Smith, Penetrating the Homunculus – near-infrared adaptive optics images of Eta Carinae. AJ. **141**, 202 (2011)
3. J.A. Baldwin, G.J. Ferland, K.T. Korista, F. Hamann, A. LaCluyzé, The origin of Fe II emission in active galactic nuclei. ApJ. **615**, 610–624 (2004)
4. M.A. Bautista, T.R. Gull, K. Ishibashi, H. Hartman, K. Davidson, Excitation of Srii lines in Eta Carinae. MNRAS. **331**, 875–879 (2002)
5. M.A. Bautista, H. Hartman, T.R. Gull, N. Smith, K. Lodders, [Ti II] and [Ni II] emission from the strontium filament of η Carinae. MNRAS. **370**, 1991–2003 (2006)
6. M.A. Bautista, C. Ballance, T.R. Gull, H. Hartman, K. Lodders, M. Martínez, M. Meléndez, Scandium and chromium in the strontium filament in the Homunculus of η Carinae. MNRAS. **393**, 1503–1512 (2009)
7. J.A. Cardelli, G.C. Clayton, J.S. Mathis, The relationship between infrared, optical, and ultraviolet extinction. ApJ. **345**, 245–256 (1989)
8. K.G. Carpenter, J.E. Pesce, R.E. Stencel, A. Brown, S. Johansson, R.F. Wing, The ultraviolet spectrum of noncoronal late-type stars – The Gamma Crucis (M3.4 III) reference spectrum. ApJS. **68**, 345–369 (1988)
9. O. Chesneau et al., The sub-arcsecond dusty environment of Eta Carinae. A&A. **435**, 1043–1061 (2005)
10. M.F. Corcoran, X-Ray monitoring of η Carinae: variations on a theme. AJ. **129**, 2018–2025 (2005)
11. M.F. Corcoran, K. Ishibashi, J.H. Swank, R. Petre, The X-Ray light curve of η Carinae: refinement of the orbit and evidence for phase-Dependent mass loss. ApJ. **547**, 1034–1039 (2001)
12. A. Dalgarno, R.A. McCray, Heating and ionization of HI regions. ARAA. **10**, 375 (1972)
13. A. Damineli, The 5.52 year cycle of Eta Carinae. ApJ. **460**, L49–52 (1996).
14. A. Damineli, O. Stahl, A. Kaufer, B. Wolf, G. Quast, D.F. Lopes, Long-term spectroscopy of η Carinae. I. The high and low excitation phases. A&AS. **133**, 299–316 (1998)
15. A. Damineli, A. Kaufer, B. Wolf, O. Stahl, D.F. Lopes, F.X. de Araújo, η Carinae: binarity confirmed. ApJ. **528**, L101–L104 (2000)
16. A. Damineli, D.J. Hillier, M.F. Corcoran et al., A multispectral view of the periodic events in η Carinae. MNRAS. **386**, 2330–2344 (2008)
17. A. Damineli, D.J. Hillier, M.F. Corcoran et al., The periodicity of the η Carinae events. MNRAS. **384**, 1649–1656 (2008).
18. K. Davidson, On the nature of Eta Carinae. MNRAS. **154**, 415–427 (1971)
19. K. Davidson, Why the binary hypothesis isn't a panacea, in *Eta Carinae at the Millennium*, ed. by J.A. Morse, R.M. Humphreys, A. Damineli. ASP Conference Series, vol. 179 (ASP, San Francisco, 1999), pp. 304–315
20. K. Davidson, Unique spectroscopic problems related to Eta Carinae, in *Eta Carinae & Other Mysterious Stars*, ed. by T.R. Gull, S. Johansson, K. Davidson. ASP Conference Series, vol. 242 (ASP, San Francisco, 2001), pp. 3–13
21. K. Davidson, R.M. Humphreys, Eta Carinae and its environment. ARAA. **35**, 1–32 (1997)
22. K. Davidson, H. Netzer, The emission lines of quasars and similar objects. Revs. Mod. Phys. **51**, 715–766 (1979)

23. K. Davidson, R.J. Dufour, N.R. Walborn, T.R. Gull, Ultraviolet and visual wavelength spectroscopy of gas around η Carinae. ApJ. **305**, 867–879 (1986)
24. K. Davidson, D. Ebbets, G. Weigelt et al., HST/FOS spectroscopy of Eta Carinae: the star itself, and ejecta within 0.3 arcsec. AJ. **109**, 1784–1796 (1995)
25. K. Davidson, D. Ebbets, S. Johansson et al., HST/GHRS observations of the compact slow ejecta of Eta Carinae. AJ. **113**, 335–345 (1997)
26. K. Davidson, K. Ishibashi, T.R. Gull, R.M. Humphreys, HST/STIS observations of the star during its recent event, in *Eta Carinae at the Millennium*, ed. by J.A. Morse, R.M. Humphreys, A. Damineli. ASP Conference Series, vol. 179 (ASP, San Francisco, 1999), pp. 227–235
27. K. Davidson, N. Smith, T.R. Gull, K. Ishibashi, D.J. Hillier, The shape and orientation of the Homunculus nebula based on spectroscopic velocities. AJ. **121**, 1569–1577 (2001)
28. K. Davidson et al., A change in the physical state of η Carinae? AJ. **129**, 900–906 (2005)
29. B.N. Dorland, D.G. Currie, A.R. Hajian, Did Eta Carinae's weigelt blobs originate circa 1941? AJ. **127**, 1052–1058 (2004)
30. R.J. Dufour, T.W. Glover, J.J. Hester et al., New HST results on the outer nebula of η Carinae, in *Luminous Blue Variables: Massive Stars in Transition*, ed. by A. Nota, H.J.G.L.M. Lamers. ASP Conference Series, vol. 120 (ASP, San Francisco, 1997), p. 255
31. D. Ebbets, H. Garner, R. White et al., HST images of η Carinae, in *Circumstellar Media in Late Stages of Stellar Evolution*, ed. by R.E.S. Clegg, I.R. Stevens, W.P.S. Meikle. 34th Herstmonceux Conference (Cambridge University Press, Cambridge, 1994), pp. 95–97
32. M. Elitzur, G.J. Ferland, Radiation pressure and emission clouds around active galactic nuclei. ApJ. **305**, 35–44 (1986).
33. M. Elitzur, H. Netzer, Line fluorescence in astrophysics. ApJ. **291**, 464–467 (1985)
34. M. Feast, P. Whitelock, F. Marang, Variability of η Carinae – III. MNRAS. **322**, 741–748 (2001)
35. G. Ferland, H. Netzer, Application of line transfer calculations to active nuclei and novae. ApJ. **229**, 274–290 (1979)
36. E. Gaviola, Eta Carinae. II. The spectrum. ApJ. **118**, 234–251 (1953)
37. S.A. Grandi, O I λ8446 emission in Seyfert 1 galaxies. ApJ. **238**, 10–16 (1980)
38. T.R. Gull, K. Ishibashi, The three-dimensional and time-variant structures of ejecta around Eta Carinae as detected by the STIS, in *Eta Carinae & Other Mysterious Stars*, ed. by T. Gull, S. Johansson, K. Davidson. ASP Conference Series, vol. 242 (ASP, San Francisco, 2001), pp. 59–70
39. T.R. Gull, G. Viera, F. Bruhweiler et al., The absorption spectrum of high-density stellar ejecta in the line of sight to η Carinae. ApJ. **620**, 442–449 (2005)
40. T.R. Gull, G. Viera Kober, K.E. Nielsen, Eta Carinae across the 2003.5 minimum: the character and variability of the ejecta absorption in the near-ultraviolet. ApJS. **163**, 173–183 (2006)
41. P.B. Hall, A quasar with broad absorption in the Balmer lines. AJ. **133**, 1271–1274 (2007)
42. F. Hamann, Emission-line studies of young stars. 4: the optical forbidden lines. ApJS. **93**, 485–518 (1994)
43. F. Hamann, The HST–Eta Carinae Treasury Team, HST spectroscopy of Eta Car's inner ejecta, in *The Fate of the Most Massive Stars*, ed. by R.M. Humphreys, K.Z. Stanek. ASP Conference Series, vol. 332 (ASP, San Francisco, 2005), pp. 283–293
44. F. Hamann, S.E. Persson, The similar emission-line spectra of the young star LkH-alpha 101 and the hypergiant MWC 300. ApJS. **71**, 931–949 (1989)
45. F. Hamann, M. Simon, 7500–9300 A spectroscopy of MWC 349A. ApJ. 327, 876–893 (1988)
46. F. Hamann, D.L. Depoy, S. Johansson, J. Elias, High-resolution 6450-24500 Å spectra of η Carinae. ApJ. **422**, 626–641 (1994)
47. F. Hamann, K. Davidson, K. Ishibashi, T.R. Gull, Preliminary analysis of HST-STIS spectra of compact ejecta from Eta Carinae, in *Eta Carinae at The Millennium*, ed. by J.A. Morse, R.M. Humphreys, A. Damineli. ASP Conference Series, vol. 179 (ASP, San Francisco, 1999), pp. 116–122
48. H. Hartman, S. Johansson, Ultraviolet fluorescence lines of Fe II observed in satellite spectra of the symbiotic star RR Telescopii. A&A. **359**, 627–634 (2000)

49. H. Hartman, T. Gull, S. Johansson, N. Smith, HST Eta Carinae Treasury Project Team, Identification of emission lines in the low-ionization strontium filament near η Carinae. A&A. **419**, 215–224 (2004)
50. H. Hartman, A. Damineli, S. Johansson, V.S. Letokhov, Time variations of the narrow Fe II and H I spectral emission lines from the close vicinity of η Carinae during the spectral event of 2003. A&A. **436**, 945–952 (2005)
51. D.B. Henley, M.F. Corcoran, J.M. Pittard et al., Chandra X-Ray grating spectrometry of η Carinae near X-Ray minimum. I. Variability of the sulfur and silicon emission lines. ApJ. **680**, 705–727 (2008)
52. D.J. Hillier, D.A. Allen, A spectroscopic investigation of η Carinae and the Homunculus nebula. I – overview of the spectra. A&A. **262**, 153–170 (1992)
53. D.J. Hillier, P.A. Crowther, F. Najarro, A.W. Fullerton, An optical and near-IR spectroscopic study of the extreme P Cygni-type supergiant HDE 316285. A&A. **340**, 483–496 (1998)
54. D.J. Hillier, K. Davidson, K. Ishibashi, T. Gull, On the nature of the central source in η Carinae. ApJ. **553**, 837–860 (2001)
55. D.J. Hillier, T. Gull, K. Nielsen et al., The UV scattering halo of the central source associated with η Carinae. ApJ. **642**, 1098–1116 (2006).
56. K.-H. Hofmann, G. Weigelt, Speckle masking observation of η Carinae. A&A. **203**, L21–L22 (1988)
57. R.M. Humphreys, K. Davidson, M. Koppelman, The early spectra of η Carinae 1892 to 1941 and the onset of its high excitation emission spectrum. AJ. **135**, 1249–1263 (2008)
58. J.B. Hutchings, D.M. Crenshaw, S.B. Kraemer et al., Balmer and He I absorption in the nuclear spectrum of NGC 4151. AJ. **124**, 2543–2547 (2002)
59. K. Ishibashi, Historical eruptions of Eta Carinae: looking through the Homunculus, in *The Fate of the Most Massive Stars*, ed. by R. Humphreys, K. Stanek. ASP Conference Series, vol. 332 (ASP, San Francisco, 2005), pp. 131–136
60. K. Ishibashi, M.F. Corcoran, K. Davidson et al., Recurrent X-Ray emission variations of η Carinae and the binary hypothesis. ApJ. **524**, 983–987 (1999)
61. K. Ishibashi, T.R. Gull, K. Davidson et al., Discovery of a little Homunculus within the Homunculus nebula of η Carinae. AJ. **125**, 3222–3236 (2003)
62. S. Johansson, Forbidden transitions of Fe II. Phys. Scr. **15**, 183 (1977)
63. S. Johansson, Strong Fe II fluorescence lines in RR Tel and V1016 CYG excited by D IV in a Bowen mechanism. MNRAS. **205**, 71P–75P (1983)
64. S. Johansson, F.W. Hamann, Fluorescence lines in ultraviolet spectra of stars. Phys. Scr. **T47**, 157 (1993)
65. S. Johansson, C. Jordan, Selective excitation of Fe II in the laboratory and late-type stellar atmospheres MNRAS. **210**, 239–256 (1984)
66. S. Johansson, V.S. Letokhov, Astrophysical lasers operating in optical Fe II lines in stellar ejecta of η Carinae. A&A. **428**, 497–509 (2004)
67. S. Johansson, T. Zethson, Atomic physics aspects on previously and newly identified iron lines in the HST spectrum of η Carinae, in *Eta Carinae at The Millennium*, ed. by J.A. Morse, R.M. Humphreys, A. Damineli. ASP Conference Series, vol. 179 (ASP, San Francisco, 1999), pp. 171–183
68. S. Johansson, G. Wallerstein, K.K. Gilroy, A. Joueizadeh, Fluorescence lines of MnII in the red spectrum of η Carinae. A&A. **300**, 521–524 (1995)
69. S. Johansson, D.S. Leckrone, K. Davidson, The impact of GHRS on atomic physics – a fruitful collaboration between laboratory and stellar spectroscopy, in *The Scientific Impact of the Goddard High Resolution Spectrograph*, ed. by J.C. Brandt, T.B. Ake, C.C. Petersen. ASP Conference Series, vol. 143 (ASP, San Francisco, 1998), p. 155
70. S. Johansson, T. Zethson, H. Hartman et al., New forbidden and fluorescent Fe III lines identified in HST spectra of η Carinae. A&A. **361**, 977–981 (2000)
71. S. Johansson, T.R. Gull, H. Hartman, V.S. Letokhov, Metastable hydrogen absorption in ejecta close to η Carinae. A&A. **435**, 183–189 (2005)

72. J.C. Martin, M.D. Koppelman, η Carinae's brightness variations since 1998: Hubble space telescope observations of the central star. AJ. **127**, 2352–2361 (2004)
73. J.C. Martin, K. Davidson, R.M. Humphreys, D.J. Hillier, K. Ishibashi, On the He II Emission in Eta Carinae and the origin of its spectroscopic events. ApJ. **640**, 474–490 (2006)
74. J.C. Martin, K. Davidson, M.D. Koppelman, The chrysalis opens? Photometry from the η Carinae Hubble space telescope treasury project, 2002–2006. AJ. **132**, 2717–2728 (2006)
75. A. Mehner, K. Davidson, G.J. Ferland, R.M. Humphreys, High-excitation emission lines near η Carinae. ApJ. **710**, 729–742 (2010)
76. A. Mehner, K. Davidson, R.M. Humphreys, A sea change in η Carinae. ApJ. **717**, L22–L25 (2010)
77. A. Mehner, K. Davidson, J.C. Martin et al., Critical differences and clues in Eta Car's 2009 event. ApJ. **740**, 2 (2011)
78. J.A. Morse, K. Davidson, J. Bally et al., Hubble space telescope wide field planetary camera 2 observations of η Carinae. AJ. **116**, 2443–2461 (1998)
79. H. Netzer, B.J. Wills, Broad emission features in QSOs and active galactic nuclei. I – new calculations of Fe II line strengths. ApJ. **275**, 445–460 (1983)
80. K.E. Nielsen, S. Ivarsson, T.R. Gull, Eta Carinae across the 2003.5 minimum: deciphering the spectrum toward weigelt D. ApJS. **168**, 289–296 (2007)
81. D.E. Osterbrock, G.J. Ferland, *Astrophysics of Gaseous Nebulae and Active Galactic Nuclei*, 2nd edn. (University Science Books, Sausalito, 2006)
82. M.V. Penston, P. Benvenuti, A. Cassatella, IUE and other new observations of the slow nova RR Tel. MNRAS. **202**, 833–857 (1983)
83. J.M. Pittard, M.F. Corcoran, In hot pursuit of the hidden companion of η Carinae: an X-ray determination of the wind parameters. A&A. **383**, 636–647 (2002)
84. N.D. Richardson, D.R. Gies, T.J. Henry, E. Fernández-Lajás, A.T. Okazaki, The Hα variations of η Carinae during the 2009.0 spectroscopic event. AJ. **139**, 1534–1541 (2010)
85. R.J. Rudy, S. Mazuk, R.C. Puetter, F. Hamann, The 1 micron Fe II lines of the seyfert galaxy I Zw 1. ApJ. **539**, 166–171 (2000)
86. B.D. Savage, K.R. Sembach, Interstellar abundances from absorption-line observations with the Hubble space telescope. ARAA. **34**, 279–330 (1996)
87. T.A.A. Sigut, A.K. Pradhan, Lyα flourescent excitation of Fe II in active galactic nuclei. ApJ. **499**, L139–L142 (1998)
88. N. Smith, J.A. Morse, Nitrogen and oxygen abundance variations in the outer ejecta of η Carinae: evidence for recent chemical enrichment. ApJ. **605**, 854–863 (2004)
89. N. Smith, K. Davidson, T.R. Gull, K. Ishibashi, D.J. Hillier, Latitude-dependent effects in the stellar wind of η Carinae. ApJ. **586**, 432–450 (2003)
90. N. Smith, R.D. Gehrz, P.M. Hinz et al., Mass and kinetic energy of the Homunculus nebula around η Carinae. AJ. **125**, 1458–1466 (2003)
91. N. Smith, J.A. Morse, T.R. Gull et al., Kinematics and ultraviolet to infrared morphology of the inner Homunculus of η Carinae. ApJ. **605**, 405–424 (2004)
92. N. Smith, J.A. Morse, N.R. Collins, T.R. Gull, The purple haze of η Carinae: binary induced variability? ApJ. **610**, L105–L108 (2004)
93. N. Smith, J.A. Morse, J. Bally, The [O III] veil: astropause of η Carinae's wind? AJ. **130**, 1778–1783 (2005)
94. O. Stahl, K. Weis, D.J. Bomans et al., A spectroscopic event of η Car viewed from different directions: the data and first results. A&A. **435**, 303–312 (2005)
95. A.D. Thackeray, Identifications in the spectra of η Carinae and RR Telescopii. MNRAS. **113**, 211–236 (1953)
96. E.M. Verner, D.A. Verner, K.T. Korista et al., Numerical simulations of Fe II emission spectra. ApJS. **120**, 101–112 (1999)
97. E.M. Verner, D.A. Verner, J.A. Baldwin, G.J. Ferland, P.G. Martin, Continuum pumping of [Fe II] in the orion nebula. ApJ. **543**, 831–839 (2000)
98. E.M. Verner, T.R. Gull, F. Bruhweiler et al., The origin of Fe II and [Fe II] emission lines in the 4000–10000 Å range in the BD weigelt blobs of η Carinae. ApJ. **581**, 1154–1167 (2002)

99. E. Verner, F. Bruhweiler, T. Gull, The binarity of η Carinae revealed from photoionization modeling of the spectral variability of the weigelt blobs B and D. ApJ. **624**, 973–982 (2005)
100. R. Viotti, L. Rossi, A. Cassatella, A. Altamore, G.B. Baratta, The ultraviolet spectrum of η Carinae. ApJS. **71**, 983–1009 (1989)
101. R. Viotti, C. Rossi, G.B. Baratta, Self-absorption curve analysis of [Fe II] and Fe II emission lines of η Carinae, in *Eta Carinae at the Millennium*, ed. by J.A. Morse, R.M. Humphreys, A. Damineli. ASP Conference Series, vol. 179 (ASP, San Francisco, 1999), pp. 184–191
102. G. Wallerstein, K.K. Gilroy, T. Zethson, S. Johansson, F. Hamann, Line identifications in the spectrum of η Carinae as observed in 1990–1991 with CCD detectors. PASP. **113**, 1210–1214 (2001)
103. G. Weigelt, J. Ebersberger, Eta Carinae resolved by speckle interferometry. A&A. **163**, L5–L6 (1986)
104. G. Weigelt, R. Albrecht, C. Barbieri et al., HST FOC Observations of Eta Carinae, in *The Eta Carinae Region: A Laboratory of Stellar Evolution*, ed. by V. Niemela, N. Morrell, A. Feinstein. Revista Mexicana de Astronomia y Astrofisica Serie de Conferencias, Vol. 2 (Instituto de Astronomía, Universidad Nacional Autónoma de México, México, D.F., 1995), pp. 11–16
105. K. Weis, W.J. Duschl, The LBV nebula around η Car and its remarkable features, in *Eta Carinae at The Millennium*, ed. by J.A. Morse, R.M. Humphreys, A. Damineli. ASP Conference Series, vol. 179 (ASP, San Francisco, 1999), pp. 155–158
106. P.A. Whitelock, M.W. Feast, C. Koen, G. Roberts, B.S. Carter, Variability of η Carinae. MNRAS. **270**, 364–372 (1994)
107. B.J. Wills, H. Netzer, D. Wills, Broad emission features in QSOs and active galactic nuclei. II – new observations and theory of Fe II and H I emission. ApJ. **288**, 94–116 (1985)
108. R. Zanella, B. Wolf, O. Stahl, Spectroscopy of the shell episode of η Car, 1981–1983. A&A. **137**, 79–84 (1984)
109. T. Zethson, S. Johansson, K. Davidson et al., Strange velocities in the equatorial ejecta of η Carinae. A&A. **344**, 211–220 (1999)
110. T. Zethson, T.R. Gull, H. Hartman et al., Sr II and [Sr II] emission in the ejecta of η Carinae. AJ. **122**, 322–326 (2001)
111. T. Zethson, H. Hartman, S. Johansson et al., Cr II fluorescence in η Carinae due to H Lyman alpha pumping, in *Eta Carinae and Other Mysterious Stars: The Hidden Opportunities of Emission Spectroscopy*, ed. by T. Gull, S. Johansson, K. Davidson. ASP Conference Series, vol. 242 (ASP, San Francisco, 2001), pp. 97–101

Chapter 6
High-Resolution Studies of Eta Carinae's Ejecta and Stellar Wind

Gerd Weigelt and Stefan Kraus

Abstract Hubble Space Telescope observations and interferometric measurements provide unique insights into the small-scale structure of η Car's circumstellar ejecta and its extreme, optically thick, aspherical stellar wind. We discuss speckle interferometric imaging and HST images of η Car's ejecta in the inner 1 arcsec. We also review the first infrared long-baseline spectro-interferometry with high spatial resolution of 5 mas and high spectral resolution. Infrared interferometry allows us, for the first time, to study the strong wavelength dependence of η Car's non-spherical wind structure within emission lines.

6.1 Introduction

Studies of η Car's recent mass-loss history, its circumstellar ejecta, and its extreme stellar wind provide us with unique insights into the late evolutionary stages of the most massive stars. The ejecta of η Car exhibit spatial structure at all scales from 0.001″ to 20″. At the large end of this range, observations before 1980 with resolution ~1″ showed basic facts about the Homunculus Nebula: major diameter ~17″, expansion speeds of 200–700 km s^{-1}, bipolar symmetry, etc. (see Refs. in [8, 42]). In this chapter, however, we focus on much smaller size scales in the central region. Sub-arcsec structure near the star was discovered in 1983–1985, when speckle interferometry and bispectrum speckle interferometry revealed three bright objects B, C, and D in the northwest quadrant at distances of 0.1–0.2″ from the brightest component A [26, 48]. Spectroscopy and imaging in the UV with the Hubble Space Telescope (HST) later confirmed that A is the central star and showed that the knots BCD are slow-moving ejecta, not companion stars [11, 12, 50]. Recent

G. Weigelt (✉) • S. Kraus
Max-Planck Institute for Radioastronomy, Auf dem Hügel 69, Bonn, Germany
e-mail: weigelt@mpifr-bonn.mpg.de; skraus@mpifr-bonn.mpg.de

K. Davidson and R.M. Humphreys (eds.), *Eta Carinae and the Supernova Impostors*, Astrophysics and Space Science Library 384, DOI 10.1007/978-1-4614-2275-4_6,

observations with ESO's Very Large Telescope (VLT) resolved the region of the ejecta BCD with unprecedented angular resolution in the near- and mid-infrared [1]. Today these objects are still not understood and we discuss them in Sect. 6.2 below (see also [21]).

At size scales smaller than approximately 0.03″, one begins to resolve the stellar wind. After 2000, HST spectroscopy showed that η Car's opaque wind is latitude-dependent and prolate with its polar axis aligned SE–NW like the Homunculus axis [41, 43]. Infrared observations with ESO's Very Large Telescope Interferometer (VLTI) have resolved this elongated wind structure [45, 51]. The first spectro-interferometric observations of η Car with the VLTI-AMBER instrument attained a spatial resolution of 5 msa (~11.5 AU for a distance of ~2300 pc) and spectral resolution $R = \lambda/\Delta\lambda \sim 1500$ and 12000. The object's characteristic diameter was found to be 4 mas in the continuum near $\lambda \sim 2.2$ μm and much larger in emission lines, with a perceptibly non-circular shape [45, 51]. The aspherical wind can be explained by models for line-driven winds from luminous hot stars rotating near their critical speed [38, 39]. The models predict a higher wind speed and density along the polar axis than in the equatorial plane. Our VLTI results and their theoretical implications will be described in Sect. 6.3 below.

η Car exhibits a 5.5-year spectroscopic cycle, most likely regulated by a hot companion star in a highly eccentric orbit; see [2–6, 12, 20, 24, 27, 29, 30, 33, 36, 37, 40, 46, 52] and various chapters in the present volume. Details are still controversial, but in some models the shape of the opaque stellar wind should change during a "spectroscopic event" of the type seen in 1997 and 2003 [24, 33, 37, 40, 43]. Furthermore, it is likely that the putative hot companion dives into the dense wind of the primary across the periastron passage, which can lead to a trapping of the FUV. Therefore, highly excited wind lines emitted from the speckle ejecta can drop within days, while recovery takes months. The interferometry mentioned above did not sample a spectroscopic event, but, fortunately, one is expected to occur near the beginning of 2009. This will provide a unique opportunity to study how the colliding stellar winds of the two hot stars and their interface change near periastron passage.

In this chapter we outline what has been learned about η Car's circumstellar environment using observations with sub-arcsecond spatial resolution. In Sect. 6.2 we describe the multi-wavelength properties of ejecta in the innermost 0.3 arcsec, including a discussion of their kinematics and ejection date. In Sect. 6.2.2 we present evidence which suggests the presence of an obscuring, equatorial disk around the central object. Finally, in Sect. 6.3 we review how infrared interferometry has allowed astronomers to zoom-in even closer to the central star and to resolve η Car's massive wind regions on scales of a few dozen AU's.

6.2 η Car's Ejecta B, C, and D at Separations of 0.1–0.3″

Figure 6.1 shows the sub-arcsecond environment of η Car for wavelengths ranging from ultraviolet to infrared. In each image one can see condensations at separations between 0.1″ and 0.3″ from the dominant central object. They were discovered in the mid-1980s, when we applied speckle interferometric techniques to η Car [26, 48]. These are imaging techniques which can overcome image degradation caused by the turbulent atmosphere and by telescope aberrations, allowing us to reconstruct diffraction-limited images [25, 31, 47, 49]. In 1985 (Fig. 6.1e) the three bright objects B, C, and D were located at separations of 0.11″, 0.18″, and 0.21″, respectively, from A. They appeared point-like in the speckle images, but later proved to be ejecta with complex emission-line spectra, not stars [11, 12, 50].

Figure 6.1a–c are broad-band images obtained at wavelengths 190, 307, and 550 nm with the HST Faint Object Camera [50]. Figure 6.1f is a narrow-band 3.74 μm VLT NACO image [1]. The two bright blobs to the north-west have positions which are similar but not identical to the positions of blobs C and D. This image resolves the speckle blob complex in the thermal infrared for the first time and suggests that there is a large region empty of dust with a typical radius of 0.1–0.15″ (∼300 AU; see [1]).

Figure 6.1e is a diffraction-limited image obtained through a filter with a central wavelength of approximately 850 nm and a bandwidth of 40 nm. Figure 6.1d is one of our new bispectrum speckle interferometry narrow-band images obtained on 10 January 2008 with the ESO 3.6 m telescope through an Hα filter with a central wavelength of 657 nm and a filter width of 4 nm. Bispectrum speckle interferometry observations will allow us to observe η Car with unprecedented angular resolution in many additional narrow-band interference filters at visible wavelengths, thus providing information that is complementary to HST spectroscopy and long-baseline infrared interferometry.

Proper motions show that B, C, and D are moving away from A, as discussed in Sect. 6.2.1 below. Evidently, they were ejected from the star [12, 50]. Moreover, since their narrow emission lines are blue-shifted by roughly 30–40 km s^{-1} relative to the star, they must be located several 100 AU closer to us than the star and fairly close to η Car's equatorial plane [8, 12, 44].

A surprising property of the ejecta BCD is their high brightness relative to star A. In 1985 B, C, and D were each almost 10% as bright as A at $\lambda \sim 850$ nm [26,48], and later HST FOC data at visual-to-UV wavelengths showed intensities above 30% relative to A [50]. Such high values constitute a serious puzzle, as we discuss in Sect. 6.2.2.

6.2.1 *Kinematics of the Ejecta in the Inner 1 arcsec: Ejection During the Outburst in 1890?*

A comparison of the 850 nm speckle image (Fig. 6.1e, epoch 1985.05 [26]) with the 550 nm HST FOC image (Fig. 6.1c, epoch 1992.84 [50]) shows that in 7.8 years, the separations A–B, A–C, and A–D increased from 0.114″ to 0.123″, 0.177″ to 0.208″, and 0.211″ to 0.241″, respectively. The two largest separations, A–C and A–D, changed by approximately 4.0 and 3.9 mas year^{-1}, respectively, corresponding to proper motion velocities of about 43 km s^{-1} if η Car's distance from us is 2,300 pc [50].

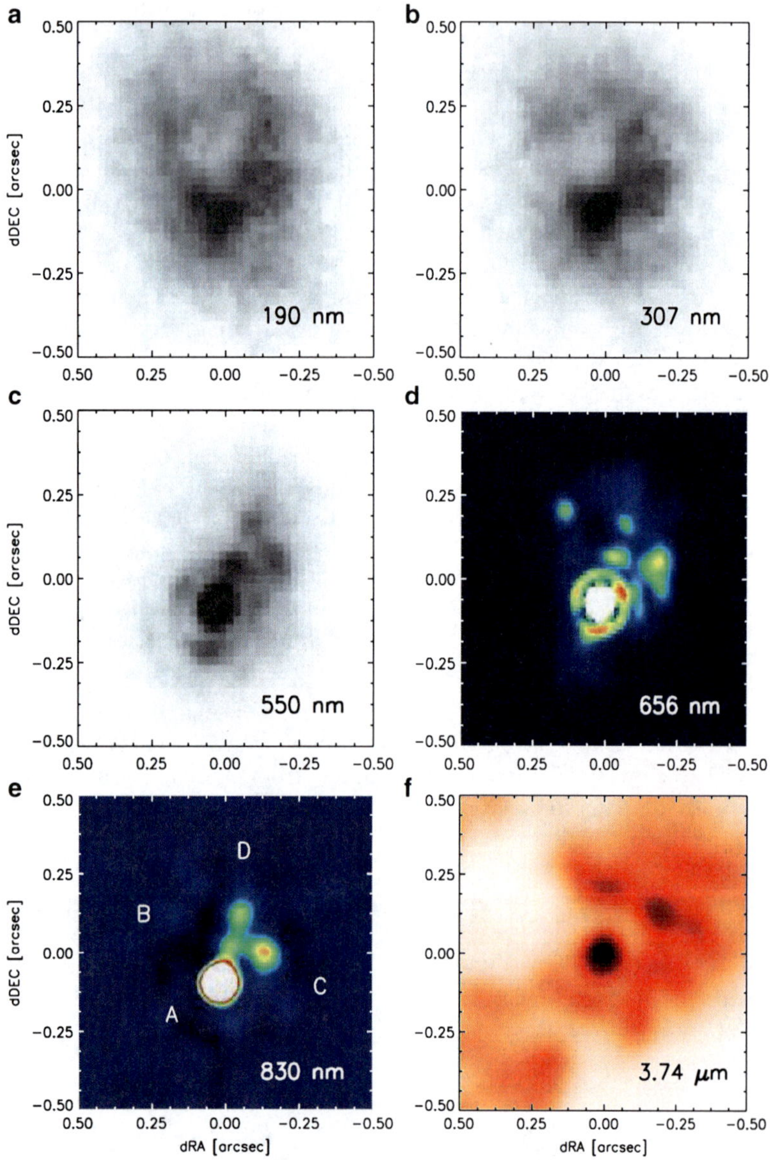

Fig. 6.1 High-resolution images of η Car's inner sub-arcsecond region at various wavelengths from the UV to the IR. (**a**), (**b**), and (**c**): Broad-band HST FOC images obtained in 1991 and 1992 near wavelengths 190, 307, and 550 nm [50]; (**d**): Hα bispectrum speckle interferometry image with central wavelength 657 nm, bandwidth 4 nm, epoch 2008.03 (the ring-like structure in panels **c** and **d** are diffraction rings); (**e**) Speckle masking image obtained through a filter with a central wavelength of 850 nm and a filter bandwidth of approximately 40 nm, epoch 1985.047 [26]; (**f**) Deconvolved VLT NACO adaptive optics image obtained through a filter with central wavelength 3.74 μm (narrow-band filter NB 3.74; bandwidth 0.02 μm; epoch 2002.877) [1]

According to these data, the separation A–D increased by about 14% from 1985.05 to 1992.84, implying an ejection date around 1930. But the uncertainty of this first age determination was large for four reasons: (1) The temporal baseline was not very long in relation to the measurement accuracy. (2) The speckle observation referred to wavelengths near 850 nm but the HST image sampled visual wavelengths around 550 nm. (3) Radiation and wind pressure from the central star may have accelerated the ejecta [12, 44]. In that case the simple linear reasoning used above tends to underestimate the age. (4) If the ejecta contain stratified ionization zones, the motion of the emission centroid may differ from the motion of the mass. However, these effects are not likely to cause a total error worse than a factor of 2. Therefore, we concluded in ref. [50]: The simplest interpretation is that components C and D were ejected from the star between 1880 and 1930. The important point is that they originated significantly later than η Car's great eruption seen in 1837–1858, which produced the large-scale Homunculus. These conclusions were later supported by the observed Doppler shifts [12]. An obvious possibility is that BCD were ejected in the secondary outburst seen around 1890.

Additional proper-motion estimates were reported after 2000. Using HST data with a temporal baseline of 5.9 years, Dorland et al. [17] derived an ejection date of $1934^{+16.0}_{-31.7}$ for blob D, while the measurement of C was less accurate. Smith et al. [44] performed new measurements based on HST image data over a longer time baseline (1992–2003) and derived an ejection date of 1907 ± 12 years. They also discussed the influence of acceleration and filter-dependent photocenter shifts on the determination of the ejection date in detail.

Figure 6.1e shows our most recent observation of η Car obtained with bispectrum speckle interferometry and the ESO 3.6 m telescope on 10 January 2008. The observations were carried out through an Hα filter with a central wavelength of 657 nm and a filter width of 4 nm (FWHM). We measured a separation of 257 mas for D. Over a temporal baseline of 23 years, the separation of A–D increased from 211 mas in 1985 to 257 mas in 2008, corresponding to a proper motion of 2 mas year^{-1}. The implied ejection date is 1880 ± 20 years if we assume a constant velocity and if the position of D does not depend on the filter used. This result is consistent with the 1890 outburst, but more observations in different filters are required to improve the accuracy. The expanding size scale will tend to be an advantage for future measurements; on the other hand, the star's rapidly increasing brightness has recently made BCD more difficult to observe at wavelengths shorter than 700 nm [34].

It is interesting to combine the proper motions with Doppler velocity measurements in order to investigate the 3-dimensional motion and the location of the speckle ejecta relative to the central star. Using UV emission lines in HST data, Davidson et al. [12] measured a heliocentric velocity of about -47 km s^{-1} in BCD. Measured velocities must be corrected for the central star's Doppler velocity, assumed to be -8 km s^{-1} heliocentric. Smith et al. [44] later found similar velocities for the same type of high-excitation lines, but smaller values of about -40 km s^{-1} for low-excitation lines, which may be better velocity tracers for these ejecta. Combining the proper motions and Doppler velocities, both sets of authors concluded that objects C and D, at least, are located within about $\pm 10^\circ$ of η Car's equatorial plane. This idea seems very reasonable, since equatorial ejecta from both the 1840 and the 1890 events are known to exist at larger distances from the star [15]. The equatorial midplane of the Homunculus nebula is inclined at about 42° from the plane of the sky [15,41].

6.2.2 HST Observations in the Visible and UV and Speckle Polarimetry: Detection of an Obscuring Edge-on, Equatorial UV Disk?

As noted above, the brightness ratios B/A, C/A, and D/A were each of the order of 0.1 in the original 1985 speckle data. Such large values were not easy to explain, since the emission from BCD is presumably excited by the central star A, but each condensation cannot intercept more than 1% of the star's energy flow (a similar argument applies to the fraction of the star's light that is reflected by dust grains in BCD). Indeed, the brightness ratios initially seemed to be hints that BCD were stars rather than ejecta [7]. This paradox was then magnified by the observations described next.

In 1991–1992 we obtained broad-band observations of BCD with the HST's pre-COSTAR Faint Object Camera (FOC); see Fig. 6.1a–c [50]. We found that the intensity ratios B/A, C/A, and D/A were each close to 0.3 at $\lambda \sim$550 nm, 0.6 at 307 nm, and 0.7 at 190 nm. Meanwhile, the first spatially resolved spectroscopy confirmed that BCD were collectively comparable to A in the same wavelength range [11]. For the reason noted above, these results were very surprising. The most plausible explanation was that the line of sight to the star includes some extra circumstellar extinction which does not apply to BCD [11,50]. Given the small $0.1''$ projected separation between A and B, this effect most likely occurs within 1000 AU of the star – but not within 150 AU, where the intense radiation field would destroy normal dust grains.

Our 1991–1992 FOC images showed a possible locale for the extra extinction. The image of the central object A was more extended than the core of the corresponding FOC point-spread function. In the 190 nm image, the resolved structure was elongated with a size of about $0.05'' \times 0.1''$ (see [50] for more details). Since this is too large to be the stellar wind per se, as we discuss in Sect. 6.3 below, instead it must be some sort of circumstellar structure. The position angle of the elongation was approximately 50°, roughly perpendicular to the Homunculus axis. It may be part of a disk-like or toroidal object which partially obscures the star but not BCD. This suggestion is also consistent with later speckle imaging polarimetry observations [18], with HST ACS/HRC observations of a somewhat larger disk-like UV structure [44], and with a related morphology traced by high-excitation [Ne III] emission in HST STIS data [35]. Altogether these structures may represent the outer parts of η Car's nonspherical stellar wind, or they may indicate a small torus related to the Little Homunculus [28], which was probably ejected in the 1890 outburst.

An unexpected development in recent years also affects this problem. From 1997 to 2007, the central star's apparent visual and UV brightness increased by a factor of 3, without a proportionate effect in BCD [14, 32, 34]. This probably indicates a substantial decrease in the special extra extinction discussed above; in other words, the hypothetical disk has very likely become less dense. In any case, the visual and UV brightness ratios B/A, C/A, and D/A are now considerably smaller than they were in the early 1990s.

It is possible to imagine a very different possible interpretation of BCD. In principle, they might be illusions caused by local minima in the circumstellar extinction, i.e., line-of-sight holes in a complex distribution of dust. This idea is motivated by the fact that images at wavelengths longer than 3 μm are not very well correlated with shorter-wavelength features (e.g., see Fig. 6.1f [1]). Thus, at wavelengths shorter than 2 μm, it is not necessarily easy to discriminate between real mass concentrations and extinction minima. For the latter, apparent proper motions do not sample the same material as the Doppler velocities. A model of this type for BCD seems unlikely to us because the standard interpretation has a good degree of consistency, but settling the question will require improved spatial resolution.

6.3 Studying the Stellar Wind Region of η Car: VLTI-AMBER Infrared Interferometry with High Spatial Resolution and Spectral Resolutions of 1,500 and 12,000

Infrared long-baseline interferometry offers fascinating new opportunities to study the inner environment around η Car with a spatial resolution of a few AU. On this spatial scale, one resolves the dense stellar wind regions around the star A. In this section we discuss new insights recently obtained with observations using the AMBER instrument at ESO's Very Large Telescope Interferometer (VLTI).

6.3.1 VLTI-AMBER Infrared Long-Baseline Interferometry

AMBER allows the measurement of visibilities, closure phases, and wavelength-differential phases, providing information about the wavelength-dependent radial intensity profile and spatial position of the emitting structures. Our VLTI-AMBER observations of η Car [51] were carried out on 2004 December 26 (phase $\phi = 0.27$ in the object's 5.5-year cycle), 2005 February 25, and 2005 February 26 ($\phi = 0.30$) with the three 8.2 m VLT Unit Telescopes UT2, UT3, and UT4 [51]. Figure 6.2

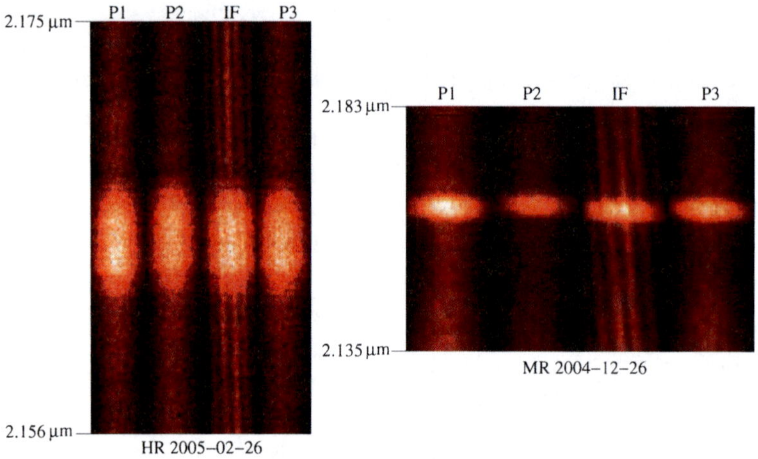

Fig. 6.2 Spectrally dispersed VLTI-AMBER Michelson interferograms of η Car [51]. The two panels show the spectrally dispersed interference fringes (*IF*) as well as the photometric calibration signals from the three telescopes (P1–P3) in high (HR mode; $R = \lambda/\Delta\lambda = 12{,}000$; *left panel*) and medium spectral resolution (MR mode; $R = \lambda/\Delta\lambda = 1{,}500$; *right panel*). In both panels, the bright regions correspond to the Brγ emission line, which is a very broad line due to the high velocities of the stellar wind. The narrow component of the Brγ line, which is caused by the inner ejecta discussed in Sect. 6.2.1, is not visible since the FOV of AMBER is only 80 mas if the 8 m telescopes are used

shows two raw AMBER interferograms taken in the wavelength range around the Brγ 2.166 μm hydrogen recombination line with spectral resolutions of 1,500 (right, MR mode) and 12,000 (left, HR mode). In the MR data sets, the Doppler-broadened Brγ line covers ~8 spectral channels, whereas in HR mode, the line is resolved by ~50 spectral channels, allowing us to measure η Car's stellar wind structure in many different channels simultaneously. High spectral resolution is useful because the resolved structures change dramatically with wavelength within the lines. Using projected baseline lengths up to 89 m, an angular resolution of 5 mas (~11 AU) was obtained in the K band.

Figure 6.3 shows our measured spectra, visibilities, wavelength-differential phases, and closure phases extracted from the VLTI-AMBER interferograms. These observations were performed in the wavelength range around the He I 2.059 μm and the Brγ 2.166 μm emission lines. Small visibilities within the emission lines indicate that the object is more extended at those wavelengths. Differential phases allow the measurement of spatial offsets of the line-emitting region with respect to the continuum emission, while closure phases indicate the degree of asymmetry. Using these measurements and comparing them to two different types of models, we can study η Car's stellar wind as discussed in the following subsections (see [51] for more details).

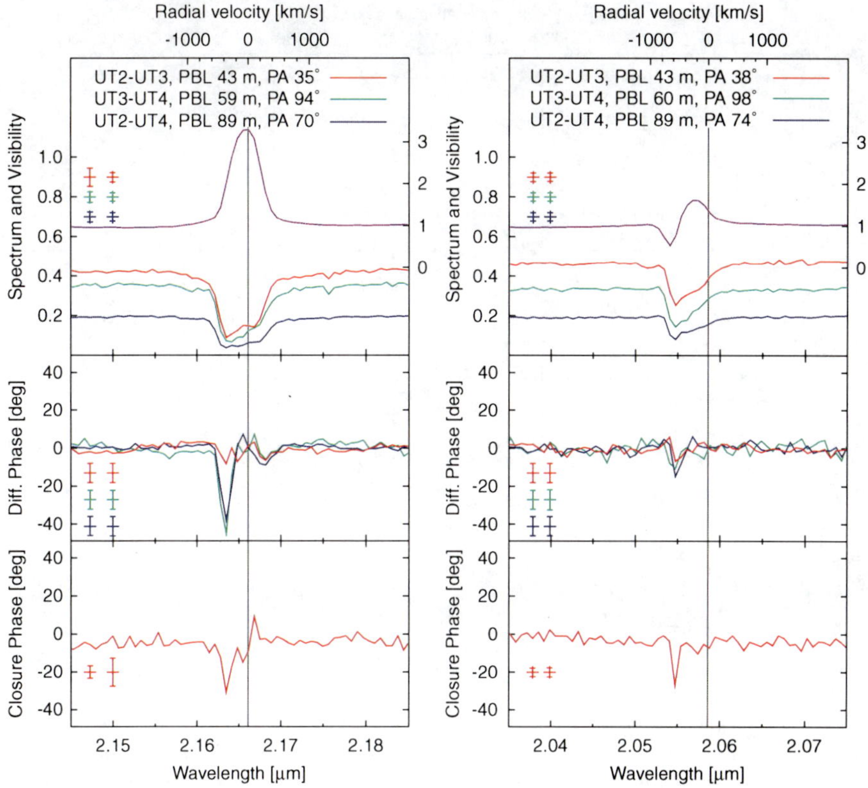

Fig. 6.3 VLTI-AMBER observations of η Car with medium spectral resolution $R = 1{,}500$ around both the Brγ (*left*) and He I line (*right*): spectra and visibilities for the three baselines described in the figures (*top*), wavelength-differential phases (*middle*), and closure phase (*bottom*) [51]. The *vertical line* marks the rest-wavelength of Brγ ($\lambda_{\text{vac}} = 2.1661\,\mu$m) and He I ($\lambda_{\text{vac}} = 2.0586\,\mu$m)

6.3.2 Comparison of VLTI-AMBER Stellar Wind Observations with the Predictions of Line Radiative Transfer Models

Physical models of density and temperature in η Car's stellar wind became feasible when HST spectra were obtained in 1991–2000 [10, 11, 13, 16, 19, 43] (previous spectroscopy was heavily contaminated by the ejecta shown in Fig. 6.1). The first NLTE model with a self-consistent treatment of continuum and emission-line radiative transfer was developed by Hillier et al. [22, 24] in order to explain HST STIS data obtained in 1998. This model assumes a spherical wind geometry; i.e., the aspherical wind structure and any contribution from a binary companion are not taken into account. Furthermore, the model is not equally suited for all orbital phases, because the 1998 data (used for modeling) were obtained soon after a spectroscopic event.

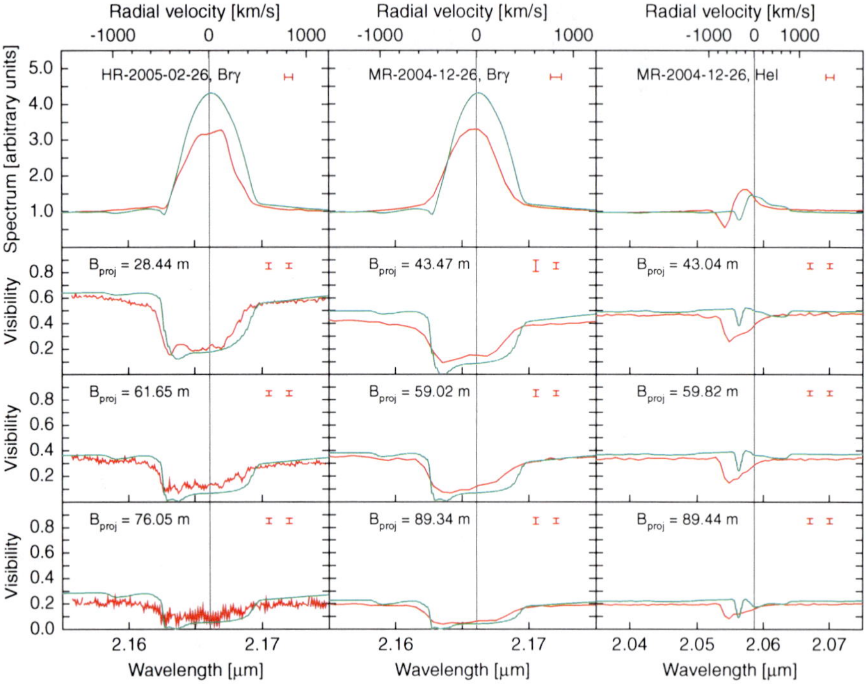

Fig. 6.4 Comparison of the AMBER η Car spectra and visibilities with the predictions of the Hillier et al. line radiative transfer model of η Car's stellar wind [22]: spectra (*upper row*) and visibilities (*bottom three rows*) of the AMBER measurements (*red/black line*) and model predictions [22] (*green/grey lines*) are shown. The data from 2004 (*middle* and *right*) was taken with medium spectral resolution (R = 1,500), whereas the 2005 Brγ measurement (*left*) was obtained with high spectral resolution (R = 12,000). The differences in the He I line (*right column*) can possibly be explained by the emission from a wind–wind collision zone (e.g., [24, 37, 51])

In Fig. 6.4 we compare the predictions of this model with the spatial information provided by our AMBER measurements. The top row displays AMBER and model spectra, while all other panels show AMBER and model visibilities for the different projected baseline lengths. As the figure shows, the NLTE model [22] approximately reproduces the AMBER continuum observations and the wavelength dependence of the visibilities inside the Brγ line. On the other hand, the visibility across the He I line shows an obvious discrepancy between the observations and the model; evidently the latter does not accurately describe the He I line-forming region. HST/STIS observations show that He I line profiles are strongly variable throughout most of the 5.54-year spectroscopic cycle (e.g., [10, 37]). It has been proposed that a large fraction of the He I emission probably originates in the outer wind, especially the wind–wind interaction zone of the binary system, excited mainly by ionizing UV from the putative hot companion star [9, 10, 23, 24, 37]. The resulting complex, asymmetrical geometry [40] obviously cannot be represented in a spherical model.

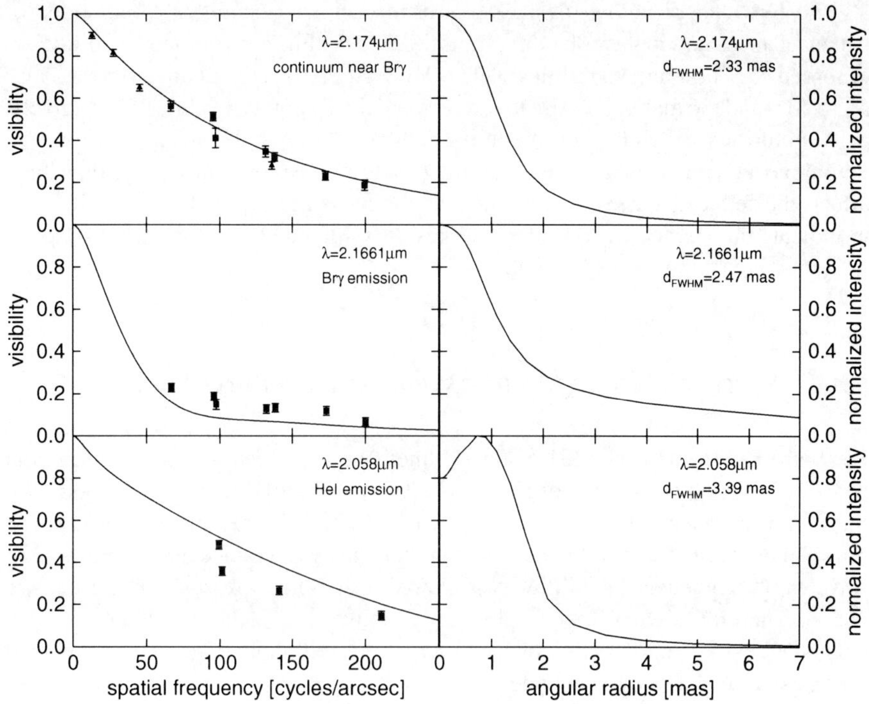

Fig. 6.5 *Left*: comparison of η Car's AMBER visibilities [51] (*squares*; baseline range 28–89 m), plotted as a function of spatial frequency, with the model predictions of [22] (*lines*) for three selected wavelengths: continuum (*upper panel*), the central wavelength of the Brγ (*middle*), and He I emission lines (*bottom*). The triangles (*top, left*) are the VLTI-VINCI K-band measurements from [45]. *Right*: center-to-limb variation of the [22] models for the wavelengths indicated by the labels

6.3.3 *Size of η Car's Stellar Wind Zone in the Continuum, He I, and Brγ Emission Line*

Figure 6.5 compares the AMBER visibilities and Hillier et al. [22] model visibilities as functions of spatial frequency or baseline length (instead of visibility versus wavelength, as shown in Fig. 6.4). Three wavelengths are shown: a continuum wavelength and the centers of the Brγ and the He I emission lines. The corresponding model center-to-limb intensity variations (CLVs) are also shown for illustration. We find good agreement between the visibilities as functions of spatial frequency measured with AMBER and those predicted by the model.

Fitting the model center-to-limb intensity profiles to the AMBER observations, we obtain a 50% encircled-energy diameter of 4.3 mas ($\sim$10 AU) for the continuum at $\lambda \sim 2.17\,\mu$m. We also find an elongation toward position angle $120 \pm 15°$ with a projected axis ratio of 1.18 ± 0.1, in agreement with previous broadband

VLTI-VINCI observations [45]. These results support theoretical predictions of enhanced mass loss in the polar directions [38,39,43]. Fits of the model visibilities to observed emission line visibilities yield 50% encircled-energy diameters of 6.5 mas ($\sim$15 AU) and 9.6 mas ($\sim$22 AU) in the He I and the Brγ emission lines, respectively.

A notorious difficulty for modeling stellar winds is the need to allow for inhomogeneous densities – the "clumping factor" or volume filling factor, which affects the deduced mass loss rate. The overall scale size, if it can be observed, helps to indicate this parameter. Based on the results outlined above, the filling factor in η Car's wind cannot be very small.

6.3.4 Direct Evidence for an Asymmetric Primary Wind

For both the Brγ and the He I emission lines, we measured non-zero differential phases (DPs) and non-zero closure phases (CPs) within the Brγ and He I emission lines, indicating a complex, asymmetric wind structure. To explain all the AMBER observables, we need a model that can simultaneously reproduce the structure of all wavelength-dependent visibilities, wavelength-differential phases, and all closure phases. Therefore we developed a physically motivated model illustrated in Fig. 6.6, which can explain all interferometric observations with a wind geometry expected for an aspherical, latitude-dependent stellar wind (see [51] for more details).

To summarize, VLTI infrared interferometry allows the study of η Car's stellar wind with unprecedented angular resolution. Comparisons of our AMBER Brγ and continuum visibilities with the radiative transfer model from Hillier et al. [22] show that a spherical radiative transfer wind model approximately agrees with most of the data even though we know that the true configuration is not spherical. Discrepancies concerning the He I line can probably be explained by supposing that this emission originates near a binary wind–wind collision interface.

From fits of model visibility profiles [22] to the AMBER visibilities, we derived 50% encircled-energy diameters of 4.2, 6.5, and 9.6 mas ($\sim$10, 15, and 22 AU) for the 2.17 μm continuum, He I, and the Brγ emission regions, respectively.

In the continuum around the Brγ line, we found an asymmetry towards position angle PA $\sim 120^\circ$ with a projected axis ratio of about 1.2. This result supports theoretical studies which predict enhanced polar winds for rotating massive stars (e.g., [38, 39]).

The measured non-zero differential phases and closure phases within the Brγ and the He I emission lines suggest an asymmetric, wavelength-dependent object structure. We developed an aspherical, latitude-dependent stellar wind model that can explain most of the available interferometric observations such as the spectrum, wavelength dependence of the visibility within the Brγ emission line (but not within the He I line), the wavelength-differential interferometric phases, and the closure phases.

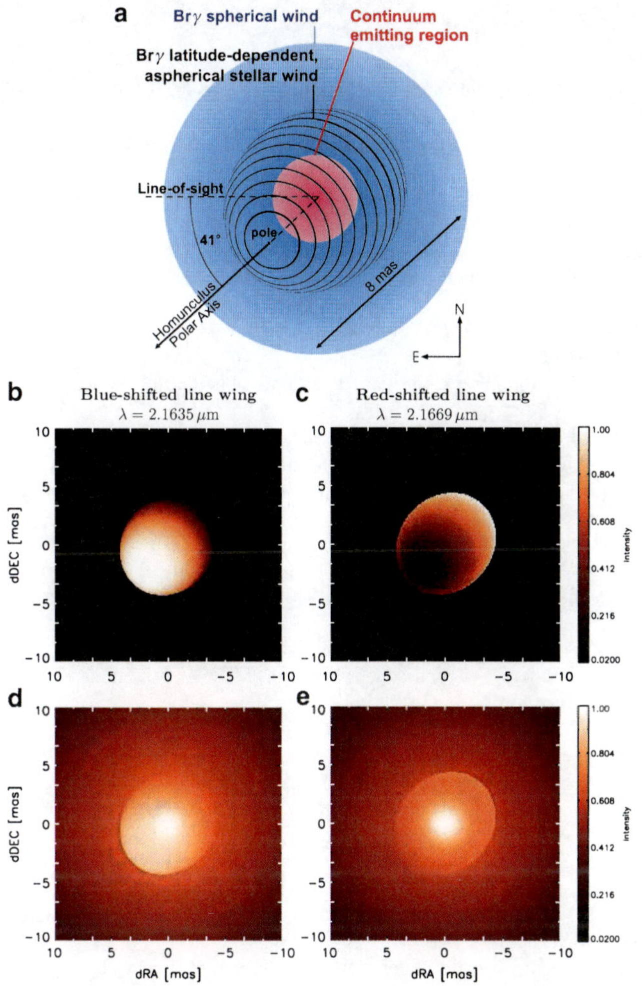

Fig. 6.6 (**a**) Illustration of the components of our geometric model for an aspherical, latitude-dependent wind (*top*; see [51] for more details). For the weak aspherical wind component, we draw the lines of latitudes to illustrate the 3D-orientation of the ellipsoid. The key point is that the inclination angle between the polar axis of η Car's stellar wind and our viewing direction is $i \sim 42°$; therefore, we are looking at the south pole of the fast, optically thick wind region. This south pole region mainly emits blue-shifted light, in contrast to the equatorial and northern region. (**b**, **c**): Brightness distribution of the modeled aspherical wind component for two wavelengths in the blue- and red-shifted wing of the Brγ emission line. The photocenter shift of, for example, the image in the blue-shifted wings of the Brγ emission line leads to the observed wavelength-differential phases. Image asymmetries cause non-zero closure phases. (**d**, **e**): Total brightness distribution after adding the contributions from the two spherical constituents of our model (From [51])

References

1. O. Chesneau, M. Min, T. Herbst, et al., The sub-arcsecond dusty environment of Eta Carinae. A&A. **435**, 1043–1061 (2005)
2. M.F. Corcoran, X-ray monitoring of η Carinae: variations on a theme. AJ. **129**, 2018–2025 (2005)
3. M.F. Corcoran, K. Ishibashi, K. Davidson, et al., Increasing X-ray emissions and periodic outbursts from the massive star Eta Carinae. Nat. **390**, 587–589 (1997)
4. A. Damineli, The 5.52 year cycle of Eta Carinae. ApJ. **460**, L49–L52 (1996)
5. A. Damineli, P.S. Conti, D.F. Lopes, Eta Carinae: a long period binary? New Astron. **2**, 107–117 (1997)
6. A. Damineli, D.J. Hillier, M.F. Corcoran, et al., The periodicity of the η Carinae events. MNRAS. **384**, 1649–1656 (2008)
7. K. Davidson, R.M. Humphreys, Eta Carinae as a possible multiple system. A&A. **164**, L7–L9 (1986)
8. K. Davidson, R.M. Humphreys, Eta Carinae and its environment. ARA&A. **35**, 1–32 (1997)
9. K. Davidson, Why the binary hypothesis isn't a panacea. in *Eta Carinae at the Millennium*, ed. by J.A. Morse, R.M. Humphreys, A. Damineli. ASP Confefence Series, vol. 179 (ASP, San Francisco, 1999), pp. 304–315
10. K. Davidson, Unique spectroscopic problems related to Eta Carinae. in *Eta Carinae and Other Mysterious Stars*, ed. by T. Gull, S. Johansson, K. Davidson. ASP Conference Series, vol. 242 (ASP, San Francisco, 2001), pp. 3–14
11. K. Davidson, D. Ebbets, G. Weigelt, et al., HST/FOS spectroscopy of Eta Carinae: the star itself, and ejecta within 0.3 arcsec. AJ. **109**, 1784–1796 (1995)
12. K. Davidson, D. Ebbets, S. Johansson, et al., HST/GHRS observations of the compact slow ejecta of Eta Carinae. AJ. **113**, 335–345 (1997)
13. K. Davidson, K. Ishibashi, T.R. Gull, R.M. Humphreys, HST/STIS observations of the star during its recent event. in *Eta Carinae at the Millennium*, ed. by J.A. Morse, R.M. Humphreys, A. Damineli. ASP Conference Series, vol. 179 (ASP, San Francisco, 1999), pp. 227–235
14. K. Davidson, T.R. Gull, R.M. Humphreys, et al., An unusual brightening of Eta Carinae. AJ. **118**, 1777–1783 (1999)
15. K. Davidson, N. Smith, T.R. Gull, et al., The shape and orientation of the Homunculus Nebula based on spectroscopic velocities. AJ. **121**, 1569–1577 (2001)
16. K. Davidson, J. Martin, R.M. Humphreys, et al., A change in the physical state of η Carinae? AJ. **129**, 900–906 (2005)
17. B.N. Dorland, D. Currie, A.R. Hajian, Did η Carinae's Weigelt blobs originate circa 1941? ApJ. **127**, 1052–1058 (2004)
18. H. Falcke, K. Davidson, K.-H. Hofmann, et al., Speckle-masking imaging polarimetry of η Carinae: evidence for an equatorial disk. A&A. **306**, L17–L20 (1996)
19. T.R. Gull, K. Ishibashi, K. Davidson, The Cycle 7 STIS Go Team: first observations of η Carinae with the space telescope imaging spectrograph. in *Eta Carinae at the Millennium*, ed. by J.A. Morse, R.M. Humphreys, A. Damineli. ASP Conference Series, vol. 179 (ASP, San Francisco, 1999), pp. 144–154
20. T.R. Gull, S. Johansson, K. Davidson (eds.), Eta Carinae and other mysterious stars: the hidden opportunities of emission spectroscopy. in *Eta Carinae and Other Mysterious Stars*, ASP Conference Series, vol. 242 (ASP, San Francisco, 2001)
21. F. Hamann, this volume, 95–128 (2012)
22. D.J. Hillier, K. Davidson, K. Ishibashi, T. Gull, On the nature of the central source in η Carinae. ApJ. **553**, 837–860 (2001)
23. D.J. Hillier, K. Davidson, K. Ishibashi, T. Gull, Eta Carinae: the central star. in *Eta Carinae and Other Mysterious Stars*, ASP Conference Series, vol. 242 (ASP, San Francisco, 2001), pp. 15–28

24. D.J. Hillier, T. Gull, K. Nielsen, et al., The UV scattering halo of the central source associated with η Carinae. ApJ. **642**, 1098–1116 (2006)
25. K.-H. Hofmann, G. Weigelt, Speckle masking observation of the central object in the giant H II region NGC 3603. A&A. **167**, L15–L16 (1986)
26. K.-H. Hofmann, G. Weigelt, Speckle masking observation of Eta Carinae. A&A. **203**, L21–L22 (1988)
27. R.M. Humphreys, K.Z. Stanek (eds.), The fate of the most massive stars. in *The Fate of the Most Massive Stars*, ASP Conference Series, vol. 332 (ASP, San Francisco, 2005)
28. K. Ishibashi, T.R. Gull, K. Davidson, Discovery of a little Homunculus within the Homunculus Nebula of η Carinae. AJ. **125**, 3222–3236 (2003)
29. W. Kundt, C. Hillemanns, ETA CARINAE – an evolved triple-star system? Chin. J. Astron. Astrophys. Suppl. **3**, 349–360 (2003)
30. M. Livio, J. E. Pringle, Can Eta Carinae be a triple system? MNRAS. **295**, L59–L60 (1998)
31. A.W. Lohmann, G. Weigelt, B. Wirnitzer, Speckle masking in astronomy – triple correlation theory and applications. Appl. Opt. **22**, 4028–4037 (1983)
32. J.C. Martin, M.D. Koppelman, η Carinae's brightness variations since 1998: Hubble space telescope observations of the central star. AJ. **127**, 2352–2361 (2004)
33. J.C. Martin, K. Davidson, R.M. Humphreys, D.J. Hillier, K. Ishibashi, On the He II emission in η Carinae and the origin of its spectroscopic events. ApJ. **640**, 474–490 (2006)
34. J.C. Martin, K. Davidson, M.D. Koppelman, The chrysalis opens? Photometry from the η Carinae Hubble space telescope treasury project, 2002–2006. AJ. **132**, 2717–2728 (2006)
35. A. Mehner, K. Davidson, G.J. Ferland, R.M. Humphreys, High-excitation emission lines near Eta Carinae, and its likely companion star. ApJ. **710**, 729–742 (2010)
36. J.A. Morse, R.M. Humphreys, A. Daminelli (eds.), Eta Carinae at the millennium. in *Eta Carinae at the Millennium*, ASP Conference Series, vol. 179 (ASP, San Francisco, 1999)
37. K.E. Nielsen, M.F. Corcoran, T. Gull, et al., η Carinae across the 2003.5 minimum: spectroscopic evidence for massive binary interactions. ApJ. **660**, 669–686 (2007)
38. S.P. Owocki, S.R. Cranmer, K.G. Gayley, Inhibition FO wind compressed disk formation by nonradial line-forces in rotating hot-star winds. ApJ. **472**, L115–L118 (1996)
39. S.P. Owocki, S.R. Cranmer, K.G. Gayley, Mass loss from rotating hot-stars: inhibition of wind compressed disks by nonradial line-forces. Ap&SS. **260**, 149–159 (1998)
40. J.M. Pittard, M.F. Corcoran, In hot pursuit of the hidden companion of Eta Carinae: an X-ray determination of the wind parameters. A&A. **383**, 636–647 (2002)
41. N. Smith, Dissecting the Homunculus Nebula around Eta Carinae with spatially resolved near-infrared spectroscopy. MNRAS. **337**, 1252–1268 (2002)
42. N. Smith, this volume, 145–169 (2012)
43. N. Smith, K. Davidson, T.R. Gull, et al., Latitude-dependent effects in the stellar wind of η Carinae. ApJ. **586**, 432–450 (2003)
44. N. Smith, J.A. Morse, T.R. Gull, et al., Kinematics and ultraviolet to infrared morphology of the inner Homunculus of η Carinae. ApJ. **605**, 405–424 (2004)
45. R. van Boekel, P. Kervella, M. Schöller, et al., Direct measurement of the size and shape of the present-day stellar wind of Eta Carinae. A&A. **410**, L37–L40 (2003)
46. A.M. van Genderen, C. Sterken, W.H. Allen, et al., An overview of the photometric events, trends and brightenings of Eta Carinae. JAD. **12**, 1–41 (2006)
47. G. Weigelt, Modified astronomical speckle interferometry 'speckle masking'. Opt. Commun. **21**, 55–59 (1977)
48. G. Weigelt, J. Ebersberger, Eta Carinae resolved by speckle interferometry. A&A. **163**, L5–L6 (1986)
49. G. Weigelt, B. Wirnitzer, B., Image reconstruction by the speckle-masking method. Opt. Lett. **8**, 389–391 (1983)

50. G. Weigelt, R. Albrecht, C. Barbieri, et al., HST FOC observations of Eta Carinae. in *The Eta Carinae Region: A Laboratory of Stellar Evolution*, ed. by V. Niemela, N. Morrell, A. Feinstein. Revista Mexicana de Astronomia y Astrofisica Serie de Conferencias, Vol. 2 (Instituto de Astronomia, Universidad Nacional Autonoma de Mexico, Mexico, 1995), pp. 11–16
51. G. Weigelt, S. Kraus, T. Driebe, et al., Near-infrared interferometry of η Carinae with spectral resolutions of 1,500 and 12,000 using AMBER/VLTI. A&A. **464**, 87–106 (2007)
52. P.A. Whitelock, M.W. Feast, F. Marang, et al., The 2003 shell event in η Carinae. MNRAS. **352**, 447–456 (2004)

Chapter 7
All Things Homunculus

Nathan Smith

Abstract The "Homunculus" nebula around Eta Carinae is one of our most valuable tools for understanding the extreme nature of episodic pre-supernova mass loss in the most massive stars, perhaps even more valuable than the historical light curve of η Car. As a young nebula that is still in free expansion, it bears the imprint of its ejection physics, making it a prototype for understanding the bipolar mass loss that is so common in astrophysics. The high mass and kinetic energy of the nebula provide a sobering example of the extreme nature of stellar eruptions in massive stars near the Eddington limit. The historical ejection event was observed, and current parameters are easily measured due to its impressive flux at all wavelengths, so the Homunculus is also a unique laboratory for studying rapid dust formation and molecular chemistry, unusual ISM abundances, and spectroscopy of dense gas. Since it is relatively nearby and bright and is expanding rapidly, its 3-D geometry, kinematics, and detailed structure can be measured accurately, providing unusually good quantitative contraints on the physics that created these structures. In this chapter I review the considerable recent history of observational and theoretical study of the Homunculus nebula, and I provide an up-to-date summary of our current understanding, as well as areas that need work.

7.1 Early Observations and Proper Motions

Eta Carinae was first recognized as a spatially extended object in 1900–1930 by Innes, van den Bos, and Voûte, who noted visual companion objects less than 2$''$ from the star (see [42, 96, 112, 137]). Around 1950 the structure was recognized as nebular by Thackeray and by Gaviola [42, 137]. Gaviola named it

N. Smith (✉)
Steward Observatory, University of Arizona, 933 N. Cherry Ave., Tucson, AZ 85721, USA
e-mail: nathans@as.arizona.edu

K. Davidson and R.M. Humphreys (eds.), *Eta Carinae and the Supernova Impostors*, Astrophysics and Space Science Library 384, DOI 10.1007/978-1-4614-2275-4_7,

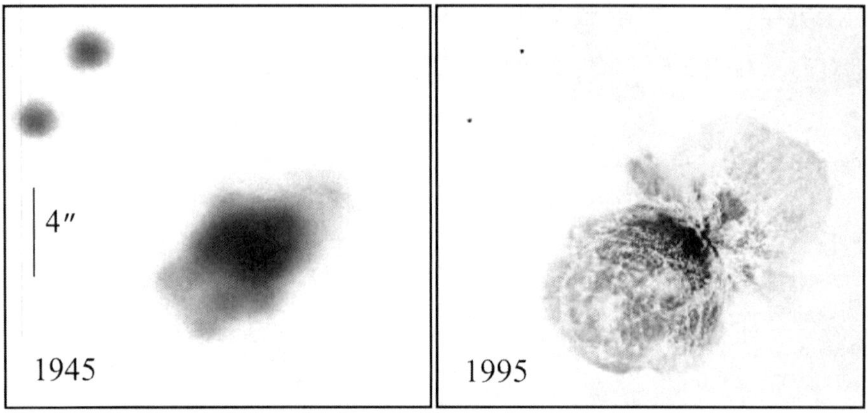

Fig. 7.1 Two visual images of the Homunculus, printed with the same spatial scale. *Left*: Gaviola's blue photograph, from an original photographic plate that was digitally scanned by the author [42, 112]. *Right*: A red-wavelength image obtained 50 year later with the *HST* [81]

"the Homunculus" because on photographic plates it resembled a small plump man (Fig. 7.1). The earliest visible features were mostly in the equatorial parts of the bipolar configuration. The polar lobes had a diameter of 5–10″ in 1900–1940, but the relatively low surface brightness of their outer extremities made them inconspicuous in the glare of the 3″ central region (see [21, 42]). During the 1940s the lobes may have brightened more than the central core region did [28, 138]. This brightening is probably an important clue to η Car's recovery from the 1843 eruption, but it is difficult to infer physical information from the scant historical record.

Early measurements of the Homunculus' expansion showed that it was ejected in the mid-to-late nineteenth century [42, 43, 96]. The obvious implication, that it was created in the Great Eruption, was later verified by measurements using data from *HST* and with higher spatial resolution or longer temporal baselines [17, 82, 112]. Thus we know reliably that the bulk of the Homunculus nebula was ejected in a relatively short time interval during the 1840s. (Smith and Frew [111] discuss the historical light curve with newly uncovered data from critical times in the 1840s.) It therefore contains some of our most valuable information about the Eruption. This doesn't necessarily mean, however, that *all* the ejecta came from that event. A number of features appear to have been ejected later, perhaps in the 1890 secondary eruption. These are discussed later.

7.2 The Brightest Object in the Mid-infrared Sky

One of the most important developments in understanding η Car was the recognition that it is extremely luminous at infrared (IR) wavelengths. In 1968–1969 Neugebauer and Westphal discovered a strong near-IR excess at 1–3 μm, and then, more important, extremely luminous mid-IR radiation at 5–20 μm [83, 150]. This led to our present understanding of the energy budget as that of a very luminous star enshrouded by circumstellar dust, which reprocesses the UV/visual luminosity into thermal IR radiation [19,66,91,92,99].

In subsequent decades, the Homunculus was a favorite target for mid-IR astronomers because it was bright and spatially extended. Early measurements of the spectral energy distribution (SED) established a range of dust temperatures, 200–400 K, a large gas mass of at least a few $M_\odot$, strong and broad silicate emission at 9.7 μm, unusually large grains with radii of roughly 1 μm, and a lack of strong fine-structure emission lines [1,5,6,44,65,76,97,135]. The reasoning for large grains was based mainly on dust temperatures near the blackbody values for the projected radii of extended features in mid-IR maps, and on the broad silicate feature [77,98], but this also agreed with the unusually low reddening/extinction ratio at shorter wavelengths (e.g., [99]). The dusty Homunculus acts as a calorimeter, providing us with an estimate of the star's total luminosity $L \approx 5 \times 10^6 L_\odot$. This is more robust than for most other massive stars, because it does not require a model-dependent spectral extrapolation to UV wavelengths.

The Homunculus showed complex spatial structure in even the earliest IR drift scans and raster maps that used single-element detectors with 1–2″ spatial resolution. Several authors noted an elongated, double-peaked, or torus-like structure in the warm inner core, plus a more extended cooler halo [3,8,12,60,78,101].

In the mid-1980s Hackwell, Gehrz, and Grasdalen made a mid-IR raster map of the Homunculus, which first clearly delineated the limb-brightened structure of the two hollow polar lobes (described as "osculating spheres"), the outer boundary of the mid-IR emitting structure, and the complex multi-temperature and multi-peaked structure of the core region [52].

The advantage of mid-IR imaging of the lobes, compared to reflected light in visual-wavelength images, is that the thermal dust radiation in the mid-IR is optically thin, so we can see that the lobes are hollow and that condensations exist in the obscured core region. Subsequent imaging with array detectors on 4m-class telescopes improved upon the spatial resolution (especially in the core), the sensitivity, and the wavelength coverage, but confirmed the same basic structure [56, 80, 93, 94, 112, 118, 120].[1] So far, the only high-quality mid-IR polarimetric imaging was presented by Aitken et al. ([2]; see also [11]), which revealed suggestive evidence of grain alignment from a possible swept-up ambient magnetic field, or a strong original field in the star's atmosphere before the eruption. Our mid-IR view of the Homunculus improved greatly with the advent of 8m-class telescopes [13, 122, 123], because at long wavelengths like 10–20 μm, the better diffraction limit of larger telescopes is critical. These modern IR observations will be discussed along with our modern view of the structure of the nebula below.

[1] Note, however, that some of the morphological interpretations in these studies were contradicted by later observations.

7.3 Mass and Kinetic Energy of the Homunculus

As noted earlier, the Homunculus is potentially our most valuable probe of the nineteenth century eruption and this class of stellar outbursts, because with the knowledge from proper motion studies that it was indeed ejected in that event, the Homunculus provides us with two essential quantities: the ejected mass and its kinetic energy.

The usual way of measuring this mass is to estimate the *dust* mass that is needed to produce the IR emission, and to then multiply by a reasonable gas/dust mass ratio. With conventional silicate grain emissivities and gas/dust = 100, the observed 2–12 μm SED yields a Homunculus mass of 2–3 $M_\odot$, representing the ~200 K material in the polar lobes [15, 120]. This is consistent with the circumstellar extinction [20]. However, dense clumps which are opaque at visual wavelengths [34, 81] may contain additional mass that is under-represented in either type of estimate. For this reason, it can be problematic to measure both the stellar luminosity and the ejecta mass from the same IR luminosity, because at least one of those parameters could be underestimated [20].

Indeed, the mass estimate of 2–3 $M_\odot$ that was adopted for many years is now known to be too small. More recent estimates from dust emission place the ejecta mass (assuming the same gas/dust mass ratio of 100) at *much higher* values of 12.5, 15, or even 20 $M_\odot$ [80, 110, 123]. This upward revision followed chiefly from high-quality measurements at far-IR wavelengths, which revealed a large mass of dust much cooler than 200 K [80, 123] (Fig. 7.2).

An important distinction must be noted regarding the location of this cool material. Morris et al. [80] fit the far-IR spectrum with dust at 110 K, and proposed

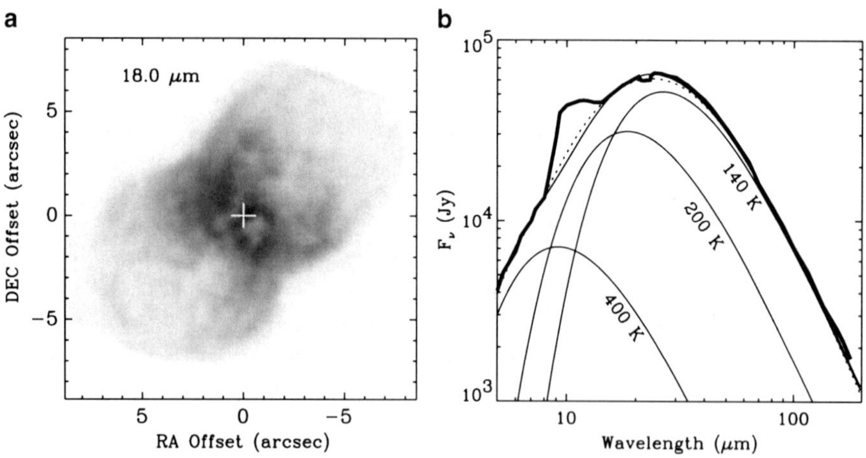

Fig. 7.2 (**a**) An 18 μm image of the Homunculus taken with the 6.5-m Magellan telescope [123]. (**b**) the mid-IR SED of η Car with the 140 K, 200 K, and warmer component fits used to derive the mass in the polar lobes [123]

that it resides in a *pre-existing* massive disk or torus that pinched the waist of the Homunculus when it was formed in the Great Eruption. But Davidson and Smith [22] noted that any thermal source at 110 K must cover a much larger projected area in order to produce the observed IR flux, and that the torus-like structure was already known to be much warmer than 110 K. After discovering a thin outer shell of 140 K dust in high-resolution mid-IR images at 18–25 μm, Smith et al. [123] performed an independent analysis of the same far-IR data and found that it could be fit equally well with 140 K dust. They suggested instead that the large mass of more than $10\,M_\odot$ exists in the polar lobes of the Homunculus, and that there is very little mass near the equator. This view had two critical implications: (1) the 1840s eruption ejected a large mass in an intrinsically bipolar flow, i.e., it was not shaped by pre-existing equatorial material; and (2) the $>10\,M_\odot$ of material is moving very fast, with a kinetic energy of $10^{49.6}$–10^{50} erg [123].

If the mass and kinetic energy values noted above are incorrect, they are probably *underestimates* because the dominant sources of error tend toward higher mass. The assumed gas/dust mass ratio of 100 is probably too low given that the ejecta are known to be C and O poor [23, 110]. Potential optical depth effects, such as shielded cold grains in dense clumps, can hide additional mass. Recent observations at sub-mm wavelengths that are sensitive to the coldest material indicate 0.3–$0.7\,M_\odot$ of dust [45], implying 30–$70\,M_\odot$ of gas if gas/dust = 100 (although some of this emission may come from a more extended region). Independent of IR dust measurements, Smith and Ferland [110] estimated a large *gas* mass of 15–$35\,M_\odot$ for nebular models that allow H_2 molecules to survive near such a luminous star (though values near the lower end of the range are favored). So far this is the only estimate of the ejecta mass that does not rely on an assumed gas/dust ratio.

If η Car ejected as much as 20–$30\,M_\odot$ in a single event that lasted just a few years, this begins to strain the star's available mass reservoir if the upper mass limit to stars really is around $150\,M_\odot$, especially if other events like the Great Eruption have happened once or twice in the past few thousand years. Such extreme values of the mass and kinetic energy underscore the profound influence of such high mass-loss events in the evolution of massive stars [58, 116], not to mention their potential role as precursors to the most luminous supernova explosions [129, 152]. The fact that they still have no theoretical explanation makes this one of the most pressing mysteries in astrophysics.

7.4 The Bi-polar Lobes

Since the Homunculus is the major product of the 1843 eruption, its shape, structure, mass, and energy are vital clues for the outburst mechanism and the origin of the bipolar shape that is so pervasive in astrophysics. This object has consistently been a favorite target of imaging and spectroscopic studies with ever-increasing resolution. But it has not yielded its secrets easily, and there has been a great deal of controversy over its true structure. Recent observations have helped.

7.4.1 *Imaging Studies of the Homunculus as a Reflection Nebula*

Although early drawings and photographs showed the polar lobes of the Homunculus [42, 43, 137], the familiar structure we recognize today was not clearly apparent. Some authors specifically described it as "bipolar" based on velocities and polarization [73, 148]. The first images to show detailed features of the polar lobes and especially the thin equatorial skirt were ground-based images shown by Duschl et al. [31]. Later Hubble Space Telescope images obtained with a carefully-chosen set of filters and dithered WFPC2 exposures provided a spectacularly sharp view of the Homunculus [81], and became among the most familiar and oft-reproduced images made with HST.[2]

The bipolar lobes appear remarkably symmetric in these images, although Morse et al. [81] remarked that a slight axial asymmetry can be perceived by simply rotating one of the familiar HST images by 180°. Multiple epochs of images with WFPC2 and HST's ACS instrument have enabled detailed investigations of the proper motions and light variations in the nebula [17, 34, 70, 82, 121, 125, 126].

Some of the most notable results from these imaging data are: (1) intricate, mottled structure on the surfaces of the polar lobes [81], resembling convective solar granulation or vegetable-inspired comparisons, (2) the ragged, streaked appearance of material in the equatorial skirt [20, 31, 81, 125], some of which seem to connect to more distant nebular features, and (3) the blue glow or Purple Haze [81, 126, 153], a near-UV fluorescent nebulosity in the core of the Homunculus, perhaps associated with the Little Homunculus or equatorial ejecta [106]. HST also revealed remarkable features inside and outside the lobes, reviewed in other chapters.

Although the Homunculus has structures similar to those seen in planetary nebulae, it is not a hot photoionized emission-line nebula. Instead it contains cool low-ionization gas, molecules, and dust grains. Its visual spectrum is a complex mixture of reflected starlight and intrinsic emission from low-ionization species such as [Fe II]; the former dominates visual-wavelength radiation at most locations [18, 54, 73, 124, 153]. The scattered light is linearly polarized, with a level of 20–40% at visual wavelengths [102, 103, 139, 140, 142, 148] and 15–30% in the near-IR [147]. Even the strong Hα emission is mainly Hα from the stellar wind, reflected by grains [74, 124]. The surface brightness distribution is consistent with scattered starlight with moderately large optical depths [21]. The Homunculus has even been detected as a Thomson-scattering nebula at hard X-ray wavelengths when the central source fades[14].

---

[2][Ed. comment:] Ironically, the lobes are now easily recognizable as such when viewed through an ordinary telescope. Eta Car's visual appearance has changed dramatically in the past 20 years. See the introductory Chap. 1 in this volume.

Light scattered by the Homunculus affords us a unique opportunity to view the star from a range of directions. The UV/visual spectrum in the middle of the approaching SE polar lobe gives a nearly pure (albeit spectrally blurred) reflected spectrum of the star [24, 153]. Smith et al. [124] exploited this fact to derive a 3-D view of the latitude dependence in the stellar wind once the 3-D shape and orientation of the nebula had been determined [18, 104]. Other than the Sun, this is perhaps the only opportunity in astrophysics where the spectrum of a star has been viewed from a range of known latitudes. Likely effects of rapid rotation and a companion star make this of significant interest.

7.4.2 Shape and Orientation of the Polar Lobes

Structures seen in images are only part of the picture. Kinematic information is needed to understand the true 3-D structure. Pioneering studies by Thackeray, Meaburn, and others combined proper motions with Doppler velocities to establish that the southeast lobe is tilted toward us and expanding away from the star with maximum speeds of roughly 650 km s^{-1}, and that the distance to η Car based on expansion is 2.0–2.5 kpc [4, 54, 73, 74, 140]. Hillier and Allen [4, 54] used long-slit spectra to probe the shape and expansion properties. They noted the distinction between narrow lines like [Ni II] and [Fe II] formed in the expanding nebula, as opposed to dust-scattered components of broad stellar-wind lines or narrow features arising near the star (see also [73,74]). Together these studies reduced the kinematic ambiguities, and confirmed that the Homunculus is basically a Hubble-like flow with a definite age. Today, with higher spatial resolution studies using HST/STIS spectra and high-resolution near-IR spectra, we have more accurate estimates of the inclination $i = 42^\circ$ ([18, 107]; the tilt angle of the polar axis from our line of sight), the distance of 2.3 kpc [18, 72, 104, 107], and the lobe shapes [107].

For several years, the shape of the polar lobes was discussed in the context of three alternative models called the "double bubble" (two hollow osculating spheroids), "bipolar caps," and a "double flask" [53]. Each found some justification in various aspects of the imperfect observational data. The hollow double-bubble model matched optical and mid-IR images that showed round, limb-brightened lobes [52, 74]. The double cap model emphasized that the polar regions appear optically thick while the side walls are more transparent [4]. The flask model had some advantages concerning polarization and kinematic properties [18]. In the end, better data made the distinctions between them largely irrelevant.

The Space Telescope Imaging Spectrograph (HST/STIS), with long-slit spectroscopy at high spatial resolution, provided the opportunity to improve upon the ground-based results. Davidson et al. [18] analyzed some of the same emission and reflection features as Hillier and Allen [54], obtaining a much better estimate of the shape and orientation of the nebula. A curious discrepancy appeared, however:

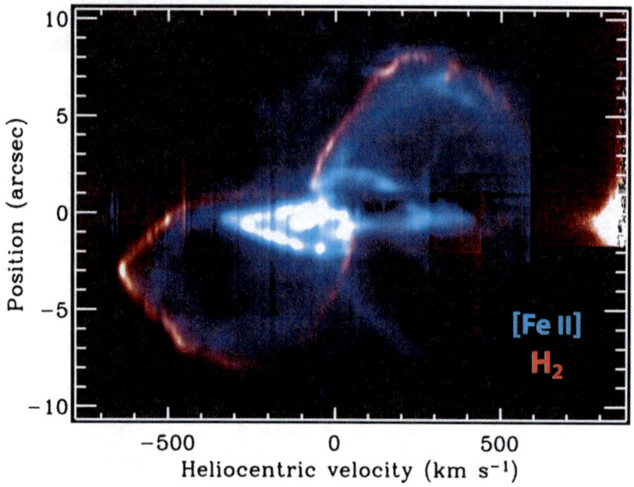

Fig. 7.3 Long-slit IR spectra of the Homunculus [107], showing the shape and detailed double-shell structure in H_2 (*r*ed; thin outer shell) and [Fe II] 1.644 μm (*b*lue, thicker inner shell). [Fe II] emission from the Little Homunculus can also be seen [106]

polar lobe shapes derived from [Ni II] emission lines in STIS data were too narrow and short compared to those seen in images. This discrepancy was solved when the first long-slit near-IR spectra were taken into account [104], showing that images in scattered light corresponded to a thin outer shell seen in H_2 emission, while the [Ni II] emission represented a layer interior to that. This double-shell structure is discussed below.

Currently the best estimate of the 3-D shape, structure, and orientation of the polar lobes is provided by near-IR, long-slit echelle ($R = 60{,}000$) spectra of H_2 emission [107]. The H_2 traces the dense outer dust shell (see Fig. 7.3) where most of the mass resides, and which is seen as the reflection nebula in images. The shape and expansion speed are now known with a precision of roughly 2–3% at all latitudes in the polar lobes. The limiting factor in this study is not the observational precision, but intrinsic variations and modulations in the lobes themselves, as well as the assumption that the nebula is axisymmetric. In other words, the present model for the shape [107] is as definitive as the configuration will allow. Ground-based IR spectra were even better suited to the task than STIS for three reasons: (1) their spatial resolution is almost as good as STIS, but the velocity resolution of the long-slit echelle spectra is much higher; (2) they trace unique IR molecular hydrogen emission from the outer shell instead of atomic gas in the inner shell (Fig. 7.3), and (3) at near-IR wavelengths one can see all the way through to the far walls of the polar lobes, mitigating optical-depth effects. Thus we now know the 3-D shape of the nebula to high enough precision to constrain physical models for its formation.

7.4.3 *Density and Ionization Structure of the Polar Lobes*

In addition to the overall shape, recent observations from large ground-based telescopes and *HST* have significantly advanced our understanding of the detailed density and ionization structure of the Homunculus. Until recently, the walls of the polar lobes were generally assumed to have a thickness of ∼10% of their radius [52, 120], and aside from the clumpy structure in *HST* images, not much more was known about their detailed structure.

Now we know that the walls actually have a well-defined double-shell structure (Fig. 7.3; [107]), where most of the mass is in a denser and geometrically thin outer molecular shell (traced most clearly by near-IR H_2 emission), and about 10% of the mass is in a geometrically thicker but optically thinner inner shell of partially-ionized atomic gas (traced by near-IR and optical lines of [Fe II] and [Ni II]). This double-shell structure was first proposed based on H_2 and [Fe II] structures seen in the first available long-slit near-IR spectra of the Homunculus [104], although subsequent spectra of the same lines with higher resolution provided a more definitive picture [107].

The double-shell structure is verified by 8–25 μm imaging on large ground based telescopes with sub-arcsecond resolution [123], revealing a thin outer shell of cool dust at 140 K, and a thicker inner shell of warmer dust at 200 K in the polar lobes (Fig. 7.2). The thicker and warmer inner shell dominated previous mid-IR imaging at shorter wavelengths near 10 μm; Smith et al. [120] noted that the warm dust emitting at 10–12 μm was ∼10% less extended than the scattered light seen at visual wavelengths, but did not articulate the nature of the double-shell structure. Fits to the IR SED imply that the outer 140 K shell contains about 11 $M_\odot$ of material, while the inner 200 K shell has only 1.5 $M_\odot$ [123].[3]

The effects of this double-shell structure are also clearly seen in UV echelle spectra obtained with STIS, tracing absorption along our line of sight to the central star. These spectra show narrow absorption lines from Fe I and diatomic molecules in the thin outer shell at $-530\,\mathrm{km\,s^{-1}}$, and a more clumpy and broader distribution of absorption lines from singly-ionized metals at slower blueshifted speeds [50, 51, 84, 141]. For the thin outer shell, these same studies suggest that the gas temperature is ∼760 K, significantly warmer than the 140 K dust in the same location [123].

The double-shell structure is explained quantitatively in model calculations of the nebula's ionization structure [110]. The inner shell is a warm (neutral-H) photodissociation region where metals like Fe and Ni are singly ionized by strong Balmer continuum radiation, while the outer thin and cool molecular shell exists where a thin layer of H_2 has absorbed much of the radiation in the Lyman-Werner bands. The required neutral H densities are roughly $10^{5.5}\,\mathrm{cm^{-3}}$ in the inner shell (or more if the material is clumpy), and $(0.5–1)\times10^{7}\,\mathrm{cm^{-3}}$ in the outer shell in

[3]Pantin and Le Mignant [93] proposed an "onion-like" structure for the polar lobes, with a smaller bubble inside a larger one. This structure was not verified in mid-IR images with higher resolution, and their proposed structure was not the double-shell structure described here.

order to allow H_2 to survive and for Fe^+ to recombine to Fe^0 [110]. UV absorption features also suggest densities of $10^7\,cm^{-3}$ in the thin outer molecular shell along our line of sight [50]. With densities this high, the observed volume of the cool shell would indicate high masses for the Homunculus of $\sim 20\,M_\odot$ [107, 110].

What gives rise to this double-shell structure? On the one hand, the presence of near-IR H_2 and [Fe II] emission features remind one of warm shocked gas [109], in two zones behind the forward and reverse shocks, respectively. This would be somewhat misleading, however: comparing radiative energy heating to shock heating indicates that radiative heating dominates the energy balance by a factor of 250 or more [110]. Thus, the near-IR lines in the Homunculus do not trace current shock excitation as in a supernova remnant, for example, but instead trace different zones of ionization in a dense photodissociation region [110]. In near-IR lines of H_2 and [Fe II], this double-shell structure of the Homunculus is qualitatively identical to that seen in the planetary nebula M 2–9 [57, 127], where UV excitation is implied by the H_2 line ratios. If the density structure was imprinted by a shock, that shock interaction must have happened long ago [107]. Further work on this is needed.

Another unsolved issue is the origin of the clumpy, mottled structure apparent on the surfaces of the polar lobes. The true nature of the dark lanes vs. bright clumps, and of the obvious "hole" at the polar axis in the SE lobe have been topics of debate because different wavelengths or different techniques (imaging, spectroscopy, polarimetry) yield conflicting answers [81, 103, 107, 120, 123]. Recent spectra of H_2 have shown that the "hole" really is an absence of material [107], and IFU spectra of [Fe II] emission confirm its presence in the inner shell of the SE lobe as well as the receding NW polar lobe [136]. The nature of the cells and dark lanes remains an open question, however, and the answer potentially holds important clues to whether they originated in Rayleigh-Taylor instabilities, thermal instabilities, convective cells, or porosity [20, 90, 107, 120].

7.4.4 *What Caused the Bipolar Shape?*

Most theoretical work on the Homunculus has attempted to answer this question. The answer, which is not yet clear, is likely to be intimately connected to the cause of the Great Eruption, the star's behavior before and after, and the potential roles of rotation and/or binarity in the system.

In some early models, the bipolar form of the Homunculus was said to arise from the interaction of winds of differing speed and density, shaping the outflow after ejection [32, 39, 67]. These borrowed from models for the ring and bipolar nebula of SN 1987A and planetary nebulae [10, 37, 38, 69]. The key requirement was a pre-existing, slow and dense equatorial disk or torus to pinch the waist of the bipolar structure. Observations, however, then showed that the equatorial skirt is no older than the polar lobes and some parts are younger [18, 82, 112]. We now know that there is not enough material near the equator to shape the overall morphology; instead, most of the mass is located at high latitudes in the polar lobes [107].

The next models to explain the morphology assumed that "intrinsic" shaping mechanisms had occurred in the ejection process. Acknowledging the apparent absence of any pre-existing "doughnut," Frank et al. [40] suggested that the wind during the Great Eruption was fast and highly aspherical. This hypothesis was extended to the shaping of the Little Homunculus as well [46, 47]. While they yield bipolar nebulae resembling the Homunculus, these hydrodynamic models did not explain why the eruption wind was aspherical to begin with.

Smith et al. [124] showed that η Car's present-day wind is in fact bipolar with more mass and higher speeds at the poles, and noted that if the present-day wind were given a mass-loss rate 1,000 times stronger, it would look like the ballistically expanding Homunculus. IR observations [104, 107, 123] also showed that the shape of the lobes approximately agrees with what one expects for the latitudinal variation in escape speed on the surface of a rotating star, that more mass is located at the poles of the Homunculus, and that the thickness of the lobes compared to their radius is the same as the duration of the eruption divided by its age. The intrinsic bipolar mass ejection must presumably then be related to either rotation or to a companion star (or rotation induced by a companion star). In the context of aspherical line-driven winds, Owocki and coworkers [87–89] proposed that gravity darkening and velocity-dependent forces on a rotating star inhibit equatorial mass loss and *enhance the mass loss toward the poles*. Thus it was suggested that the Homunculus lobes arose from a bipolar continuum-driven wind with a higher polar mass-loss rate because of the latitudinal variation in escape speed [33, 68, 85–87]. This would require an enhanced wind for a brief time during the eruption, ejected directly from the surface of a rotating luminous star.

An intrinsically bipolar wind from a rotating star has several advantages over a pre-existing torus that pinches the waist, but there was still no explanation for the origin of the thin equatorial skirt described in Sect. 7.7 below. Smith and Townsend [117] noted that since disk-inhibition mechanism arise because of velocity-dependent forces in a line-driven wind [88], a wind-compressed disk [9] could still be created in a continuum-driven wind. In that case the disk is formed by hyperbolic orbits crossing the equator after being launched from the surface of a rotating star, while the enhanced polar mass flux arises because of gravity darkening. Smith and Townsend showed that one can thereby account for both the shape of the polar lobes and a thin fast equatorial ejecta skirt.

In all the models mentioned so far, mass ejection by the primary star is the principle agent for shaping the nebula. (This includes a scenario where the rotating primary has increased angular momentum through interaction with a nearby companion [124]). Binary models have also been proposed to explain the bipolar shape. Soker [130–133] advocates an analytic accretion model, where a close companion star accretes matter from the primary wind through Bondi-Hoyle accretion, and then launches bipolar jets that shape the polar lobes. Alternatively, the idea of a binary merger (i.e. a merger of a close binary in a hierarchical triple system) has been suggested more than once to explain the Great Eruption and the origin of the bipolar nebula [41,61,79], but it is difficult to understand the formation of the thin disk or the occurrence of previous giant eruptions in this type of scenario.

7.5 Dust, Molecules and Chemistry

By virtue of its phenomenal IR luminosity, the Homunculus has long been appreciated as a unique laboratory for the study of dust formation and survival in harsh conditions. The grains in the Homunculus are unusual compared to grains in the ISM and AGB stars in two important ways:

1. They seem to be unusually large, $a \sim 1\,\mu$m or more, based on the following clues. The grains are near equilibrium blackbody temperatures [52,77,78,110,120,123], their UV-to-IR extinction ratio is extraordinarily gray [5, 54, 55, 99, 100, 151], and they scatter efficiently even at $\sim 2\,\mu$m [113]. (Polarization properties of the nebula indicate that some small grains are also present [103].)
2. They may have unusual composition, probably as a result of C and O depletion in the ejecta. While η Car shows what appears to be a strong 9.7 μm silicate feature, it is unusually broad and shifted to longer wavelengths. This has prompted suggestions that alternative grain mineralogy such as corrundum (Al_2O_3) and olivine ($MgFeSiO_4$) might be present [13, 76]. In particular, Chesneau et al. [13] demonstrated that a combination of olivine and corrundum can account for the observed shape of the spatially dependent "silicate" feature in the core of the Homunculus. Since corrundum forms at higher temperatures than normal silicates, this type of dust might dominate the composition of grains that form today in the colliding-wind binary system[108]. The unusual composition of the grains may suggest that current mass estimates of the Homunculus of 10–15 $M_{\odot}$ may be underestimates [27] (see above for other reasons why the mass estimates are conservative).

Our understanding of how these large and unusual grains nucleated rapidly and survived in the outflow of the Great Eruption is far from complete, and the puzzle deserves further investigation both theoretically and observationally. An additional mystery concerns the fact that gas-phase Fe has basically solar abundance in such a dusty region as the Homunculus [110]. Since we know when the material was ejected and when the dust formed from the historical record, and η Car is bright enough in the IR to provide high-quality data today, the Homunculus is a rare and valuable laboratory to address these questions.

In addition to unusual dust grains, the Homunculus harbors a surprisingly large mass of molecular gas, less than 0.1 pc from one of the most luminous hot supergiants known (i.e., η Car itself). Molecules were first discovered there via near-IR H_2 emission lines [104, 109], and were later shown to reside in the cool outer skin of each polar lobe [104, 107]. Given the presence of molecular hydrogen in a presumably nitrogen-rich environment, one might expect species like NH_3, which indeed was the first polyatomic molecule detected in the Homunculus [128]. This was also the first detection of a polyatomic molecule around any LBV. UV absorption studies with STIS have revealed diatomic molecules such as CH and OH in the thin outer shell [141]. These UV absorption observations and radio studies have found no evidence for CO, presumably because most CNO in η Car's ejecta is N rather than C and O. While the outer ejecta and the inner Wiegelt Knots are known to be N-rich from their atomic emission-line spectra [23, 24, 115], the presence of ammonia gives our first reliable verification that the Homunculus too is N-rich [128]. The formation and survival of these molecules is far from understood, and more theoretical work on the molecular chemistry of the Homunculus would be useful [36].

7.6 The Little Homunculus

The Little Homunculus (LH) is a smaller bipolar nebula nested inside the main Homunculus, oriented along the same bipolar axis [64]. It is a relatively new addition to η Car's family of known morphological features. The LH was briefly called the "integral-sign filament" (referring to the shape of its emission lines in *HST*/STIS spectrograms) and the "Matryoshka Nebula" [49, 63].

Its discovery by Ishibashi et al. was based on its Doppler-shift morphology in visual-wavelength spectra [64]; bright reflection nebulosity and extinction in the Homunculus make its structure very difficult to perceive in visual-wavelength images. The LH's 3-D structure is best seen in near-IR emission lines of [Fe II], because these lines are bright and they can penetrate the foreground dust screen [104, 106]. This structure is clearly evident in Fig. 7.3. The LH spans 4–5$''$ along the polar axis of the Homunculus [64, 106]. It does not align with any features in visual scattered light, nor does it match the inner IR torus [106]. Since it was not detected at 10–20 μm, it probably does not contain a substantial amount of dust [123].

It does, however, match up quite well with the "Purple Haze" [81, 126] and the inner radio continuum nebula [30]. Smith [106] proposed that periodic and direction-dependent UV illumination of the LH is responsible for the extended radio continuum nebula and its associated changes through the 5.5 year cycle. This could offer a powerful test of models for the 5.5 year variability. Because it is seen in the radio continuum, the LH apparently absorbs whatever last remaining ionizing photons escape the dense wind of the central system.

Most indications suggest that the LH was ejected during η Car's 1890 eruption [64, 106]. The LH has an expansion speed of 300 km s^{-1} at the poles [62, 64, 106] and slower values at lower latitudes, with an expansion speed of $\sim$140 km s^{-1} along our line of sight [106]. Thus, the LH is responsible for the -146 km s^{-1} absorption feature seen in high-resolution UV spectra [50]. Blueshifted absorption features around 200–300 km s^{-1} in the spectrum observed during the 1890's eruption [59, 146] support the association of the LH with this event. However, nebular kinematics [106] and proper motions in STIS spectra [62] both imply an ejection date around 1910, but this assumes constant linear expansion. The LH's polar expansion speed of 300 km s^{-1} is much slower than the 500–600 km s^{-1} stellar wind of η Car that pushes behind it, and Smith [106] showed that the LH would indeed be accelerated to its present value if it had been ejected in 1890, and that this acceleration leads us to infer an incorrect and more recent ejection date. The same may be true for the kinematics of the Weigelt knots [125]. Smith [106] proposed that the Weigelt knots are simply the equatorial pinched waist of the LH, analogous to the IR torus of the larger Homunculus (see below).[4]

Estimates of densities in the LH range from a few times $10^4\,cm^{-3}$ [104] to 10^6 or even $10^7\,cm^{-3}$ [64]. Combining the observed 3-D volume of the LH with an uncertain ionization fraction, the mass is estimated as $0.1\,M_\odot$ [106], which is in rough agreement with other estimates based on upper limits to the dust mass and the kinematics. The kinetic energy of the 1890 event is then roughly $10^{46.9}$ erg [106]. Given that the mass and kinetic energy of the 1890 event were orders of magnitude less than the Great Eruption [123], the fact that the two mass ejections shared the same basic geometry is thought provoking, to say the least. From a similar study of infrared [Fe II] emission from P Cygni's shell, Smith and Hartigan [114] concluded that the 1600 AD giant eruption of P Cygni was nearly identical to the 1890 lesser eruption of η Car in terms of its ejected mass, speed, and kinetic energy, although it was not bipolar.[5]

7.7 Equatorial Structures

7.7.1 The Equatorial Skirt

While bipolar lobes often occur in planetary nebulae and other astronomical objects, the ragged debris disk in η Carinae's equatorial plane is an uncommon and somewhat mysterious sight. This equatorial "skirt" or "debris", as it is sometimes called to distinguish it from a smooth (and especially a Keplerian) disk, was first seen in high-resolution ground-based images [31]. Images with *HST* showed more detail in the radial streaks and other features [81]. Images alone were inadequate, however, because some "equatorial features" later proved illusory.[6] The geometrical thinness, planar form, and velocity structure of the skirt were especially apparent in HST spectral data [26], and Doppler velocities remain necessary to establish whether a feature is equatorial or not.

Though prominent in scattered-light images with brightness comparable to the polar lobes, the equatorial skirt is not seen clearly in thermal-IR maps [52, 94, 120, 123], suggesting that it contains little mass or that it is not heated efficiently.

[4] Also see Chap. 6.

[5] See the chapter on the outer ejecta and other LBV-associated nebulae by Weis.

[6] For instance, a bright fan-shaped structure 4″ northwest of the star was often cited in the 1990s as an especially obvious equatorial feature. But Doppler velocities and IR measurements show that it is really a less-obscured patch of the NW polar lobe, despite appearances [18, 120, 123].

If the dust were cooler and substantially massive, however, one would expect the equatorial features to become more prominent at longer wavelengths or to cause severe extinction, but they are as hard to see at 25 μm as at 10 μm [123]. Smith et al. [123] have discussed the illumination of the equatorial ejecta in detail.

The age of the equatorial ejecta inspired a lively debate which is still not entirely settled [16, 17, 82, 112]. Proper motions at low spatial resolution with a time baseline of 50 years yielded an ejection date around 1885 for the equatorial features, suggesting that they came from the 1890 event [112]. Gaviola noted hints toward the same conclusion [42, 96]. Doppler velocities indicate post-1860 material in the skirt [25, 26, 104, 106] and in the roughly-equatorial Weigelt Knots closer to the star [25, 29, 125, 149]. Proper motions measured in high-resolution *HST* data [16, 82], however, suggested that the skirt features were older, perhaps coeval with the Homunculus. Subsequent spectra showed emission lines in the equatorial material with *both* ages [18]. Thus, the resolution of the disagreement may be that the early blue plates (1945) and color slides (1972) were contaminated by line emission from younger diffuse gas originating in the 1890 event, whereas the small clumps measured in *HST* images were dust condensations tracing most of the mass. At least one of the features overlapped with the "Purple Haze" [126], a region of line emission that may be partly due to the younger Little Homunculus.

Only two models so far have attempted to explain the simultaneous origin of the polar lobes and this equatorial skirt, and in both cases the origin of the asymmetry was an inherently aspherical ejection by the star. Matt and Balick [71] presented a magnetohydrodynamic model for the present-day bipolar wind and disk, and conjectured that if this occurred at much higher mass-loss rate it may also explain the Homunculus. Smith and Townsend [117] presented a model of ejection from the surface of a rotating star that would produce a structure like the skirt via a compressed disk, along with the polar lobes. This type of mechanism is intuitively appealing because the required "splashing" at the equator may partly account for the ragged streaked appearance of the skirt [20, 25].

Some prominent features in the skirt share apparent connections to more distant and perplexing features in the outer nebulosity. For instance, the bright equatorial structure located $\sim 4''$ north-east of the star seems to connect to a nebular feature, sometimes termed the NN Condensation, north-east "jet", or "NN jet" [75, 144, 145]. This structure extends radially in the same direction, driving an apparent bow-shock structure in the outer ejecta that is seen in nebular emission-line images [73, 75, 81]. Similarly, the streaked features in the skirt located west of the star appear to connect to the S Condensation. These perceived connections may be due to either common hydrodynamic flows or illusions caused by beams of starlight escaping in preferred directions, or both. These collimated protrusions in the equatorial plane have few parallels in astrophysical objects. These and additional structures in the outer ejecta are discussed further in Chap. 8.

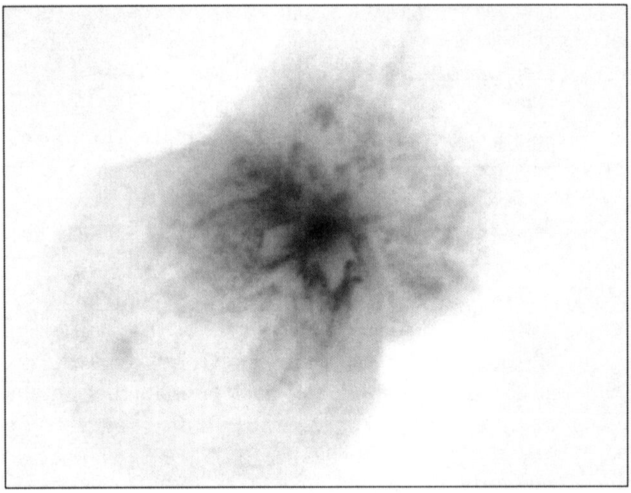

Fig. 7.4 A near-IR adaptive-optics image of the core of the Homunculus obtained with the VLT. This figure was prepared by the author from a K-band image kindly provided by O. Chesneau (see [13])

7.7.2 The Obscured Inner Torus

As noted earlier, IR imaging during the 1970–1990s revealed a complex, multi-peaked structure in the core of the Homunculus. Several near-IR and mid-IR studies with 2–4-m class telescopes, each of which improved on previous work, unveiled these inner structures at high resolution, most often interpreted as some variation of a dusty torus [35, 52, 56, 80, 94, 95, 113, 118, 120]. Access to mid-IR imaging on 6–8 m class telescopes improved our view of the torus [122], and we now have stunning adaptive optics images in the near-IR (Fig. 7.4) obtained with the VLT and Gemini-South [7, 13].

Like the larger equatorial skirt, this compact and complex structure has sparked interesting debates. Morris et al. [80] proposed that this structure was a pre-existing $15\,M_{\odot}$ torus, including the 110 K dust that domainted the far-IR flux, and was the agent responsible for pinching the waist of the Homunculus. Davidson and Smith [22] argued against this interpretation, and pointed out that the torus was actually quite warm. Hony et al. [56] proposed that the inner structure was a pair of polar rings like those around SN 1987A, and that the polar axis of the binary system had precessed by 37–58° caused by a close interaction of a triple system. Smith et al. [122] presented higher resolution images showing that these rings were an artifact of lower spatial resolution, resembling a fragmented torus instead. Chesneau et al. [13] presented even higher-resolution images showing surprisingly complex structure (Fig. 7.4), and proposed that these unprecedented features were part of an inner bipolar nebula that they named the "butterfly nebula".

Examination H_2 velocities in near-IR long-slit echelle spectra showed that the dusty features in question were in the mid-plane of the Homunculus where the two polar lobes meet at the pinched waist of the nebula, and were not related to polar features of the Little Homunculus [106, 107]. The expected dust temperatures at this location agree with the observed warm color temperatures of the dust of 250–400 K [123]. The strange and irregularly corrugated structures seen in the VLT images [13] probably result from the strong post-eruption wind of η Car pushing out the relatively low density regions in between clumps in the torus [122]. In some cases, perhaps these protrusions have broken all the way through in the mid plane, allowing the stars visual light to escape to large radii, producing radial streaks seen in images of reflected light.[7] This provides only a partial explanation for outer features like the NN jet and the S Condensation, which still evade our understanding.[8]

The clumpy torus seen in near- and mid-IR images has no counterpart in visual-wavelength images in scattered light, but it can be seen by virtue of its emission-line variability caused by ionization changes during the 5.5 year cycle [121], and it can be recognized by its Doppler shifted narrow line emission in long-slit STIS spectra of the inner Homunculus [48].

7.8 Summary: Fast Facts

The Homunculus of η Car is a spectacular and complex object that is a rich laboratory for studying many phenomena associated with ISM processes and the physics of stellar mass loss. It is seen in scattered starlight at visual and near-IR wavelengths, allowing us to see the star from multiple directions [124], but it also has intrinsic emission features that tell us about the kinematics and excitation of the nebula [18, 54, 107, 110]. Its intrinsic spectrum is that of a low-ionization warm photodissociation region, but with peculiar gas and dust abundances. Extended emission has been detected at all wavelengths from scattered X-rays [14], the UV [125], visual and near-IR wavelengths, thermal-IR, and even low-level radio continuum emission from the polar lobes [30]. It is rivaled by few other objects in the sky, but its complexity often defies the simplest interpretation. Our understanding of the Homunculus will no-doubt continue to improve, but here I list current best estimates for several observed properties of the nebula:

[7] [Editor's note:] After this chapter was complete, Artigau et al. [7] described new near-infrared adaptive optics images. Proper motions measured from the NICI and NaCO [13] images combined with the radial velocities [107], show that the knots and filaments that define the bright rims of the "butterfly" were ejected at two different epochs corresponding approximately to the Great Eruption and the second eruption. Most of the material is spatially distributed above and below the equatorial plane apparently behind the Little Homunculus and the larger SE lobe. Artigau at al. conclude that the butterfly is not a coherent physical structure or equatorial torus but spatially separate clumps and filaments ejected at different times.

[8] See relevant Chaps. 8 and 6.

- Proper motions measured with post-refurbishment *HST* images give an ejection date for the Homunculus of 1847 ($\pm$4–6 year) [82]. This is consistent with the 1843 eruption peak within the uncertainty, but hints that there may have been some post-ejection acceleration.
- The total IR luminosity integrated from 2 to 200 μm is $4.3 \times 10^6\, L_\odot$ [123]. The total stellar luminosity is 10–20% higher, since UV/visual light escapes.
- The total mass of the Homunculus (gas + dust) is at least 12–15 $M_\odot$ [80, 110, 123], but could be substantially more as noted above. The total kinetic energy is then at least $10^{49.7}$ erg [123].
- The shape of the polar lobes is traced best by near-IR H_2 emission [107] and is shown in Fig. 7.3. Smith [107] also gives the expansion speed, mass, and kinetic energy as a function of latitude. Most of the mass and kinetic energy is at high latitudes.
- The heliocentric distance to the Homunculus, derived from its proper motion and Doppler shifts assuming axisymmetry, is 2.3 ± 0.05 kpc [18, 72, 104, 107].
- The inclination angle of the Homunculus (the angle that the polar axis is tilted from the line-of-sight) is $i = 42^\circ \pm 1^\circ$ [18, 107].
- The systemic velocity of η Car is $-8.1(\pm 1)$ km s^{-1} heliocentric, or $-19.7(\pm 1)$ km s^{-1} LSR, measured from narrow H_2 lines [105].
- The Homunculus has a well-defined double-shell structure. The thin outer shell has about 90% of the total mass, is primarily molecular gas, Fe is neutral, the particle density is $\sim 10^7$ cm^{-3}, and the dust temperature is about 140 K. The inner shell is thicker and less massive, the dust is warmer at about 200 K, metals like Fe and Ni are singly ionized, and H is mostly neutral [107, 110].
- The dust grains that dominate thermal emission and extinction are large, with $a \simeq 1$ μm. Judging by the 10 μm emission feature, the chemical makeup of the dust appears to be unusual compared to other evolved stars, with significant amounts of corrundum and olivine.
- The detection of NH_3 and lack of CO verifies that the neutral/molecular Homunculus is N-rich like the outer ejecta and Weigelt knots [128]. The molecules that have been detected so far are H_2 [104], CH and OH [141], and NH_3 [128]. Verner et al. suggest that CH^+, CH_2^+, CH_2, and NH may also be detectable. One might also expect to find N_2H^+ [128]. In other words, the Homunculus is a valuable laboratory for studying N-rich molecular chemistry.
- Most of the dense clumpy structures in the equatorial skirt seen in optical images were ejected at the same time as the polar lobes. There is also younger material from the 1890 eruption intermixed with these equatorial ejecta, seen mainly in emission lines. The equatorial skirt is *not* a pre-existing disk that pinched the waist of the nebula.
- IR images show a bright IR torus inside the Homunculus that is not seen in visual-wavelength images. It is bright in the IR because it contains some of the warmest dust in the Homunculus, where the thin walls of the two polar lobes meet at the equator. The mass in this torus is not nearly enough to have pinched the waist of the Homunculus.

7.9 The Future for the Homunculus

Eta Carinae is among the most bizarre extended objects in the sky, with some of the most spectacularly complex and puzzling circumstellar structures known. It is wise to keep this in perspective, however, remembering that many of the reasons the Homunculus seems so peculiar arise *because it is so young*. It is young enough that it is still in free expansion, and so the observed geometry still bears the imprint of the initial ejection – before its signature is erased by deceleration from the surrounding medium. This makes η Car unique among massive stars, and this is why it is such a valuable tool for constraining the physics of episodic mass loss. It is therefore instructive to end by briefly considering the future fate of the Homunculus:

In another 500–1,000 years, the Homunculus will have expanded to $\sim$5 times its current size, and will plow into the outer ejecta. Its very clean bipolar and thin disk geometry will probably be diluted or erased. It will no longer obscure the central star (and will therefore be harder to observe), and it will not be one of the brightest 10–20 μm objects in the sky because the dust will absorb a smaller fraction of the total luminosity and will be much cooler. At lower densities it will probably be largely ionized (if not by η Car, then by the remaining $\sim$60 O-type stars in the Carina nebula), and the thin outer molecular shell will probably not survive.

In other words, it will probably appear as a fairly ill-defined, hollow ellipsoidal ionized gas shell with unclear kinematics, much like η Car's current outer ejecta from a previous eruption or the nebula around AG Car [119, 134, 143]. In context, then, the Homunculus is perhaps not so bizarre after all, since shell nebulae with masses of $\sim 10\,M_\odot$ are not unusual for LBVs with luminosities above $10^6\,L_\odot$ [116]. This epitaph underscores how lucky we are to observe the Homunculus in our lifetimes with available technology at an optimal time when $\tau = 1$ throughout much of the nebula. Until the distant (or not so distant?) future when fuel has been exhausted in the star's core, the Homunculus will continue to disperse, leaving the star unobscured when it explodes ... unless another giant eruption happens first.

Acknowledgments I am indebted to numerous people with whom I have collaborated on studies of the Homunculus and discussed related topics, especially Kris Davidson, Gary Ferland, Bob Gehrz, John Hillier, Jon Morse, Stan Owocki, and Rich Townsend.

References

1. D.K. Aitken, B. Jones, The infrared spectrum and structure of Eta Carinae. MNRAS. **172**, 141–147 (1975)
2. D.K. Aitken, C.H. Smith, T.J.T. Moore, P.F. Roche, Mid-infrared studies of eta Carinae -II. Polarimetric imaging at 12.5 micron and the magnetic field structure. MNRAS. **273**, 359–366 (1995)
3. D.A. Allen, The hollow, clumpy outflow around Eta Carinae. MNRAS. **241**, 195–207 (1989)

4. D.A. Allen, D.J. Hillier, The shape of the Homunculus Nebula around Eta Carinae. PASA. **10**, 338–341 (1993)
5. C.D. Andriesse, B.D. Donn, R. Viotti, The condensation of dust around Eta Carinae. MNRAS. **185**, 771–788 (1978)
6. J.P. Apruzese, Radiative transfer in spherical circumstellar dust envelopes. II. Is the infrared continuum of Eta Carinae produced by thermal dust emission? ApJ. **196**, 753–760 (1975)
7. É. Artigau, J.C. Martin, R.M. Humphreys, K. Davidson, O. Chesneau, N. Smith, Penetrating the Homunculus – Near-infrared adaptive optics images of Eta Carinae. AJ. **141**, 202 (2011)
8. S. Bensammar, N. Letourneur, F. Perrier, M. Friedjung, R. Viotti, Multiplex imagery of the infrared core of Eta Carinae. A&A. **146**, L1–L2 (1985)
9. J.F. Bjorkmann, J.P. Cassinelli, Equatorial disk formation around rotating stars due to ram pressure confinement by the stellar wind. ApJ. **409**, 429–449 (1993)
10. J. Blondin, P. Lundqvist, Formation of the circumstellar shell around SN 1987A. ApJ. **405**, 337–352 (1993)
11. G.R. Briggs, D.K. Aitken, Middle infrared spectropolarimetry of Eta Carinae. PASA. **6**, 145–147 (1985)
12. A. Chelli, C. Perrier, Y.G. Biraud, One-dimensional high resolution image reconstruction on Eta Carinae at 4.6 micron with Speckle Data. A&A, **117**, 199–204 (1983)
13. O. Chesneau et al. The sub-arcsecond dusty environment of Eta Carinae. A&A. **435**, 1043–1061 (2005)
14. M.F. Corcoran et al., Waiting in the wings: reflected X-ray emission from the Homunculus Nebula. ApJ. **613**, 381–386 (2004)
15. P. Cox et al., Millimeter emission of Eta Carinae and its surroundings. A&A. **297**, 168–174 (1995)
16. D.G. Currie, D.M. Dowling, Astrometric motion and Doppler velocity, in *Eta Carinae at the Millenium*, ed. by J.A. Morse, R.M. Humphreys, A. Damineli. ASP Conference Series, vol. 179 (ASP, San Francisco, 1999), pp. 72–82
17. D.G. Currie et al., Astrometric analysis of the Homunculus of Eta Carinae with the Hubble Space Telescope. AJ. **112**, 1115–1127 (1996a)
18. D.G. Currie et al., Structure of the bipolar dust shell of Eta Carinae, in *The Role of Dust in the Formation of Stars*, ed. by H.U. Käufl, R. Siebenmorgen (Springer, Berlin, 1996b), pp. 89–94 3-D
19. K. Davidson, On the nature of Eta Carinae. MNRAS. **154**, 415–427 (1971)
20. K. Davidson, R.M. Humphreys, Eta Carinae and its environment. ARA&A **35**, 1 (1997)
21. K. Davidson, M.T. Ruiz, Scattering by dust and the photographic appearance of Eta Carinae. ApJ. **202**, 421–424 (1975)
22. K. Davidson, N. Smith, A massive cool dust torus around Eta Carinae? Nature **405**, 532 (2000)
23. K. Davidson, R.J. Dufour, N.R. Walborn, T.R. Gull, Ultraviolet and visual wavelength spectroscopy of gas around Eta Carinae. ApH. **305**, 867–879 (1986)
24. K. Davidson, D. Ebbets, G. Weigelt, R.M. Humphreys, A. Hajian, N.R. Walborn, M. Rosa, HST/FOS spectroscopy of Eta Carinae: the star itself, and ejecta within 0.3 Arcsec. AJ. **109**, 1784–1796 (1995)
25. K. Davidson, D. Ebbets, S. Johansson, J.A. Morse, F.W. Hamann, B. Balick, R.M. Humphreys, G. Weigelt, A. Frank, HST/GHRS observations of the compact slow ejecta of Eta Carinae. AJ. **113**, 335–345 (1997)
26. K. Davidson, N. Smith, T.R. Gull, K. Ishibashi, D.J. Hillier, The shape and orientation of the Homunculus Nebula based on spectroscopic velocities. AJ. **121**, 1569–1577 (2001)
27. A. de Koter, M. Min, R. van Boekel, O. Chesneau, The solid-state composition and mass of the Homunculus of Eta Carinae, in *The Fate of the Most Massive Stars*, ed. by R. Humphreys, K. Stanek. ASP Conference Series, vol. 332 (ASP, San Francisco, 2005), pp. 313–316
28. G. de Vaucouleurs, O.C. Eggen, The brightening of Eta Carinae. PASP. **64**, 185–190 (1952)
29. B.N. Dorland, D.G. Currie, A.R. Hajian, Did Eta Carinae's Weigelt Blobs Originate Circa 1941? AJ. **127**, 1052–1058 (2004)

30. R.A. Duncan, S.M. White, J. Lim, Evolution of the radio outburst from the supermassive star Eta Carinae from 1992 to 1996. MNRAS, **290**, 680–688 (1997)
31. W.J. Duschl, K.H. Hofman, F. Rigaut, G. Weigelt, Morphology and kinematics of Eta Carinae. Rev. Mex. AA, Ser. Conf. **2**, 17 (1995)
32. V.V. Dwarkadas, B. Balick, On the formation of the Homunculus Nebula around Eta Carinae. AJ. **116**, 829–839 (1998)
33. V.V. Dwarkadas, S.P. Owocki, Radiatively driven winds and the shaping of bipolar luminous blue variable nebulae. ApJ. **581**, 1337–1343 (2002)
34. D. Ebbets, H. Garner, R. White, K. Davidson, E. Malumuth, N. Walborn, HST images of Eta Carinae, in *Circumstellar Media in Late Stages of Stellar Evolution*, ed. by R.E.S. Clegg, I.R. Stevens, W.P.S. Meikle. 34th Herstmonceux Conference (Cambridge University Press, Cambridge, 1994), pp. 95–97
35. H. Falcke, K. Davidson, K.H. Hofmann, G. Weigelt, Speckly-masking imaging polarimetry of Eta Carinae: evidence for an equatorial disk. A&A. **306**, L17–L20 (1996)
36. G.J. Ferland, N. Abel, K. Davidson, N. Smith, Physical conditions in the Homunculus, in *The Fate of the Most Massive Stars*, ed. by R. Humphreys, K. Stanek. ASP Conference Series, vol. 332 (ASP, San Francisco, 2005), pp. 294–301
37. A. Frank, Bipolar outflows and the evolution of stars. New Astron. Rev. **43**, 31–65 (1999)
38. A. Frank, G. Mellema, The radiation gas dynamics of planetary nebulae. IV. From the Owl to the Eskimo. ApJ. **430**, 800–813 (1994)
39. A. Frank, B. Balick, K. Davidson, The Homunculus of Eta Carinae: an interacting stellar winds paradigm. ApJ. **441**, L77–L80 (1995)
40. A. Frank, D. Ryu, K. Davidson, Where is the doughnut? Luminous blue variable bubbles and aspherical fast winds. ApJ. **500**, 291–301 (1998)
41. J.S. Gallagher, Close binary models for luminous blue variable stars, in *Physics of Luminous Blue Variables*, ed. by K. Davidson, A.F.J. Moffat, H.J.G.L.M. Lamers (Kluwer, Dordrecht, 1989), pp. 185–194.
42. E. Gaviola, I. Eta Carinae, The nebulosity. ApJ. **111**, 408–413 (1950)
43. R.D. Gehrz, E.P. Ney, The core of Eta Carinae. Sky Telesc. **44**, 4–5 (1972)
44. R.D. Gehrz, E.P. Ney, E.E. Becklin, G. Neugebauer, The infrared spectrum and angular size of Eta Carinae. Astrophys. Lett. **13**, 89–93 (1973)
45. H.L. Gomez, L. Dunne, S.A. Eales, M.G. Edmunds, Submillimetre emission from Eta Carinae. MNRAS. **372**, 1133–1139 (2006)
46. R. Gonzalez, E.M. de Gouveia Dal Pino, A.C. Raga, P.F. Velazquez, Gasdynamical simulations of the large and little Homunculus Nebulae of Eta Carinae. ApJ. **600**, L59–L62 (2004a)
47. R. Gonzalez, E.M. de Gouveia Dal Pino, A.C. Raga, P.F. Velazquez, Numerical modeling of Eta Carinae bipolar outflows. ApJ. **616**, 976–987 (2004b)
48. T.R. Gull, K. Ishibashi, The three-dimensional and time-variant structures of ejecta around Eta Carinae as detected by the STIS, in *Eta Carinae & Other Mysterious Stars*, ed. by T. Gull, S. Johansson, K. Davidson. ASP Conference Series, vol. 242 (ASP, San Francisco, 2001), pp. 59–70
49. T.R. Gull, K. Ishibashi, K. Davidson, the Cycle 7 STIS GTO Team, First observations of Eta Carinae with the space telescope imaging spectrograph, in *Eta Carinae at the Millenium*, ed. by J.A. Morse, R.M. Humphreys, A. Damineli. ASP Conference Series, vol. 179 (ASP, San Francisco, 1999), pp. 144–154
50. T.R. Gull, G. Viera, F. Bruhweiler, K.E. Nielsen, E. Verner, A. Danks, The absorption spectrum of high-density stellar ejecta in the line of sight to Eta Carinae. ApJ. **620**, 442–449 (2005)
51. T.R. Gull, G. Viera Kober, K.E. Nielsen, Eta Carinae across the 2003.5 Minimum: the character and variability of the ejecta absorption in the near-ultraviolet. ApJS. **163**, 173–183 (2006)
52. J.A. Hackwell, R.D. Gehrz, G.L. Grasdalen, The internal structure of the dust shell of Eta Carinae deduced from six channel 8–13 micron mapping. ApJ. **311**, 380–399 (1986)

53. D.J. Hillier, Models for Eta Carinae, in *Luminous Blue Variables: Massive Stars in Transition*, ed. A. Nota, H.J.G.L.M. Lamers. ASP Conference Series, vol. 120 (ASP, San Francisco, 1997), pp. 287–293
54. D.J. Hillier, D.A. Allen, A spectroscopic investigation of Eta Carinae and the Homunculus Nebula. I. Overview of the spectra. A&A. **262**, 153–170 (1992)
55. D.J. Hillier, K. Davidson, K. Ishibashi, T.R. Gull, On the nature of the central source in Eta Carinae. ApJ. **553**, 837–860 (2001)
56. S. Hony et al., Discovery of a double ring in the core of Eta Carinae. A&A. **377**, L1–L4 (2001)
57. J.L. Hora, W.B. Latter, The near-infrared structure and spectra of the bipolar Nebulae M2-9 and AFGL 2688: the role of ultraviolet pumping and shocks in molecular hydrogen excitation. ApJ. **437**, 281–295 (1994)
58. R.M. Humphreys, K. Davidson, The luminous blue variables: astrophysical geysers. PASP. **106**, 1025 (1994)
59. R.M. Humphreys, K. Davidson, M. Koppelman, The early spectra of Eta Carinae 1892 to 1941 and the onset of its high excitation emission spectrum. AJ. **135**, 1249–1263 (2008)
60. A.R. Hyland, G. Robinson, R.M. Mitchell, J.A. Thomas, E.E. Becklin, The spectral and spatial distribution of radiation from Eta Carinae. II. High-resolution infrared maps of the Homunculus. ApJ. **233**, 145–153 (1979)
61. I. Iben, The effects of possible binary and tertiary companions on the behavior of Eta Carinae, in *Eta Carinae at the Millenium*, ed. by J.A. Morse, R.M. Humphreys, A. Damineli. ASP Conference Series, vol. 179 (ASP, San Francisco, 1999), pp. 367–372
62. K. Ishibashi, Historical eruptions of Eta Carinae: looking through the Homunculus, in *The Fate of the Most Massive Stars*, ed. by R. Humphreys, K. Stanek. ASP Conference Series, vol. 332 (ASP, San Francisco, 2005), pp. 131–136
63. K. Ishibashi, T.R. Gull, K. Davidson, The HST/STIS mapping of the Eta Carinae Nebula, in *Eta Carinae & Other Mysterious Stars*, ed. by T. Gull, S. Johansson, K. Davidson. ASP Conference Series, vol. 242 (ASP, San Francisco, 2001), pp. 71–79
64. K. Ishibashi et al., Discovery of a little Homunculus within the Homunculus Nebula of Eta Carinae. AJ. **125**, 3222–3236 (2003)
65. R.R. Joyce, The infrared spectrum of Eta Carinae: 3-14 microns. PASP. **87**, 917–921 (1975)
66. K.S. Krishna Swamy, On the infra-red radiation from Eta Carinae. Observatory **91**, 120–123 (1971)
67. N. Langer, G. Garcia-Segura, M.M. Mac Low, Giant outbursts of luminous blue variables and the formation of the Homunculus Nebula around Eta Carinae. ApJ. **520**, L49–L53 (1999)
68. A. Maeder, V. Desjacues, The shape of Eta Carinae and LBV Nebulae. A&A. **372**, L9–L12 (2001)
69. C.L. Martin, W.D. Arnett, The origin of the rings around SN 1987A: an evaluation of the interacting-winds model. ApJ. **447**, 378–390 (1985)
70. J.C. Martin, K. Davidson, M.D. Koppelman, The Chrysalis opens? Photometry from the Eta Carinae Hubble Space Telescope Treasury Project, 2002–2006. AJ. **132**, 2717–2728 (2006)
71. S. Matt, B. Balick, Simultaneous production of disk and lobes: a single-wind MHD model for the Eta Carinae Nebula, ApJ. **615**, 921–933 (2004)
72. J. Meaburn, An updated proper-motion/spectropolarimetric distance to η Carinae, in *Eta Carinae at the Millennium*, ed. by J.A. Morse, R.M Humphreys, A. Damineli, ASP Conference Series, vol. 179 (ASP, San Francisco, 1999), pp. 89–91
73. J. Meaburn, R.D. Wolstencroft, J.R. Walsh, Echelle and spectropolarimetric observations of the Eta Carinae Nebulosity. A&A. **181**, 333–342 (1987)
74. J. Meaburn, J.R. Walsh, R.D. Wolstencroft, The outflowing dust around Eta Carinae, A&A. **268**, 283–293 (1993)
75. J. Meaburn, G. Gehring, J.R. Walsh, J.W. Palmer, J.A. Lopez, M. Bryce, A.C. Raga, An episodic jet from Eta Carinae. A&A. **276**, L21–L24 (1993)
76. R.M. Mitchell, G. Robinson, The spectral and spatial distribution of radiation from Eta Carinae. I. A spherical dust shell model approach. ApJ. **220**, 841–852 (1978)

77. R.M. Mitchell, G. Robinson, Grain size and geometrical effects in the Eta Carinae dust shell. MNRAS. **222**, 347–355 (1986)
78. R.M. Mitchell, G. Robinson, A.R. Hyland, T.J. Jones, The spectral and spatial distribution of radiation from Eta Carinae. III. A high resolution 2.2 micron map and morphological considerations of the evolutionary status. ApJ. **271**, 133–142 (1983)
79. T. Morris, P. Podsiadlowski, Anisotropic mass ejection in binary mergers. MNRAS. **365**, 2–10 (2006)
80. P.W. Morris et al., Discovery of a massive equatorial torus in the Eta Carinae stellar system. Nature **402**, 502–504 (1999)
81. J.A. Morse, K. Davidson, J. Bally, D. Ebbets, B. Balick, A. Frank, Hubble Space Telescope Wide Field Planetary Camera 2 observations of Eta Carinae. AJ. **116**, 2443–2461 (1998)
82. J.A. Morse, J.R. Kellogg, J. Bally, K. Davidson, B. Balick, D. Ebbets, Hubble Space Telescope proper-motion measurements of the Eta Carinae Nebula. ApJ. **548**, L207–L211 (2001)
83. G. Neugebauer, J.A. Westphal, Infrared observations of Eta Carinae. ApJ. **152**, L89–L94 (1968)
84. K.E. Nielsen, T.R. Gull, G. Viera Kober, The ultraviolet spectrum of Eta Carinae: investigation of the ejecta absorption. ApJS. **157**, 138–146 (2005)
85. S.P. Owocki, Instabilities in massive stars. in *A Massive Star Odyssey: From Main Sequence to Supernova*, ed. by K. van der Hucht, A. Herrero, C. Esteban. IAU Symposium, vol. 212 (ASP, San Francisco, 2003), pp. 281–290.
86. S.P. Owocki, Radiatively driven stellar winds and aspherical mass loss. in *The Fate of the Most Massive Stars*, ed. by R. Humphreys, K. Stanek. ASP Conference Series, vol. 332 (ASP, San Francisco, 2005), pp. 169–179
87. S.P. Owocki, K.G. Gayley, The physics of stellar winds near the Eddington limit, in *Luminous Blue Variables: Massive Stars in Transition*, ed. by A. Nota, H.J.G.L.M. Lamers. ASP Conference Series, vol. 120 (ASP, San Francisco, 1997), pp. 121–127.
88. S.P. Owocki, S.R. Cranmer, K. Gayley, Inhibition FO wind compressed disk formation by nonradial line-forces in rotating hot-star winds. ApJ. **472**, L115–L118 (1996)
89. S.P. Owocki, S.R. Cranmer, K. Gayley, Mass loss from rotating hot-stars: inhibition of wind compressed disks by nonradial line-forces. Ap&SS. **260**, 149–159 (1998)
90. S.P. Owocki, K.G. Gayley, N.J. Shaviv, A porosity-length formalism for photon-tiring-limited mass loss from stars above the Eddington limit. ApJ. **616**, 525–541 (2004)
91. B.E.J. Pagel, Intrinsic reddening of Eta Carinae. Nature **221**, 325–327 (1969a)
92. B.E.J. Pagel, Energy budget for the infrared radiation from Eta Carinae. Astrophys. Lett. **4**, 221–224 (1969b)
93. E. Pantin, D. Le Mignant, 17 micron imaging of Eta Carinae: an onion-like structure for the Lobes? A&A. **355**, 155–164 (2000)
94. E. Polomski, C.M. Telesco, R.K. Pina, R.S. Fisher, Complex structure of Eta Carinae in the mid-infrared. AJ. **118**, 2369–2377 (1999)
95. F. Rigaut, G. Gehring, The inner core of Eta Carinae in the near infrared. Rev. Mex. AA, Ser. Conf. **2**, 27–35 (1995)
96. A.E. Ringuelet, Note on the nebulosity around Eta Carinae. Z. Astrophys. **46**, 276–278 (1958)
97. G. Robinson, A.R. Hyland, J.A. Thomas, Observation and interpretation of the infra-red spectrum of Eta Carinae. MNRAS. **161**, 281–292 (1973)
98. G. Robinson, R.M. Mitchell, D.K. Aitken, G.P. Briggs, P.F. Roche, Infrared studies of Eta Carinae – I. Spectroscopy and a Composite Dust Model. MNRAS. **227**, 535–542 (1987)
99. A.W. Rodgers, The reddening of Eta Carinae. ApJ. **165**, 665–667 (1971)
100. A.W. Rodgers, L. Searle, Spectrophotometry of the object Eta Carinae. MNRAS. **135**, 99–119 (1967)
101. R.W. Russel et al., Airborne spectrophotometry of Eta Carinae from 4.5 to 7.5 microns and a model for source morphology. ApJ. **321**, 937–942 (1987)
102. R.E. Schulte-Ladbeck et al., The structure of Eta Carinae's Homunculus from imaging polarimetry and spectropolarimetry, in *Luminous Blue Variables: Massive Stars in Transition*,

ed. by A. Nota, H.J.G.L.M. Lamers. ASP Conference Series, vol. 120 (ASP, San Francisco, 1997), pp. 260–266
103. R.E. Schulte-Ladbeck et al. Hubble Space Telescope imaging polarimetry of Eta Carinae. AJ. **118**, 1320–1337 (1999)
104. N. Smith, Dissecting the Homunculus Nebula around Eta Carinae with spatially resolved near-infrared spectroscopy. MNRAS. **337**, 1252–1268 (2002)
105. N. Smith, The systemic velocity of Eta Carinae. MNRAS. **351**, L15–L18 (2004)
106. N. Smith, Doppler tomography of the little Homunculus: high-resolution spectra of [Fe II] 16435 around Eta Carinae. MNRAS. **357**, 1330–1336 (2005)
107. N. Smith, The structure of the Homunculus. I. Shape and latitude dependence from H_2 and [Fe II] velocity maps of Eta Carinae. ApJ. **644**, 1151–1163 (2006)
108. N. Smith, Episodic post-shock dust formation in the colliding winds of Eta Carinae. MNRAS. **402**, 145–151 (2010)
109. N. Smith, K. Davidson, The shocking near-infrared spectrum of the Homunculus nebula surrounding Eta Carinae. ApJ. **551**, L101–L104 (2001)
110. N. Smith, G.J. Ferland, The structure of the Homunculus. II. Modeling the physical conditions in Eta Carinae's molecular shell. ApJ. **655**, 911–919 (2007)
111. N. Smith, D.J. Frew, A revised historical light curve of Eta Carinae and the timing of close periastron encounters. MNRAS. **415**, 2009–2019 (2011)
112. N. Smith, R.D. Gehrz, Proper motions in the ejecta of Eta Carinae with a 50 year baseline. AJ. **116**, 823–828 (1998)
113. N. Smith, R.D. Gehrz, Recent changes in the near-infrared struture of Eta Carinae. ApJ. **529**, L99–L102 (2000)
114. N. Smith, P. Hartigan, Infrared [Fe II] emission from P Cygni's Nebula: atomic data, mass, kinematics, and the 1600 AD outburst. ApJ. **638**, 1045–1055 (2006)
115. N. Smith, J.A. Morse, Nitrogen and oxygen abundance variations in the outer ejecta of Eta Carinae: evidence for recent chemical enrichment. ApJ. **605**, 854–863 (2004)
116. N. Smith, S.P. Owocki, On the role of continuum-driven eruptions in the evolution of very massive stars and population III stars. ApJ. **645**, L45–L48 (2006)
117. N. Smith, R.H.D. Townsend, The structure of the Homunculus. III. Forming a disk and bipolar lobes in a rotating surface explosion. ApJ. **666**, 967–975 (2007)
118. C.H. Smith, D.K. Aitken, T.J.T. Moore, P.F. Roche, R.C. Puetter, R.K. Pina, Mid-infrared studies of Eta Carinae-I. Subarcsecond imaging of 12.5 and 17 micron. MNRAS. **273**, 354–358 (1995)
119. L.J. Smith, M.P. Stroud, C. Esteban, M. Vilchez, MNRAS. **290**, 265–275 (1997)
120. N. Smith, R.D. Gehrz, J. Krautter, The infrared morphology of Eta Carinae. AJ. **116**, 1332–1345 (1998)
121. N. Smith, J.A. Morse, K. Davidson, R.M. Humphreys, Recent changes in the near-ultraviolet and optical structure of Eta Carinae. AJ. **120**, 920–934 (2000)
122. N. Smith, R.D. Gehrz, P.M. Hinz, W.F. Hoffmann, E.E. Mamajek, M.R. Meyer, J.L. Hora, A disrupted circumstellar torus inside Eta Carinae's Homunculus Nebula. ApJ. **567**, L77–L80 (2002)
123. N. Smith, R.D. Gehrz, P.M. Hinz, W.F. Hoffmann, J.L. Hora, E.E. Mamajek, M.R. Meyer, Mass and kinetic energy of the Homunculus Nebula around Eta Carinae. AJ. **125**, 1458–1466 (2003a)
124. N. Smith, K. Davidson, T.R. Gull, K. Ishibashi, D.J. Hillier, Latitude-dependent effects in the stellar wind of Eta Carinae. ApJ. **586**, 432–450 (2003b)
125. N. Smith et al., Kinematics and ultraviolet to infrared morphology of the inner Homunuclus of Eta Carinae. ApJ. **605**, 405–424 (2004a)
126. N. Smith, J.A. Morse, N.R. Collins, T.R. Gull, The purple haze of Eta Carinae: binary induced variability? ApJ. **610**, L105–L108 (2004b)
127. N. Smith, B. Balick, R.D. Gehrz, Kinematic structure of H_2 and [Fe II] in the bipolar planetary Nebula M2-9. AJ. **130**, 853–861 (2005)

128. N. Smith, K.J. Brooks, B. Koribalski, J. Bally, cleaning up Eta Carinae: detection of ammonia in the Homunculus Nebula. ApJ. **645**, L41–L44 (2006)
129. N. Smith, et al. SN 2006gy: discovery of the most luminous supernova ever recorded, powered by the death of an extremely massive star like Eta Carinae. ApJ. **666**, 1116–1128 (2007)
130. N. Soker, The departure of Eta Carinae from axisymmetry and the binary hypothesis. MNRAS. **325**, 584–588 (2001)
131. N. Soker, Why a single-star model cannot explain the bipolar Nebula of Eta Carinae. ApJ. **612**, 1060–1064 (2004)
132. N. Soker, The binarity of Eta Carinae and its similarity to related astrophysical objects. ApJ. **619**, 1064–1071 (2005)
133. N. Soker, Comparing Eta Carinae with the red rectangle. ApJ. **661**, 490–495 (2007)
134. O. Stahl, Direct imagery of circumstellar shells around Ofpe/WN9 stars in the galaxy and in the LMC. A&A. **182**, 229–236 (1987)
135. E. Sutton, E.E. Becklin, G. Neugebauer, 34-micron observations of Eta Carinae, G333.6-0.2, and the galactic center. ApJ. **190**, L69–L70 (1974)
136. M. Teodoro et al., Near-infrared integral field spectroscopy of the Homunculus nebula around Carinae using Gemini/CIRPASS. MNRAS. **387**, 564–576 (2008)
137. A.D. Thackeray, Nebulosity surrounding Eta Carinae. Observatory **69**, 31–33 (1949)
138. A.D. Thackeray, Note on the brightening of Eta Carinae. MNRAS. **113**, 237–238 (1953)
139. A.D. Thackeray, Polarization of Eta Carinae. Observatory **76**, 154–155 (1956)
140. A.D. Thackeray, Spectra of the polarized halo around Eta Carinae. Observatory **81**, 99–104 (1961)
141. E. Verner, F. Bruhweiler, K.E. Nielsen, T.R. Gull, G. Viera kober, M.F. Corcoran, Discovery of CH and OH in the $-513\,\mathrm{km\,s^{-1}}$ ejecta of Eta Carinae. ApJ. **629**, 1034–1039 (2005)
142. N. Visvanathan, Polarization in Eta Carinae. MNRAS. **135**, 275–286 (1967)
143. R.H.M. Voors, et al. Infrared imaging and spectroscopy of the Luminous Blue Variables Wra 751 and AG Car. A&A. **356**, 501–516 (2000)
144. N.R. Walborn, The complex outer shell of Eta Carinae. ApJ. **204**, L17–L19 (1973)
145. N.R. Walborn, B.M. Blanco, A.D. Thackeray, Proper motions in the outer shell of Eta Carinae. ApJ. **219**, 498–503 (1978)
146. N.R. Walborn, M. Liller, The earliest spectroscopic observations of Eta Carinae and its interaction with the Carina Nebula. ApJ. **211**, 181–183 (1977)
147. J.R. Walsh, N. Ageorges, High resolution near-infrared polarimetry of Eta Carinae and the Homunculus Nebula. A&A. **357**, 255–267 (2000)
148. R.F. Warren-Smith, S.M. Scarrott, P. Murdin, R.G. Bingham, Optical polarization map of Eta Carinae and the nature of its outburst. MNRAS. **187**, 761–768 (1979)
149. G. Weigelt et al., HST FOC observations of Eta Carinae. Rev. Mex. AA, Ser. Conf. **2**, 11–16 (1995)
150. J.A. Westphal, G. Neugebauer, Infrared observations of Eta Carinae to 20 microns. ApJ. **156**, L45–L48 (1969)
151. P.A. Whitelock, M.W. Feast, C. Koen, G. Roberts, B.S. Carter, Variability of Eta Carinae. MNRAS. **270**, 364–372 (1994)
152. S.F. Woosley, S. Blinnikov, A. Heger, Pulsational pair instability as an explanation for the most luminous supernovae. Nature **450**, 390–392 (2007)
153. T. Zethson et al., Strange velocities in the equatorial ejecta of Eta Carinae, A&A. **344**, 211–220 (1999)

Chapter 8
The Outer Ejecta

Kerstin Weis

Abstract η Carinae is surrounded by a complex circumstellar nebula ejected during more than one eruption, the great eruption in the 1840s and the second or lesser eruption in the 1890s. Beyond the well-defined edges of its famous bipolar nebula are additional nebulous features and ejecta referred to as the outer ejecta. The outer ejecta includes a variety of structures of very different sizes and morphologies distributed in a region 0.67 pc in diameter with a mass of >2–4 $M_{\odot}$. Some individual features in the outer ejecta are moving extremely fast, up to 3,200 km/s, with most of the expansion velocities between 400–900 km/s. As a consequence of these high velocities, structures in the outer ejecta interact with the surrounding medium and with each other. The strong shocks that arise from these interactions give rise to soft X-ray emission. The global expansion pattern of the outer ejecta reveals an overall bipolar distribution, giving a symmetric structure to its morphologically more irregular appearance. The long, highly collimated filaments, called strings, are particularly unusual. The material in the strings follow a Hubble-flow and appear to originate at the central star. The properties of the nebulae associated with other LBVs also are described and compared with η Car. HR Car and AG Car show similar bipolar morphologies but are much older; HR Car's nebula may be η Car's older twin. The larger, extended nebulae detected around the giant eruption LBV P Cygni, and the extended nebulosity associated with AG Car and HR Car could be either from previous eruptions or facsimiles to η Car's outer ejecta.

K. Weis (✉)
Astronomical Institute, Ruhr-University Bochum, Germany
e-mail: kweis@astro.rub.de

K. Davidson and R.M. Humphreys (eds.), *Eta Carinae and the Supernova Impostors*, Astrophysics and Space Science Library 384, DOI 10.1007/978-1-4614-2275-4_8,

8.1 The History of the Homunculus and the Outer Ejecta

The earliest reference to the presence of extended nebular emission around η Carinae is from van den Bos [54, 55] who suggested that some ‘companions’ he detected around the η Argus system might be “nebular nuclei”. Shortly after that, Gaviola [17, 18] published a series of images of a coherent nebular structure, which he called the Manikin or the *Homunculus* due to its human-like appearance. Gaviola also identified individual bright regions around the Homunculus, some of which had also been recognized by van den Bos. At about the same time Thackeray [52,53] also observed the nebula. His sketch showed emission outside the bright Homunculus region and he speculated about the presence of a fainter outer nebula or shell. Since then the term Homunculus has been applied only to the bright, coherent bipolar nebular structure around η Car.

Walborn's [56] ground based images were the first to show the detailed structure of the outer nebula, now called the *outer ejecta*. He recognized and named various individual structures which are identified in Fig. 8.1 on a more recent HST image. The *skirt* (or equatorial disk of *streamers*) and the NN bow, which marks the larger, jet-like region in which the NN and NS condensations are embedded are also marked in Fig. 8.1 (for details see Sect. 8.2.2). Using many very short ground based exposures and a shift-add technique Duschl et al. [10] produced an image in 1985 that clearly showed for the first time the bipolar nature of the Homunculus with its equatorial disk or skirt of streamers. The bipolar lobes of the Homunculus were later made famous in some of the first images from HST [13, 23].

Gaviola was the first to note that the Homunculus nebula is expanding at a rate of several arcseconds per century based on proper motions [18]. More detailed analysis of the proper motions [22, 44] yielded expansion velocities as high as 300 km/s which at the currently accepted distance to η Car, translate into motions of about 500 km/s. Similar values for the expansion of the Homunculus have been obtained from more recent multi-epoch data from HST [3, 39, 48]. Walborn and collaborators [56–58], however, derived velocities, from proper motions, between 280 and 1,360 km/s for structures in the outer ejecta.

In his spectral analysis Gaviola [19] identified sections of the outer ejecta (his region k, in Walborn's notation the S condensation) which move with a speed in excess of 1,000 km/s. Meaburn's analysis with higher spectral resolution and long-slit spectroscopy [30–33] yielded the motions of many more structures and regions in the outer ejecta, with radial velocities spanning values between +300 and −1200 km/s. Other studies, [5, 24], including Fabry-Perot measurements [59], support these results. In the course of these early kinematic studies several knots were also found in the outer ejecta which expand much faster than the Homunculus including a feature in the S ridge moving at +2,150 km/s [8]!

Several chemical composition analyses of the outer ejecta revealed an overabundance of nitrogen and helium as expected from CNO processed material presumably

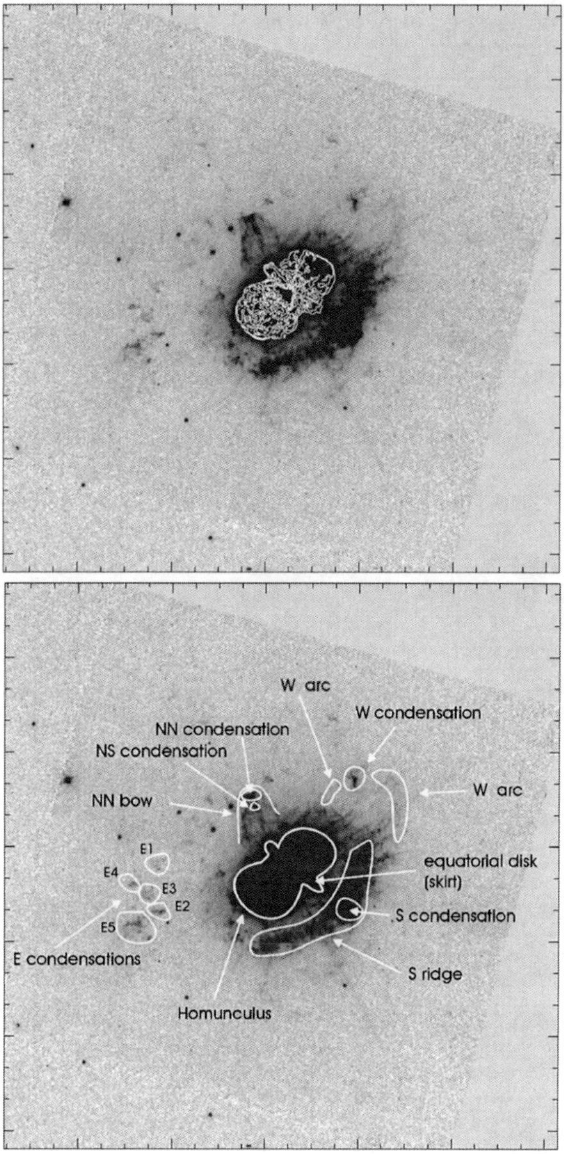

Fig. 8.1 This figure shows an 1$'.5 \times$ 1$'.5$ HST image of η Carinae taken with the F658N filter, north is up, east to the left. Contours in the upper panel highlight the *brightest regions* and reveal the bipolar Homunculus with its equatorial skirt. In the lower panel a sketch outlines several distinct features identified by Walborn [56, 58]. The NN bow and skirt markings have been added (Figure from Weis [63])

ejected from the evolved central star [6–8]. More recent measurements [50] of the abundances at several position in the outer ejecta confirm that the ejecta is CNO processed, and suggest that more distant clumps may have a lower nitrogen overabundance.

8.2 The Present Optical Appearance

8.2.1 The Global View

8.2.1.1 Size and Shape

The outer ejecta extends to at least 30″ (0.34 pc) from the central star based on deep HST WFPC2 images in the F656N (Hα) and F658N ([N II]) filters.[1] The central bipolar structure, the Homunculus, has a diameter (long axis) of about 18″ (0.2 pc). In the left panel of Fig. 8.1 one clearly sees that the Homunculus is by far the brightest structure in η Car's associated circumstellar nebulosity. In the outer ejecta, the NN, NS, E, W, S condensations and the S ridge have the highest surface brightnesses and were therefore the first to be recognized. Numerous lower surface brightness emission features from small filaments and knots were detected later in deep HST images. Unlike the Homunculus, the outer ejecta at first does not look symmetric, or like a coherent structure. Individual structures can be described as filaments, jets, arcs, bullets or knots (see Sect. 8.2.2), and *strings* (Sect. 8.2.3). While the Homunculus with its two lobes (and equatorial ejecta) has a near-perfect bipolar morphology, the outer ejecta appears much more asymmetric, composed of many irregularly shaped structures in an elliptically shaped region around the Homunculus.

8.2.1.2 Mass Estimate

The electron density in the outer ejecta is $\sim 10^4\,\mathrm{cm}^{-3}$ [63] based on the [S II] ratio from HST/STIS observations with an assumed typical electron temperature of 14,000 K [9]. This density estimate is quite constant for the structures detected in four different spectra across the outer ejecta (including the strings, see Sect. 8.2.3) and is similar to the value obtained for the S ridge [9] using the [Si III] line diagnostics. With this density, an estimate of the volume filling factor,[2] and maximum diameter of 0.67 pc ejecta, we derive a total mass of 2–4 $M_{\odot}$. The S ridge, W arc and NN bow contribute roughly 85% of the total mass. Given that the density measurements are close to the high density limit, the very faint diffuse emission is not included, and we have no knowledge of possible structures hidden by the Homunculus or projected onto it, this value has to be considered a lower limit.

[1] Due to the fast motion of the structures in the outer ejecta (see Sect. 8.3) emission detected in the Hα filter can be the result of blueshifted [N II] emission and vice versa.

[2] The much higher filling factor for the S ridge, W arc and NN bow has been taken into account.

8.2.1.3 Origin

The nebulosity associated with many LBVs is generally considered to be the result of episodes of high mass loss in the LBV phase, and possible giant eruptions, like that of η Car in 1843. Models for the formation of η Car's Homunculus and the outer ejecta are based on hydrodynamic simulations that include the mass loss history, the interaction of various wind phases, and possible instabilities that could lead to an eruption. García-Segura et al. [16] produced one of the first simulations of a nebula around a $60\,M_\odot$ star, with a bipolar structure and the onset of instabilities that lead to fragmentation of the shell. The same group [26] also proposed a model specifically for η Carinae, assuming the nebula was created in the 1843 event, that reproduces the bipolar shape, the equatorial skirt, and filamentary structures comparable to those observed in the outer ejecta. However their model is focused on reproducing the bipolar Homunculus, not the outer ejecta. Other models (some assuming that the disk formed first) were also able to reproduce the Homunculus [11,14,15,21], and even the *Little Homunculus* [20], a smaller bipolar shaped nebula embedded in the Homunculus. A comprehensive model for the formation of the outer ejecta however is missing. González et al. [21] suggest that it might be the relic of an earlier eruptive phase before the giant eruption. Various kinematic analyses have produced arguments in favor of both an earlier ejection and at the same time as the 1843 eruption. Both possibilities are discussed later. The abundances of the outer ejecta nevertheless support its formation during the evolved – most likely LBV – phase of the star. The high nitrogen and helium abundances (CNO processed material), rule out its origin as the remnant of an older main-sequence wind blown bubble.

8.2.2 Knots, Arcs, Streamers and Jets

The morphology of the outer ejecta are described here starting with the historically identified structures shown in Fig. 8.1. Higher resolution HST images of some of the regions discussed below are also presented in Fig. 8.2.

8.2.2.1 The E Condensations

The E condensations consists of five discrete knots (E1, E2, E3, E4, E5) described by Walborn [56]. With the much higher resolution of the HST images each of these knots breaks up into several substructures (Fig. 8.2, middle left, $10'' \times 20''$, 0.1×0.2 pc). There are also many smaller knots and diffuse emission in between each of the E condensations. The upper limit for the number of knots clearly depends on the detection limit in the current images.

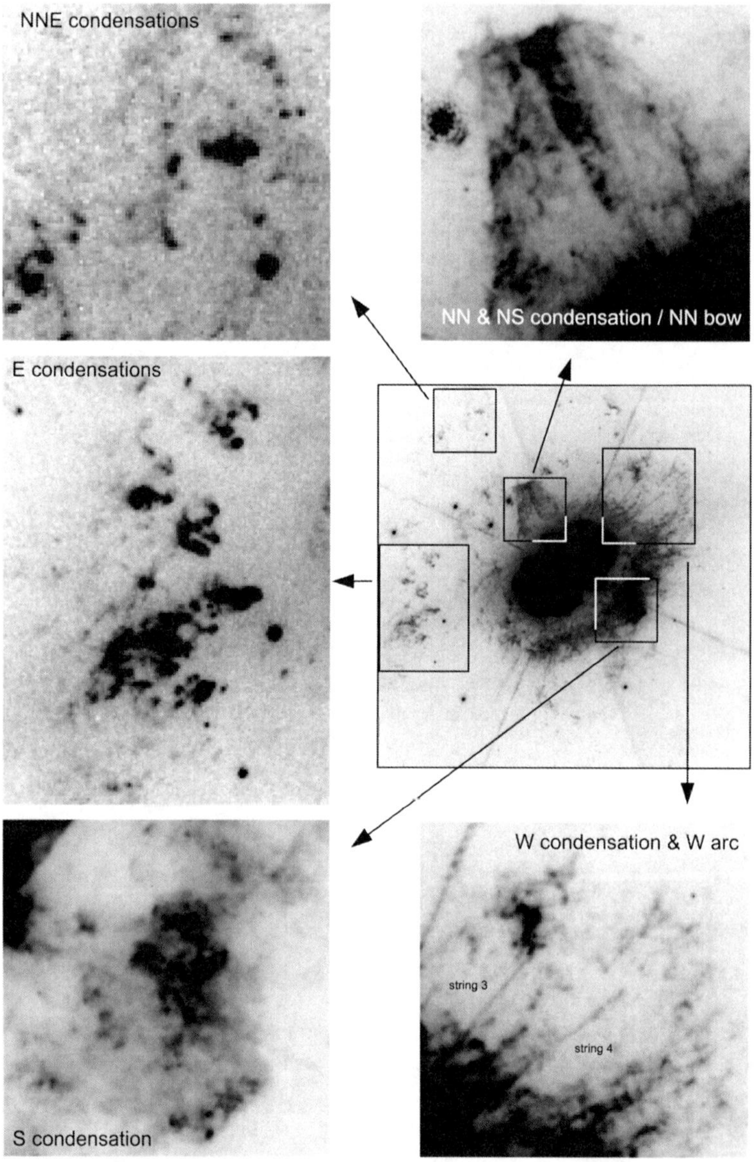

Fig. 8.2 In the *middle right* a 60″ × 60″ section of the HST image from Fig. 8.1 is shown with north up and east to the left. In several side panels enlargements of distinct regions in the outer ejecta are shown: The NN bow region, the E, S and W condensation(s) as well a region in the very north-east (designated NNE condensations) in which several knots are present in the very outskirts of the outer ejecta. In the image of the W condensation & W arc the strings 3 and 4 are also visible

8.2.2.2 The S Condensation and S Ridge

The S ridge including the brightest part, the S condensation, is, together with the N condensation the brightest and largest structure in the outer ejecta (e.g. Walborn [56]). In deeper HST images, the S ridge is visible as an arc covering 150° from the south to north. It consists of a countless number of knots, bullets, tiny filaments, arcs and finger like structures. a significant number of which seem to interact and collide with each other (see also Sect. 8.4). The S and N condensations appear to be natural extensions of the equatorial skirt or streamers [10]. A large amount of material has apparently been ejected preferentially in the equatorial direction and gives rise to the highest surface brightness in the S ridge. In Fig. 8.2 (lower left, $10'' \times 10''$, 0.1×0.1 pc) one can even see diffuse filaments connecting the bright region of the S condensation (central part of the images) with the skirt (bright bow like structure in the upper left).

8.2.2.3 The W Condensation and W Arc

The W condensation is by far the brightest emission in the western section of the outer ejecta. HST images resolve it into several much smaller clumps (Fig. 8.2, lower right, $15'' \times 15''$, 0.15×0.15 pc). It is also not as isolated as it appeared in older ground based images. Several less bright knots and filaments accompany this structure. The W arc resembles an extension of the S ridge but at lower intensity. The HST image resolves it into many smaller elongated filaments that in several cases point radially away from the Homunculus. There are two structures within the W arc and partially projected onto the W condensation now classified as strings (see Sect. 8.2.3).

8.2.2.4 The N Condensation or Jet

In his first sketch of the outer ejecta Walborn [56] identified the N condensation that marks the very northern edge or tip of a bow like structure, that was present in his sketch but unnamed. It is now called the NN bow in Fig. 8.1. In a later paper, Walborn & Blanco [58] (see also Fig. 8.1) subdivided the N condensation into two separate entities, the NN and NS condensation. The NN and NS condensation as well as the NN bow appear as a counterpart to the S condensation and parts of the S ridge. A natural interpretation of both is that they are a projected extension of the streamers defining the inner skirt [10], and represent the brightest area of a larger equatorial disk. Meaburn et al. [32] interpreted the NN condensation as the tip of what he defined as the *jet* which marks the inner region of the bow in Walborn's sketch and runs straight from the skirt to the NS condensation and then to the tip of the NN condensation. In Meaburn's interpretation, the jet is the result of an episodic ejection of material. Figure 8.2 (upper right, $10'' \times 10''$, 0.1×0.1 pc) shows the NN bow (jet) in detail and the bright NN and NS condensations. To some degree, the

NN bow is a relatively coherent structure with a lot of substructures. The NN bow has a broad top at the northern edge and does not appear tapered as it is normally the case for a jet. If the NN bow is an extension of the skirt this broader top could result from the projection of the extended skirt (streamers) on the plane of the sky, due to the different position angles of each streamer.

8.2.2.5 New Structures Detected in HST-Images

With deep, high spatial resolution HST images, smaller individual entities and diffuse emission in the outer ejecta became visible. The high spatial resolution of the deep HST images (primarily from HST/WFPC2) significantly increased the contrast of small knots and narrow filaments against the diffuse, bright, and structured background of the Carina H II region, allowing us to see the significant substructure in known features in the outer ejecta, and also led to the detection of many new small and/or low surface brightness emission-line regions. The upper left panel of Fig. 8.2 ($10'' \times 10''$, 0.1×0.1 pc) shows a region with many small clumps far from the central star, in the outskirts of the outer ejecta, even though they are much fainter than those closer to the Homunculus. By analogy with the older nomenclature, we call them the NNE condensations. Note also that they again appear spatially on a line with an even larger extension of the skirt and NN bow.

The smallest objects detected so far in the outer ejecta are those barely resolved or just below the resolution of HST which limits their size to $7.7\,10^{-4}$ pc or 160 AU at a distance of 2.3 kpc. The brightness of η Car, responsible for stray light and ghosts, and the nebular background also restrict the detectable surface brightness, and we may still be missing the lowest surface brightness ejecta.

In summary the outer ejecta is a conglomerate of knots, arcs, streamers and jet-like structures of very different sizes. The NN bow seems to be the only structure we can assume to be coherent, and therefore one of the largest with a length or height of 0.084 pc and a width of 0.081 pc, at its broadest section, close to the Homunculus. The longest single objects by far in the outer ejecta, however, are the strings described in Sect. 8.2.3.

8.2.3 ... and Strings !

The most amazing structures in the outer ejecta are long, highly collimated linear features, now called the strings (see lower right panel in Fig. 8.2 and left panel in Fig. 8.3). The first structure of this type, String 1, was identified nearly simultaneously by Malin [27] and by Meaburn et al. [30] and was called, respectively, a *spike* or *jet*. Five such strings have now been identified by visual inspection of HST images and labeled string 1, 2, 3, 4, 5 by Weis et al. [69]. Morphologically similar features can be seen on smaller scales, but have been provisionally excluded from this class as they are not as prominent and show less extreme parameters.

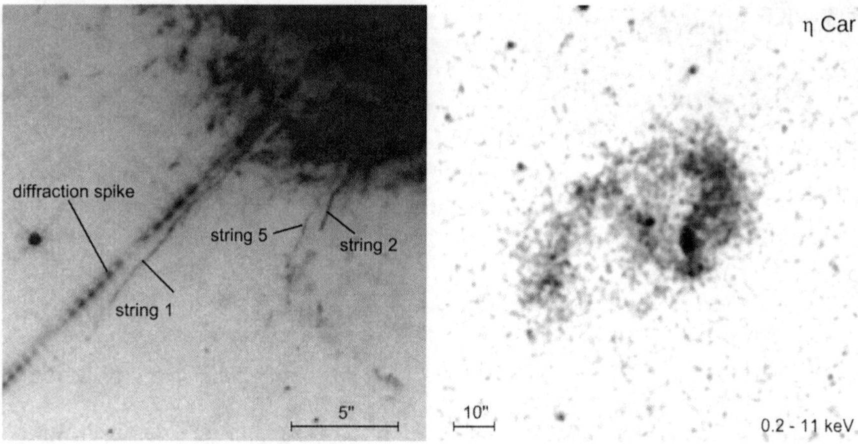

Fig. 8.3 *Left:* Three of the highly collimated structures, the strings, found in the south(-east) part of η Carinae's outer ejecta (Image from Weis et al. [69]). *Right:* An image from the CHANDRA satellite showing η Carinae's and the nebula's X-ray emission (Weis et al. [72])

Table 8.1 Parameters of the five strings in the outer ejecta (Table adapted from Weis et al. [69])

String #	Length (pc)	Width (pc)	Length-to-width ratio	v_{min} ($km\,s^{-1}$)	v_{max} ($km\,s^{-1}$)	(N II)λ 6583Å/H_α
1	0.177	0.003	70	−522	−995	3.3
2	0.044	0.002	31	−442	−591	2.7
3	0.095	0.002	42	200	300	2.3
4	0.103	0.002	68	–	–	–
5	0.058	0.002	38	−383	−565	3.0

Two of the strings were found in the north(-western) (Fig. 8.2, lower right panel) and three in the south(-eastern) (Fig. 8.3 left panel) part of the outer ejecta. None were detected in the other two quadrants. All strings appear to emerge from the rim of the Homunculus lobe either between it and the S ridge or the W arc. The length of the strings varies from the shortest (string 2) with 0.044–0.177 pc for the longest (string 1). Both are visible in Fig. 8.3. The width of the strings was resolved with HST to be 0.002–0.003 pc, revealing a length-to-width ratio between 31 and 70. Parameters for all strings are given in Table 8.1.

All of the strings point radially away from the Homunculus, and if they could be traced back further, would appear to emanate from η Car itself. Except for string 2 which shows a noticeable kink, they seem to be quite straight with only minor wiggles. String 1 however does split and merge again in a short section close to the S ridge (barely visible in Fig. 8.3, just underneath the diffraction spike). Measurements of the [S II] lines yield a density estimate of $1.6\,10^{-23}$ kg/cm^3 for string 1 and assuming the string to be a filled cylinder a total mass estimate of 3–4 $10^{-4}\,M_\odot$ [63]. Kinematic information exists for strings 1, 2, 3, and 5, and will be discussed in Sect. 8.3.

8.3 Kinematics

8.3.1 The Outer Ejecta

As already mentioned in Sect. 8.1 the expansion velocities in the outer ejecta can reach up to more than 1000 km/s. Several newer studies show velocities as high as 2000 km/s or more [50, 60, 61]. Weis's [61, 70] systematic high-resolution longslit echelle spectroscopy of the complete nebula revealed unexpected results for the kinematic behavior of the the individual structures with expansion velocities ranging from roughly −1000 to 3200 km/s, with the majority between 600 and 700 km/s.[3] Individual knots or arcs can however show a very large velocity dispersion, that also reaches up as high as several hundred km/s. The upper panel in Fig. 8.4 shows the maximum expansion velocities measured for the knots and regions in the ejecta surrounding the Homunculus. While the observed features in the outer ejecta show a more or less irregular and random distribution, the red and blueshifted velocities seem to follow a systematic bipolar expansion pattern. All of the structures more or less to the (south-)east are blueshifted, and those more to the (north-)west are redshifted. The outer ejecta thus follows the same symmetry and has the same axis as the Homunculus: the southern lobe in the Homunculus is tilted towards, the northern away from the observer. So, even though the morphology does not appear bipolar, the kinematics reveal the bipolar nature of the outer ejecta (e.g. Weis [60, 72]). An additional expanding shell [4] in front of the south-east lobe of the Homunculus also matches the global expansion pattern of the outer ejecta.

Finally what do the kinematics tell us about the the formation and age of the outer ejecta? Is the outer ejecta the result of the 1843 eruption or is it much older? From proper motion measurements, Walborn & Blanco [57] and Ebbets et al. [12] found that the NN, S, and some of the E condensations were ejected between 1797 and 1891. This makes it very likely that the outer ejecta originated in the 1843 eruption, however, the measured velocities could be affected by deceleration [57]. The proper motion measurements from the HST images [3,40] are also compatible with an 1843 origin of the outer ejecta. If deceleration is allowed for, the observed velocities for the N condensation could also match an ejection during the 1890 event, or second eruption. Smith & Gehrz [48] suggested the same origin for the formation of the equatorial features (skirt).

Meaburn et al. [30,32,33] and Weis [61] obtained similar results for the origin of the outer ejecta using radial velocity measurements. Assuming an origin of the outer ejetca around 1843, Weis determined the the inclination angle for all of the knots with known radial velocities. The spatial distribution derived is in good agreement with an outburst in bipolar directions.

---

[3] Editors note: Smith [47] has recently reported ejecta moving at 3,500 km s^{-1} and possibly as high as 6,000 km s^{-1} which he attributes to a blast wave from the 1843 eruption.

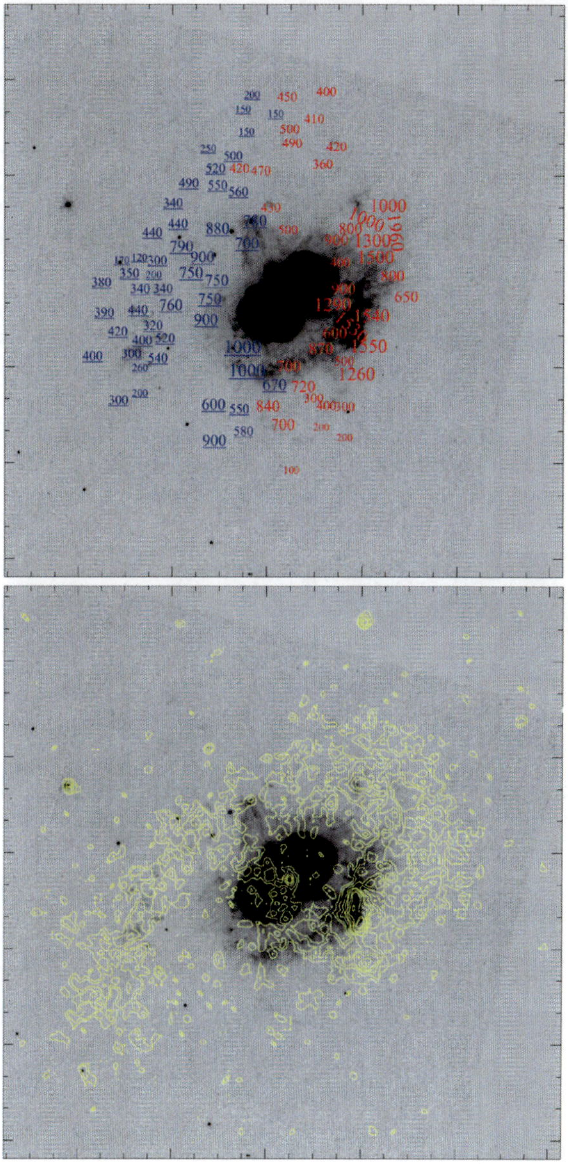

Fig. 8.4 In both figures the underlying grayscale image is the HST F658N picture. In the *upper panel* radial velocities for certain clumps (areas) measured from optical echelle spectroscopy are shown. For better visualization, the expansion velocities have been color coded: blueshifted structures in *blue* (and underlined), redshifted in *red*. The font size has also been increased with increasing velocity. The *lower panel* shows the X-ray emission measured with the CHANDRA satellite in the energy band between 0.6 and 1.2 keV superposed on the same optical image (Figure from Weis [63])

8.3.2 The Strings

Weis et al. [69] obtained high-resolution echelle spectroscopy and kinematic measurements for strings 1, 2, and 5 (see Fig. 8.3). All of these strings posses a blueshifted emission and show radial velocities between −383 and −995 km/s. They also all show an increasing radial velocity towards their tip or end furthest from the

Homunculus, a perfect Hubble-flow. Extrapolating the strings back to their origin (see Fig. 6 in Weis et al. [69]), they seem to have been expelled at the position of the central object and their radial velocities steadily decrease towards zero at this position. Their velocities are blueshifted like other structures in this region, and they thus follow the bipolar symmetry derived from kinematical analysis. An HST-STIS spectrum at higher spatial, but lower spectral resolution, indicates that string 3 shows the same kinematic behaviour as the others. As expected for its north(-west) location in the outer ejecta the radial velocity of string 3 is redshifted, fitting to the global bipolar expansion pattern. The origin of these highly collimated strings and their linearly increasing velocities is still very puzzling. Are they a single physical entity or something else? Weis et al. [69] discussed several possibilities including coherent structures analogous to a water jet, trails or wakes following a projectile at the far end or tip of the string like a comet, or numerous much smaller individual knots or bullets following the same path. Other models proposed for the strings include a decelerating flow of ablated gas from a bullet [43] or a radiative bullet [42], and an alternative model assuming shadowing effects has also been proposed [51].

8.4 X-ray Emission

8.4.1 Morphology of the X-ray Emission

The first X-ray satellite with enough spatial resolution to disentangle the nebula and the stellar emission was the *Einstein* Observatory. The Einstein images [1, 45] showed that the X-ray emission partially originates in the outer ejecta and is not only from the star. A more detailed comparison was possible with the data from the *High Resolution Imager* (HRI; spatial resolution $\sim 5''$) on board the *Röntgensatellite* (ROSAT). An overlay of the optical HST image and the ROSAT image shows a very good spatial agreement between the optical and diffuse X-ray emission, indicating that the hot X-ray emitting material was created by processes in the outer ejecta [70]. The X-ray emission around η Carinae is hook-shaped with two brighter knots roughly identified with the optical S ridge and W arc. With the launch of the CHANDRA satellite, the spatial and spectral resolution of X-ray satellites again improved and images of the outer ejecta in various energy bands became possible. Seward et al. [46] provided a short overview of the early CHANDRA observations, followed by a more detailed analysis of a more sensitive dataset and a comparison with the kinematics by Weis et al. [72].

The right panel in Fig. 8.3 shows the CHANDRA image in the energy range between 0.2 and 11 keV. The central object is visible as a bright point source. A second bright source is located towards the south-west and coincides with the S condensation. In the lower panel in Fig. 8.4 the soft X-ray emission (0.6–1.2 keV), which forms nearly the total emission of the outer ejecta, is superimposed on an HST image. Both figures show that the X-ray emission still exhibits the hook shape structure, but is composed of smaller substructures. Across the hook, from the very bright region (the S condensation) a *bridge* projects onto the Homunculus to the north-east, ending south of the N condensation. The NN bow (or jet) is extremely faint in X-ray emission, hinting at photoelectric absorption of the soft X-ray emission. Virtually no emission is found in the south-west section of the outer ejecta, where the number density of clumps is also low. In general, comparison between optical and X-ray images shows that optical counterparts are present in most of the regions were X-ray emission is present. It is not a perfect match in position and size of the structures, but the overall distribution is the same. The best agreements are seen in the S ridge, S, E and W condensations. An anti-correlation is found as already mentioned between the NN bow and the Homunculus itself. Except for the bridge, soft X-ray emission does not coincide with the Homunculus. X-ray emission from the Homunculus is detected in significantly harder X-ray bands only, being an X-ray reflection nebula [2]. The bridge therefore most likely results from clumps and structures in the outer ejecta lying in front and projected onto the Homunculus. The spatial concentration of the bridge suggests that these features may be extensions of the streamers/skirt.

8.4.2 Comparing X-ray Emission with the Kinematic Measurements

A natural origin for the X-ray emission in the outer ejecta is from shocks due to collisions of the structures with the interstellar medium and with each other. A comparison of both images in Fig. 8.4 indicates that where strong X-ray emission is detected, the (radial) expansion velocities range from several hundred to 2,000 km/s in the S ridge. In most cases the X-ray emission is highest in regions with high expansion velocities and with the highest density of clumps (as traced by the optical emission). To test this shock hypothesis, Weis [72] compared the X-ray temperatures and postshock temperatures. The temperatures of the shocks resulting from the expansion of the outer ejecta were calculated from the kinematic data (Sect. 8.3). The X-ray temperatures were measured from CHANDRA spectra selected regions: the E rim, which includes all E condensations; the S ridge, here defined as the region north of the S condensation; the SS ridge, an bright area south of the S ridge which posses only a few optical counterparts; S condensation and the W arc (see Fig. 5 in Weis et al. [72]). All spectra have been fit using a Mewe-Kaastra-Liedahl thermal plasma model, with a foreground column density of $\log N_{\mathrm{H}} = 21.3\,\mathrm{cm}^{-3}$ and a solar abundance, except for nitrogen which together with the temperature and emission measure were left as free parameters. The results for the temperature and nitrogen abundance (as the number of nuclei per hydrogen nucleus relative to solar abundances), are listed in Table 8.2. The W arc is the hottest area, with the E rim being only slightly cooler. All spectra of the S region (S ridge, SS ridge and S condensation) are very similar in temperature. The nitrogen abundances are high and, vary only slightly except for the S condensation which has a very high value. Nitrogen abundances from these model fits match well those derived from optical spectra [50].

Table 8.2 Results of the models fit to the X-ray spectra in selected regions and the measured velocities in the same area given by the measured maximum velocity and the typical velocity of the majority of the clumps. The nitrogen abundance is given as the number of nuclei per hydrogen nucleus relative to solar abundances (Table adapted from Weis et al. [72])

Area	Temperature (keV)	Temperature (K)	Nitrogen	$\|v_{postshock}\|$ (km/s)	$\|v_{max}\|$ (km/s)	$\|v_{typical}\|$ (km/s)
E rim	0.70 ± 0.03	$8.0 \pm 0.4\,10^6$	8	726	750	350–700
S ridge	0.60 ± 0.02	$7.0 \pm 0.2 10^6$	6	670	1,500	600–800
SS ridge	0.63 ± 0.04	$7.3 \pm 0.5 10^6$	6	690	1,260	800–900
S condensation	0.63 ± 0.04	$7.3 \pm 0.5 10^6$	22	690	1,550	700–1,000
W arc	0.76 ± 0.03	$8.8 \pm 0.4 10^6$	4	756	1,960	800–1,000

Assuming that the X-ray temperatures are postshock temperatures, the necessary velocities of the gas are determined using the relationship $T = v^2 (3\mu/16k)$. Here v is the postshock velocity of the gas to be calculated, T the postshock temperature taken from the X-ray temperature, k the Boltzmann constant and μ the molecular weight. Since the outer ejecta consists of CNO processed material (e.g. Davidson et al. [7]) $\mu = 0.67$ (fully ionized gas with a mixture of X = 0.6, Y = 0.4) was used. The results are also given in Table 8.2. The postshock velocities necessary to reproduce the X-ray emission are roughly 670–760 km/s.

Comparing these values with the measured expansion velocities (Fig. 8.4), we conclude that the X-ray emission is indeed due to shocks in the outer ejecta. In some regions (especially E rim = E condensations) the velocities are a bit too low, however, the plotted velocities are radial velocities, representing a lower limit to the total space velocity. For example, E5 has a tangential velocity of 600 km/s [58] and therefore a total expansion velocity of 744 km/s, matching the required velocity. Therefore one can conclude that the X-ray image of the outer ejecta displays the shock fronts of the material which interacts with the ambient medium and with itself. Interaction of the outer ejecta material with each other seems to take place particularly in the S ridge and S condensation. Very high radial velocities have been detected in the S ridge and the W arc indicative of very high postshock temperatures.

The CHANDRA images extracted in the hard X-ray band ([72] lower panel Fig. 4) show the central object and a larger diffuse halo which extends into the W arc and S ridge. This emission partly results from residuals of the PSF (from the more central area), and partly from the Homunculus X-ray reflection nebula found by Corcoran et al. [2], but the more elongated and asymmetric regions to the south-west are real hard X-ray emission sources in this area, where the highest velocities have also been detected. This again supports the shock model for the origin of the soft X-ray emission. The number of fast components is however much smaller compared to those which move slower. Therefore the X-ray emission is dominated by emission in the lower energy bands (0.6–1.2 keV, Fig. 8.5).

8.5 LBV Nebulae and Associated Ejecta

8.5.1 Other LBV Nebulae

Numerous massive stars exhibit episodes of high mass loss. Those that are hot or early-type supergiants in their quiescent phase are known as the Luminous Blue Variables (LBVs); see the contribution to this volume on "Eta Carinae and the Luminous Blue Variables" by Vink. It is not known if all massive evolved stars go through an LBV phase, but in principle, they all certainly have the potential to enter the LBV phase. Initial mass, metalicity, stellar wind, and rotation may impact the star's evolution and therefore its passage through an LBV phase (see e.g. Meynet & Maeder [38]). The new stellar evolution models – which include the treatment

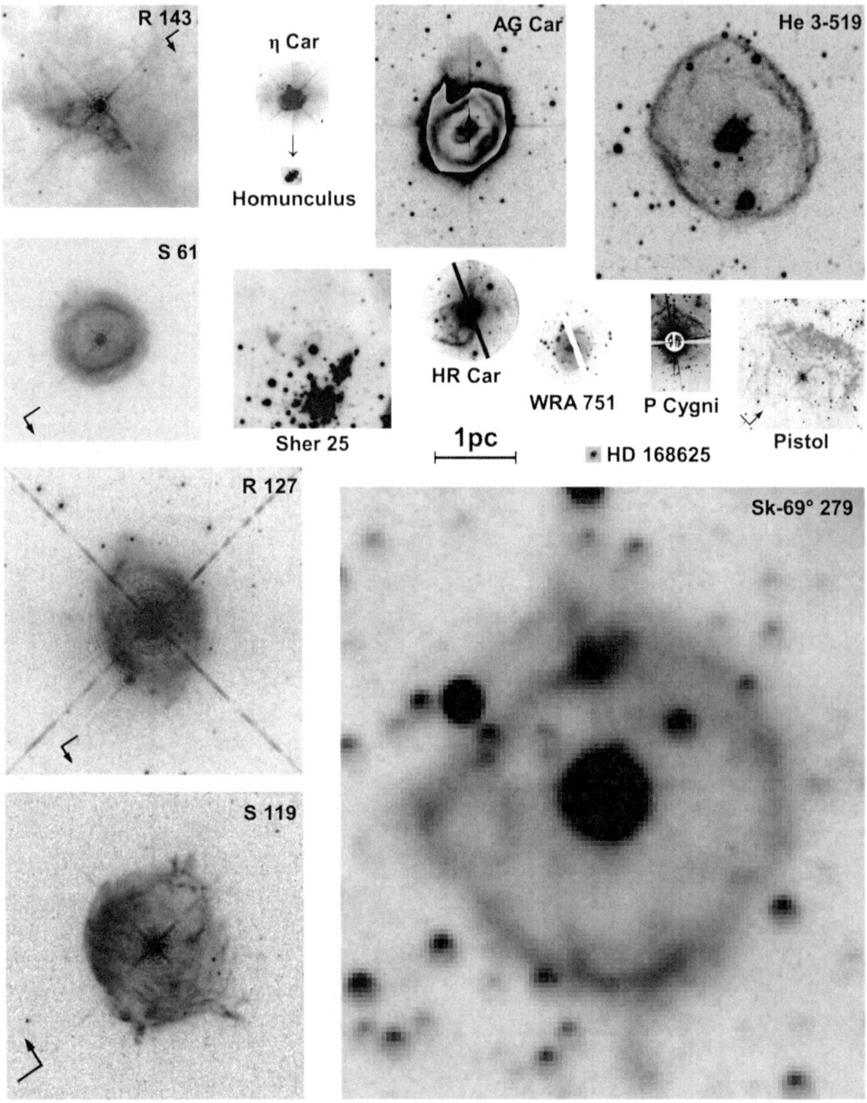

Fig. 8.5 This image shows a collage of several galactic and LMC LBV with extended nebulae. All pictures are shown to the same physical scale (Figure adapted from [64, 67])

of stellar rotation with an initial rotation rate of 300 km/s – indicate that stars with initial masses as low $22\,M_\odot$ (for $Z = 0.02$) may have LBV phases (compared to about $40\,M_\odot$ without rotation). The same models also suggest that a rotating star with very high mass might totally skip the LBV phase and directly enter the Wolf-Rayet phase. As a consequence of the high mass loss during the LBV state and possibly through giant eruptions (like η Car's 1840s eruption), many LBVs form circumstellar nebulae (like the Homunculus and outer ejecta). Figure 8.5 shows a selection of LBV nebulae detected in the optical (or infrared for the Pistol star). In this figure all of the nebulae are shown to the same physical scale. The size, the expansion velocity and morphology for several LBV nebulae in the Milky Way and the LMC are summarized in Table 8.3.[4] They display a wide range of values, and comparing their parameters with η Car's Homunculus and outer ejecta shows that: (a) the outer ejecta of η Car is by far the fastest expanding structure (b) it is one of the smallest (the Homunculus is only marginally larger than the nebula around HD 168625), and (c) it is not the only bipolar structure, but the Homunculus has the most clearly defined bipolar morphology.

Many of the LBV nebulae in the LMC are larger than their galactic counterparts, while their expansion velocities are much smaller. The LMC LBV nebulae could be older, which would make them larger and would give them time for deceleration of their expansion. In that case, however, a population of larger galactic LBV nebulae should also exist, which is not observed. The lower metalicity in the LMC though will result in weaker stellar winds, different evolutionary times, changes in the time scales of stellar and interstellar instabilities, and longer cooling times which could directly influence the parameters of the nebulae in a complex way, and could explain the disparity in observed properties of the LMC and Milky Way LBV nebulae [62]. Table 8.3 and Fig. 8.5 show that bipolarity among LBV nebulae is common. Based on existing morphological or kinematic analyses about 40% are roughly spherical, and less than 10%, (R 143) appear completely irregular. A large number of LBV nebulae are bipolar, showing either a hourglass shape like the Homunculus, or bipolar attachments (caps) as in R 127(see Fig. 8.5). In total at least 50% of all LBVs are bipolar, for the Galactic objects the fraction is 75%. On smaller scales, however, the nebulae are highly complex. We have already discussed the fine structure in η Car's outer ejecta with its variety of small and large scale objects. The HST image of S 119 reveals a similar filigree texture in the nearly spherical nebula [71]. Four bright filaments, possibly due to Vishniac instabilities, point outwards to the north and south, and are all comparable in size to the entire Homunculus. An image of the nebula around S 61 shows a bright, inner ring structure surrounded by a larger fainter ring suggestive of two different wind phases or a porous inner shell. The kinematic analyses of the nebula around R 127 indicates its bipolar nature [71]. AG Car and HR Car both reveal a bipolar character, and Weis and collaborators concluded that the nebula around HR Car may be an older twin of the Homunculus [60, 68]. It is larger in size but is expanding slower with a kinematic age of 4000–9000 years.

The characteristics and appearance of each LBV nebula therefore depend on the history of its stellar wind, eruptions, the interstellar medium, stellar rotation, and possible (magneto-)hydrodynamic instabilities.

---

[4]For nebulae with a distinct inner and outer section both values are given and are separated with a slash. The maximum size is either the largest diameter measured if spherical or the major and minor axes. For hourglass shaped bipolar nebulae, the radius and expansion velocities (marked with *) refer to just one lobe. Adapted from Weis [60, 62].

Table 8.3 Parameters of galactic and LMC LBV nebulae

LBV	Host galaxy	Maximum size (pc)	Radius (pc)	v_{exp} (km/s)	Kinematic age (10^3 years)	Morphology
η Carinae	Milky way	<0.2/0.67	<0.05/0.335	300*/10–3,200		Bipolar
AG Carinae	Milky way	$<1.4 \times 2$	<0.4	~25*	~30	Bipolar
HD 168625	Milky way	$<0.13 \times 0.17$	<0.075	30	1.8	Bipolar ?
He 3–519	Milky way	<2.1	<1.05	61	16.8	Spherical/elliptical
HR Carinae	Milky way	$<0.65 \times 1.3$	<0.325	75*	4.2	Bipolar
P Cygni	Milky way	<0.2/0.84	<0.1/0.42	110–140/185	0.7/2.1	Spherical
Pistol star	Milky way	$<0.8 \times 1.2$	<0.5	60	8.2	Spherical
Sher 25	Milky way	$<0.4 \times 1$	$<0.2 \times 0.5$	30–70	6.5–6.9	Bipolar
WRA 751	Milky way	<0.5	<0.25	26	9.4	Bipolar
R 71	LMC	<0.1 ?	<0.05 ?	20	2.5 ?	?
R 84	LMC	<0.3 ?	<0.15 ?	24 (split)	6 ?	?
R 127	LMC	<1.3	<0.77	32	23.5	Bipolar
R 143	LMC	<1.2	<0.6	24 (split)	49	Irregular
S Dor	LMC	<0.25 ?	<0.13 ?	<40 (FWHM)	3.2 ?	?
S 61	LMC	<0.82	<0.41	27	15	Spherical
S 119	LMC	<1.8	<0.9	26	33.9	Spherical/outflow
Sk −69° 279	LMC	$<4.5 \times 6.2$	<2.25	14	157	Spherical/outflow

8.5.2 *Extended Outer Ejecta*

Three LBVs or giant eruption LBVs show evidence for extended ejecta from possible earlier eruptions or nebulosity similar to η Car's outer ejecta, that is, extended over a larger region, compared with the size of their inner nebulae.

8.5.2.1 P Cygni

P Cygni is famous for its brightening in the early 1600s and may be an example of a giant eruption similar to that of η Car [25, 49]. Meaburn and his associates [28, 29, 34–37] have identified three separate extended nebulae associated with P Cyg; the inner shell with a diameter of 0.2 pc ($v = 110$–140 km/s), the outer shell at 0.84 pc ($v = 160$–180 km/s) and the giant lobe wich was traced out to $7'$corresponding to a linear extent of about 4 pc ($v = 80$–130 km/s) from P Cyg. The kinematic age of the inner shell is of the right order to match its eruption in the 1600s, assuming small deceleration. The age of the outer shell is about a factor of 3 larger, it is either an older event (ejection or wind phase) or has been significantly slowed down. Considering its dynamical age of 5 10^4 years, the giant lobe may not have been formed during the LBV phase, depending on its duration. Since the lobe is visible in [O III] images it is most likely an interstellar bubble created by P Cyg's former O star phase and not an circumstellar nebula containing CNO processed material (which would be depleted in oxygen).

8.5.2.2 HR Car

As mentioned earlier the nebula around HR Car may be an older bipolar, evolved twin of the Homunculus. It is therefore natural to ask about a possible outer ejecta around HR Car. Images of its larger environment show the presence of nebulosity to the north-west (see left panel in Fig. 8.6, Figs. 1 and 2 in Weis et al. [68], and Nota et al. [41]). The extended nebula has a conical (partially elliptical) shape and becomes extremely narrow at the north-western end where it merges with a structure called the funnel. So far no counterpart in the south-east has been detected, otherwise this outer nebula would have a bipolar structure. The orientation of the north-western cone and funnel however is perfectly aligned with the bipolar axis of the inner nebula. The image shown in Fig. 8.6 (left) is a composite of two fields (H_α) observed with the 0.9 m CTIO telescope. In the center (round subsection) an image made with ESO/NTT (and a coronograph) shows the bipolar inner nebula in more detail (the NTT image is also shown in Fig. 8.5). The inner bipolar nebula is about 1.3×0.65 pc and is expanding with 75–150 km/s. The outer nebula lies at a distance between 3 and 4.5 pc from HR Car. This is roughly the same size as the largest LBV nebula detected so far (Sk$-69°279$ see Table 8.3). Kinematic analysis of the southern most parts of the outer nebula indicate that it is slowly expanding with

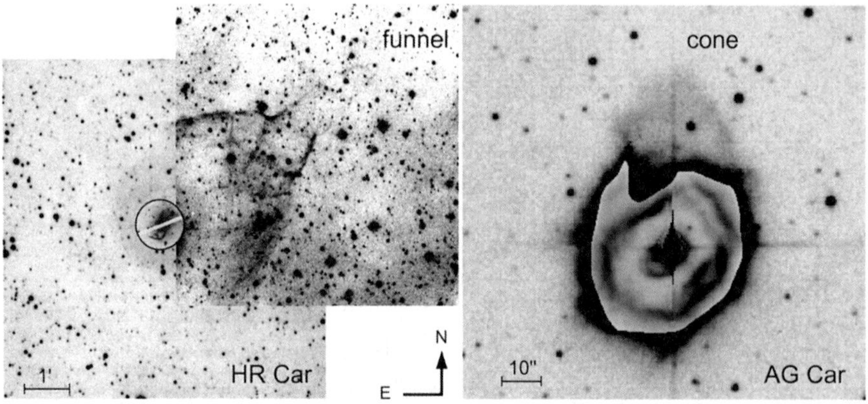

Fig. 8.6 *Left:* A composite of three images of the region around the LBV HR Carinae. In the center the (round) subset shows the bipolar LBV nebula. On larger scale an elliptically shaped nebular emission is detected to the north-west which is topped with a narrow funnel shaped structure [68]. *Right:* Two H_α images of the nebula around the LBV AG Carinae are superimposed. A short exposure shows the inner nearly boxy like nebula. A longer exposure reveals the much larger fainter nebula, extending mainly to the north(-east) (Image taken from Weis [65])

a detected line split of 20 km/s. The true deprojected expansion velocity might be much higher if the outer nebula is (as the images indicate) expanding preferentially in the plane of the sky towards the north-west. Depending on the inclination, the kinematic age could be between several 1,000 years and 10,000–20,000 years old. The [N II]λ6583Å/H_α ratio of the outer nebula is similar to that of the expanding lobes, but a factor of 2–3 lower than found in several knots of the inner nebula. The results are compatible with the outer nebula having formed either during the main sequence phase or the onset of the LBV phase.

8.5.2.3 AG Car

Images of the nebula around AG Car (Fig. 8.6) suggest an appearance not unlike that of two boxes stacked on top of each other with a size of 0.87×1.16 pc. Much deeper ground based image of AG Car however reveal a larger nebula with fainter outer emission extending to a total maximum size of 1.4×2 pc [65, 66]. A faint conical emission to the north, dubbed the *cone*, is detectable beginning from a small arm-like structure to the north-west. The outer emission, including the cone has the same [N II]λ6583Å/H_α ratio (indicative of CNO processed material) as the brighter part of the nebula, supporting a LBV phase origin. Kinematic analyses of the AG Car nebula furthermore demonstrated that the 'stacked boxes' expand in opposite directions. A redshifted and a blueshifted shell can be identified projected on top of each other with a slight shift of the blueshifed shell towards the southwest (see the rim in Fig. 8.6). The fainter emission is simply an extension of the shell and not a separate entity of the nebula. The expansion velocity (derived from the Doppler ellipses) of each of the shell is about 30 km/s. This clearly identifies the nebula as bipolar, with two shells like the Homuculus. The cone is not an outer outflow emission or nebula rather a slightly faster moving section of the redshifted shell.

8.6 Summary and Conclusions

η Car's outer ejecta is the largest structure attributed to a major ejection or eruption in its recent history. Its nitrogen enriched material is strong evidence for CNO processed elements in the outer ejecta brought to the surface and peeled off or violently ejected. The outer ejecta is more like a conglomerate of individual smaller structures, not a coherent circumstellar shell. It therefore requires either a very rapid and efficient fragmentation within the nebula or a creation mechanism which favors the ejection of clumps. Its probable origin in the early nineteenth century together with Walborn's [56] images which showed its clumpy character only 130 years later, makes formation of the outer ejecta by fragmentation of an expanding shell seem unlikely. Its morphology, high velocities and large velocity dispersion of individual structures, and the X-ray emission formed through shocks, support its creation in a more explosive event, during which several outer layers of the star's surface were ejected. The Homunculus, a more well-defined, dusty and denser nebula, would then be attributed to material from deeper layers expelled during the outburst. The highly filamentary, clumped structure is the outstanding characteristic of the outer ejecta, and is not observed in any of the other nebulae associated with LBVs.

Acknowledgements I like to thank Dominik J. Bomans for many years of stimulating discussions on all topics concerning η Carinae and beyond, as well as reading and improving the manuscript. I am grateful to Wolfgang J. Duschl my teacher, long-time collaborator, and for his contributions to this text. I thank Roberta Humphreys for helpul comments to and corrections of this chapter.

References

1. T. Chlebowski, F.D. Seward, J. Swank, A. Szymkowiak, X-rays from Eta Carinae. ApJ. **281**, 665–672 (1984)
2. M.F. Corcoran, K. Hamaguchi, T. Gull, K. Davidson, R. Petre, D.J. Hillier, N. Smith, A. Damineli, J.A. Morse, N.R. Walborn, E. Verner, N. Collins, S. White, J.M. Pittard, K. Weis, D.J. Bomans, Y. Butt, Waiting in the wings: reflected X-ray emission from the Homunculus nebula. ApJ. **613**, 381–386 (2004)
3. D.G. Currie, D.M. Dowling, E.J. Shaya, J. Hester, P. Scowen, E.J. Groth, R. Lynds, E.J. O'Neil Jr., WFPC instrument definition team astrometric analysis of the Homunculus of eta Carinae with the Hubble space telescope. AJ. **112**, 1115–1127 (1996)
4. D.G. Currie, B.N. Dorland, A. Kaufer, Discovery of a high velocity, spatially extended emission 'shell' in front of the southeast lobe of the eta Carinae Homunculus. A&A. **389**, L65–68 (2002)

5. A. Damineli Neto, R. Viotti, G.B. Baratta, F.X. de Araujo, High velocity outflow from Eta Carinae. A&A. **268**, 183–186 (1993)
6. K. Davidson, N.R. Walborn, T.R. Gull, The remarkable spectrum of some material ejected by Eta Carinae. ApJ. **254**, L47–51 (1982)
7. K. Davidson, R.J. Dufour, N.R. Walborn, T.R. Gull, Ultraviolet and visual wavelength spectroscopy of gas around ETA Carinae. ApJ. **305**, 867–876 (1986)
8. R.J. Dufour, Shell nebulae around luminous evolved stars. RevMexAA. **18**, 87–98 (1989)
9. R.J. Dufour, T.W. Glover, J.J. Hester, D.G. Currie, D. van Orsow, D.K. Walter, New HST results on the outer nebula of eta Carinae. ASP Conf. Series **120**, 255–259 (1997)
10. W.J. Duschl, K.-H. Hofmann, F. Rigaut, G. Weigelt, Morphology and kinematics of Eta Carinae. RevMexAA SdC. **2**, 17–22 (1995)
11. V.V. Dwarkadas, B. Balick, On the formation of the Homunculus nebula around eta Carinae. AJ. **116**, 829–839 (1998)
12. D. Ebbets, E. Malumuth, K. Davidson, R. White, N. Walborn, Proper motions of the N condensations of Eta Carinae. ASP Conf. Series **35**, 263–265 (1993)
13. D. Ebbets, H. Garner, R. White, K. Davidson, E. Malumuth, N. Walborn, HST images of Eta Carinae, in *Circumstellar Media in Late Stages of Stellar Evolution*, ed. by R.E.S. Clegg, I.R. Stevens, W.P.S. Meikle. 34th Herstmonceux Conference (Cambridge University Press, Cambridge, 1994), pp. 95–97
14. A. Frank, B. Balick, K. Davidson, The homunculus of Eta Carinae: an interacting stellar winds paradigm. ApJ. **441**, L77–80 (1995)
15. A. Frank, D. Ryu, K. Davidson, Where is the doughnut? Luminous blue variable bubbles and aspherical fast winds. ApJ. **500**, 291–301 (1998)
16. G. García-Segura, M.-M. Mac Low, N. Langer, The dynamical evolution of circumstellar gas around massive stars. I. The impact of the time sequence Ostar $\rightarrow$ LBV $\rightarrow$ WR star. A&A. **305**, 229–244 (1996)
17. E. Gaviola, Seventy-fifth anniversary of the Cordoba Observatory. Nature **158**, 402–403 (1946)
18. E. Gaviola, Eta Carinae. I. The nebulosity. ApJ. **111**, 408–413 (1950)
19. E. Gaviola, Eta Carinae. II. The spectrum. ApJ. **118**, 234–251 (1953)
20. R.F. González, E.M. de Gouveia Dal Pino, A.C. Raga, P.F. Velázquez, Gasdynamical simulations of the large and little Homunculus nebulae of η Carinae. ApJ. **600**, 59–62 (2004)
21. R.F. González, E.M. de Gouveia Dal Pino, A.C. Raga, P.F. Velázquez, Numerical modeling of η Carinae bipolar outflows. ApJ. **616**, 976–987 (2004)
22. L. Gratton, The problem of Eta Carinae, in *Star Evolution*, ed. by L. Gratton (Academic Press, New York, 1963), pp. 297–311
23. J.J. Hester, R.M. Light, J.A. Westphal, D.G. Currie, E.J. Groth, J.A. Holtzmann, T.R. Lauer, E.J. O'Neil Jr., Hubble space telescope imaging of Eta Carinae. AJ. **102**, 654–657 (1991)
24. D.J. Hillier, D.A. Allen, A spectroscopic investigation of Eta Carinae and the Homunculus nebula. I – Overview of the spectra. A&A. **262**, 153–170 (1992)
25. R.M. Humphreys, K. Davidson, N. Smith, Eta Carinae's second eruption and the light curves of the eta Carinae variables. PASP. **111**, 1124–1131 (1999)
26. N. Langer, G. García-Segura, M.-M. Mac Low, Giant outbursts of luminous blue variables and the formation of the Homunculus nebula around eta Carinae. ApJ. **520**, L49–53 (1999)
27. D. Malin, The splendor of Eta Carinae. Sky & Telescope **73**, 14–18 (1987)
28. J. Meaburn, The eruption history of P Cygni, in *P Cygni 2000: 400 Years of Progress*, ed. by M. de Groot, C. Sterken. ASP Conference Series, vol. 233 (Astronomical Society of the Pacific, San Francisco, 2001), pp. 253–260
29. J. Meaburn, The high-speed circumstellar shells, lobes, funnels, knots and fingers of P Cygni and Eta Carinae, in *Eta Carinae and Other Mysterious Stars: The Hidden Opportunities of Emission Spectroscopy*, ed. by T. Gull, S. Johannson, K. Davidson. ASP Conference Series, vol. 242 (Astronomical Society of the Pacific, San Francisco, 2001), pp. 141–150
30. J. Meaburn, R.D. Wolstencroft, J.R. Walsh, Echelle and spectropolarimetric observations of the Eta Carinae nebulosity. A&A. **181**, 333–342 (1987)

31. J. Meaburn, J.R. Walsh, R.D. Wolstencroft, The outflowing dust around Eta Carinae. A&A. **268**, 283–293 (1993)
32. J. Meaburn, G. Gehring, J.R. Walsh, J.W. Palmer, J.A. López, M. Bryce, A.C. Raga, An episodic jet from Eta-Carinae. A&A. **276**, L21–24 (1993)
33. J. Meaburn, P. Boumis, J.R. Walsh, W. Steffen, A.J. Holloway, R.J.R. Williams, M. Bryce, Highly supersonic motions within the outer features of the eta Carinae nebulosity. MNRAS. **282**, 1313–1320 (1996)
34. J. Meaburn, J. A. López, M. J. Barlow, J. E. Drew, The expansion of the outer circumstellar shell of P Cygni. MNRAS. **283**, L69–L71 (1996)
35. J. Meaburn, J.A. López, J.A. O'Connor, The kinematical association of a giant lobe with the luminous blue variable star P Cygni. ApJ. **516**, L29–L32 (1999)
36. J. Meaburn, J. A. O'Connor, J. A. López, M. Bryce, M.P. Redman, A. Noriega-Crespo, The ejecta from the luminous blue variable star P Cygni. MNRAS. **318**, 561–572 (2000)
37. J. Meaburn, P. Boumis, M.P. Redman, J. A. López, F. Mavromatakis, Candidates for giant lobes projecting from the LBV stars P Cygni and R 143. A&A. **422**, 603–608 (2004)
38. G. Meynet, A. Maeder, Stellar evolution with rotation. XI. Wolf-Rayet star populations at different metallicities. A&A. **429**, 581–598 (2005)
39. J.A. Morse, K. Davidson, J. Bally, D. Ebbets, B. Balick, A. Frank, Hubble space telescope Wide Field Planetary Camera 2 observations of eta Carinae. AJ. **116**, 2443–2461 (1998)
40. J.A. Morse, J.R. Kellogg, J. Bally, K. Davidson, B. Balick, D. Ebbets, Hubble space telescope proper-motion measurements of the η Carinae nebula. ApJ. **548**, L207–211 (2001)
41. A. Nota, L. Smith, A. Pasquali, M. Clampin, M. Stroud, The bipolar HR Carinae nebula: dynamics and chemical abundances. ApJ. **486**, 338–354 (1997)
42. A.Y. Poludnenko, A. Frank, S. Mitran, Strings in the η Carinae nebula: hypersonic radiative cosmic bullets. ApJ. **613**, 387–392 (2004)
43. M.P. Redman, J. Meaburn, A.J. Holloway, The origin of the strings in the outer regions of η Carinae. MNRAS. **332**, 754–758 (2002)
44. A.E. Ringuelet, Note on the nebulosity around Eta Carina. Zeitschrift für Astrophysik **46**, 276–278 (1958)
45. F.D. Seward, W.R. Forman, R. Giacconi, R.E. Griffiths, F.R. Harnden Jr., C. Jones, J.P. Pye, X-rays from Eta Carinae and the surrounding nebula. ApJ. **234**, L55–58 (1979)
46. F.D. Seward, Y.M. Butt, M. Karovska, A. Prestwich, E.M. Schlegel, M. Corcoran, Early chandra X-ray observations of η Carinae. ApJ. **553**, 832–836 (2001)
47. N. Smith, A blast wave from the 1843 eruption of η Carinae. Nature **455**, 201–203 (2008)
48. N. Smith, R.D. Gehrz, Proper motions in the ejecta of eta Carinae with a 50 year baseline. AJ. **116**, 823–828 (1998)
49. N. Smith, P. Hartigan, Infrared [Fe II] emission from P Cygni's nebula: atomic data, mass, kinematics, and the 1600 AD outburst. ApJ. **638**, 1045–1055 (2006)
50. N. Smith, J.A. Morse, Nitrogen and oxygen abundance variations in the outer ejecta of η Carinae: evidence for recent chemical enrichment. ApJ. **605**, 854–853 (2004)
51. N. Soker, A model for the strings of η Carinae. A&A. **377**, 672–676 (2001)
52. A.D. Thackeray, Nebulosity surrounding eta Carinae. The Observatory **69**, 31–33 (1949)
53. A.D. Thackeray, Some southern stars involved in nebulosity. MNRAS. **110**, 524–530 (1950)
54. W.H. van den Bos, Measures of double stars – 23rd Series Union Obs. Circ. **100**, 522 (1938)
55. W.H. van den Bos, Union Obs. Circ. **106**, 207 (1948)
56. N.R. Walborn, The complex outer shell of Eta Carinae. ApJ. **204**, L17–19 (1976)
57. N.R. Walborn, B.M. Blanco, Third-epoch proper motions in the outer shell of ETA Carinae. PASP. **100**, 797–800 (1988)
58. N.R. Walborn, B.M. Blanco, A.D. Thackeray, Proper motions in the outer shell of Eta Carinae. ApJ. **219**, 498–503 (1978)
59. N.R. Walborn, I.N. Evans, E.L. Fitzpatrick, M.M. Phillips, New observations of LBV environments, in *Wolf-Rayet Stars and Interrelations with Other Massive Stars in Galaxies*, ed. by K.A. van der Hucht, B. Hidayat. IAU Symposium, vol. 143 (Kluwer, Holland, 1991), pp. 505–512

60. K. Weis, LBV nebulae: the mass lost from the most massive stars. Rev. Mod. Astron. **14**, 261–281 (2001)
61. K. Weis, The outer ejecta of Eta Carinae, in *Eta Carinae and Other Mysterious Stars*, ed. by T. Gull, S. Johansson, K Davidson. ASP Conference Series, vol. 242 (Astronomical Society of the Pacific, San Francisco, 2001), pp. 129–140
62. K. Weis, On the structure and kinematics of nebulae around LBVs and LBV candidates in the LMC. A&A. **408**, 205–229 (2003)
63. K. Weis, The outer ejecta of η Carinae, in *The Fate of the Most Massive Stars*, ed. by R. Humphreys, K. Stanek. ASP Conference Series, vol. 332 (Astronomical Society of the Pacific, San Francisco, 2005), pp. 271–276
64. K. Weis, Wind relics: clumps, inhomogeneities and outflows in LBV nebulae, in *Clumping in Hot-Star Winds*, ed. by W.-R. Hamann, A. Feldmeier, L. Oskinova (Universitäts Verlag, Potsdam, 2008), pp. 183–186. http://nbn-resolving.de/urn:nbn:de:kobv:517-opus-13981. ISBN 978-3-940793-33-1
65. K. Weis, The AG Carinae nebula – bigger than ever? in *Mass Loss from Stars and the Evolution of Stellar Clusters*, ed. by A. Koter, L. Smith, R. Waters. ASP Conference Series, vol. 388 (Astronomical Society of the Pacific, San Francisco, 2008), pp. 231–232
66. K. Weis, Gone with the wind: nebulae around LBVs. Bulletin de la Societe Royale des Sciences de Liege **80**. 440–444 (2011)
67. K. Weis, Nebulae around luminous blue variables - large bipolar variety, in *Active OB Stars: Structure, Evolution, Mass Loss, and Critical Limits*, ed. by C. Neiner, G. Wade, G. Meynet, G. Peters. IAU Symposium, vol. 272 (Cambridge University Press, Cambridge, 2011), pp. 72–377
68. K. Weis, W.J. Duschl, D.J. Bomans, Y.-H. Chu, M.D. Joner, The bipolar structure of the LBV nebula around HR Carinae. A&A. **320**, 568–574 (1997)
69. K. Weis, W.J. Duschl, Y.-H. Chu, The nature of strings in the nebula around η Carinae. A&A. **349**, 467–474 (1999)
70. K. Weis, W.J. Duschl, D.J. Bomans, High velocity structures in and the X-ray emission from the LBV nebula around η Carinae. A&A. **367**, 566–576 (2001)
71. K. Weis, W. J. Duschl, D. Bomans, An outflow from the nebula around the LBV candidate S 119. A&A. **398**, 1041–1048 (2003)
72. K. Weis, M.F. Corcoran, D.J. Bomans, K. Davidson, A spectral and spatial analysis of η Carinae's diffuse X-ray emission using CHANDRA. A&A. **415**, 595–607 (200

Chapter 9
X-ray Variability and the Secondary Star

M.F. Corcoran and K. Ishibashi

Abstract We discuss the history of X-ray observations of the η Car system, concentrating on the periodic variability discovered in the 1990s. We discuss the interpretation of these variations, concentrating on a model of the system as a "colliding-wind" binary. This interpretation allows the physical and orbital parameters of η Car and its companion star to be constrained.

9.1 The 5.5 Year Period

Periodic variability is a key diagnostic in astrophysics. Astronomers long sought periodicities in η Car's behavior ever since its giant eruption [27]. Before 1870 Rudolf Wolf and Elias Loomis suggested photometric cycles of 46 and 67 years, respectively [45, 60, 99]; a century later Payne-Gaposchkin proposed a cycle of 15–16 years [73], and Feinstein and Marraco suspected a 3-year periodicity [35]. The complex behavior of the observed variations from η Car in broad-band photometry and in spectrographic observations made determination of the true period difficult.

The discovery of η Car's first confirmed period depended on observations spanning 20 years or more. This began with the recognition of a so-called "spectroscopic event" by Zanella, Wolf and Stahl in 1981 [95, 100]. These "events"

M.F. Corcoran (✉)
Center for Research and Exploration in Space Science & Technology, Code 662, Goddard Space Flight Center, Greenbelt, MD, 20771 USA

Universities Space Research Association, 10211 Wincopin Circle, Suite 500, Columbia, MD, 21044 USA
e-mail: Michael.F.Corcoran@nasa.gov

K. Ishibashi
Nagoya University, Department of Physics, 1 Furo-cho Chikusa-ku, Nagoya, Aichi, Japan 464-8602
e-mail: bish@u.phys.nagoya-u.ac.jp

K. Davidson and R.M. Humphreys (eds.), *Eta Carinae and the Supernova Impostors*, Astrophysics and Space Science Library 384, DOI 10.1007/978-1-4614-2275-4_9,

represent a dramatic and complex change in η Car's visible-band spectrum in which high-excitation He I and [Ne III] emission lines abruptly fade, then recover. Similar changes happened in 1948 and 1965, which suggested to them a recurrence timescale of 17 years [100]. They were on the right track, but did not know that other such events had occurred (as we deduce today) in 1953, 1959, 1970, and 1975. Near-infrared photometry from South Africa obtained by Patricia Whitelock and her collaborators [96] over 156 nights in 1975–1994 showed a significant 5–6 year cycle, but this cycle was confused with other near-infrared variations on other timescales. Meanwhile new spectroscopic events occurred in 1986 and in 1992, though the former was not widely reported.

The breakthrough was provided by a set of 27 measurements of the He I 10830Å emission line obtained by Augusto Damineli at Brazil's National Astrophysical Laboratory. These observations, augmented by four additional measures from other observers from 1981 to 1995, showed a clear, rapid weakening of this He I line every 5.5 years, lasting only for a few weeks (a duty cycle of $<5\%$). Damineli showed [16] the fading of the He I 10830 line occurred at nearly the same time as brightening of the system in the near IR, and that the 5.5 year period recovered not only the "spectroscopic event" seen by Zanella, Wolf and Stahl in 1981, but other "spectroscopic events" observed in the previous 50 years. Damineli successfully predicted the occurrence of the next minimum at the end of 1997.

Thanks to multi-wavelength observations of η Car during the subsequent minima in December 1997 and June 2003, we know that the fading occurs throughout nearly the entire electromagnetic spectrum from the radio through the IR, optical, UV and X-ray region (the situation in the Gamma-ray band is less clear). The period is almost exactly the same in every wavelength region [21]. Detailed monitoring observations since 1997 have refined the period to 2,024 days (5.54 year) with an uncertainty that appears to be no more than a week and may be as small as ± 2 days [20].

What regulates the cycle? Following Zanella et al., Damineli initially supposed that each spectroscopic event is a single-star "shell ejection", a temporary large increase in the stellar mass-loss rate, reminiscent of phenomena seen in S Dor variables and Luminous Blue Variable stars (see, e.g., [46]). But η Car's variability cycle is too long to represent a pulsation or rotation period, since those periods should be only days or weeks. A thermal/rotational recovery cycle could have a timescale of several years [26], but this seems unlikely to produce so regular a period. The He I 10830 variation seems strictly periodic and strongly suggests a simple clocklike explanation: a companion star in orbit around η Car. If the 5.54 year period is assumed to be the orbital period of the system, the high total mass (probably 150–200 $M_{\odot}$) implies a semi-major axis close to 17 AU. The short duty cycle implies that the phenomenon occurs near periastron passage in a highly eccentric orbit. Proposed orbits [17, 18, 22, 23, 28, 29, 50] are mostly comparable in size and shape to the orbit of Halley's Comet.

But the UV-to-IR emission regions include both the wind of η Car and the detritus of the eruptions from the nineteenth century and are spatially complex. Emission from one region may be absorbed by cold gas in another. Dense clouds may shadow one parcel of gas from the photospheric radiation but not its neighbor. More distant gas reacts more slowly to changes in the inner system because of light-travel time, which may be days or even weeks. These complications make interpreting the spectrum at these wavelengths difficult.

A binary system composed of luminous stars must produce significant amounts of X-ray emission resulting from the high speed collision of the massive stellar winds [5,76] which constantly flow off the stellar surfaces. This wind-wind collision will produce a "bow shock" where the wind kinetic energy is converted to thermal energy which radiates mostly in the X-ray band (as discussed below). Such emission is an important probe of massive binaries since it originates in a localized region (defined by system parameters like the separation between the two stars and the relative mass loss rates and wind speeds), and it penetrates through large columns of intervening cold material. What do X-ray observations tell us about η Car?

9.2 X-ray Emission and Variability

Studies of X-ray emission from η Car go back 30 years or more, nearly to the beginning of X-ray astronomy. The earliest data were obtained by non-imaging detectors with fields of view defined by collimators, flown on rockets or orbiting platforms [36, 39, 42, 69, 78]. They established η Car as a source of X-rays at energies $E > 5$ keV, and detected iron K-line emission at 6.7 keV. They also provided hints of variability [56], though direct comparisons of X-ray flux were hampered by differences in instrumentation (fields of view, sensitivities, etc.).

Today we know that many X-ray sources, both discrete and diffuse, exist near η Car, so it was not until the advent of imaging X-ray telescopes that X-ray emission from η Car could be cleanly disentangled from this contamination. The first X-ray images of η Car and the Carina Nebula were obtained in the early 1980s with the *HEAO-2/Einstein* satellite observatory [6, 77, 79]. Those observations showed that at energies above 3 keV, η Car was the only source visible in the field. They showed that X-ray emission from η Car consists of at least two components: an extended ($\sim 20''$) "soft" ($E < 1$ keV) component (likely produced by the collision of material from the 1840s eruption with interstellar material) surrounding a point-like, "hard" ($E > 3$ keV) heavily absorbed component centered on the optical position of η Car. This "point-like" component has a temperature of $4 - 8 \times 10^7$ K and its X-ray emission is absorbed by intervening material having a column density of $5 - 50 \times 10^{22}$ atoms cm^{-2} [6, 13, 40]. The observed luminosity of the hard, point-like source measured by *Einstein* (adjusted to the current best estimated distance of 2,300 pc) was $\approx 4 \times 10^{33}$ ergs s^{-1} in the 0.2–4 keV band; correcting for absorption, the luminosity of the source is about 2×10^{34} erg s^{-1}. While these observations were key to establishing the overall character of the high-energy X-ray emission from η Car, these relatively short "snapshots" could not really address issues of temporal variability.

Figure 9.1 shows a Hubble Space Telescope image of η Car and the Homunculus nebula superimposed on a more recent X-ray image from the Chandra X-ray

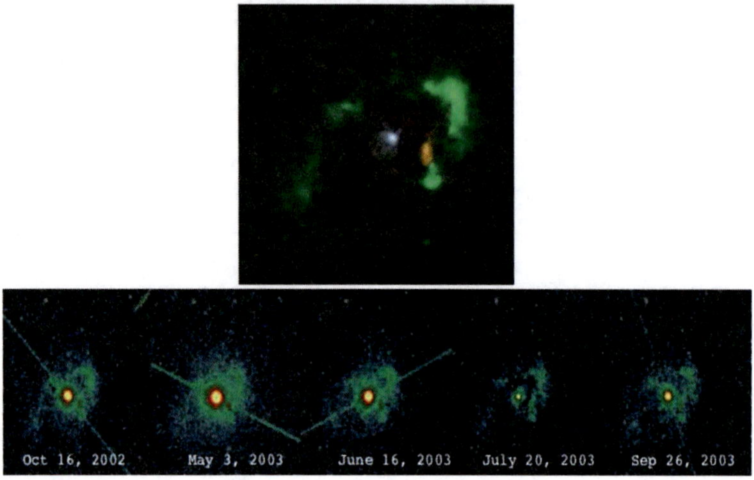

Fig. 9.1 η Car in X-rays. The *upper figure* shows an X-ray image from the Chandra X-ray observatory (in *green*) combined with an optical image from the Hubble Space Telescope (in *red*). The unresolved white point source at the center of the image, at the position of the optical star, is the source of hard X-ray emissions. The *lower figure* shows five Chandra ACIS-I images of η Car, highlighting the variation of the central pointlike source. The streaks at various angles to the central source are detector artifacts

Observatory. This image clearly shows the extended "shell" of X-ray emission first seen by *Einstein* surrounding the Homunculus along with the point-like X-ray emission coincident with the optical position of the star. The point-like, hard emission was initially interpreted [6, 77] as the collision of a fast wind from η Car with a circumstellar "shell" about 1,000 AU from the star. As we shall see, the temperature of the emission requires a wind velocity of 2,000–3,000 km s^{-1} or more if it's solely produced by conversion of the wind kinetic energy to heat. This is about four times faster than η Car's bulk wind seen in visual-wavelength spectra – a highly signficant point in retrospect, though it was not emphasized at the time of the *Einstein* observations. The X-ray absorbing column depth was thought to be consistent with absorption in a three solar mass cold shell, roughly agreeing with what was known of the mass of the Homunculus at that time.

9.2.1 Discovery of X-ray Variability

X-ray variability of η Car was first detected by the *ROSAT* X-ray observatory in short observations obtained in 1992–1993 analyzed by G. L. Rawley. She showed that the central source was not visible in June and August of 1992 at energies above 2 keV, but was clearly detected in January 1993 [8]. The flux above 2 keV varied by a factor of 2 in a 4-month interval, requiring a tripling of the mass-loss rate if the

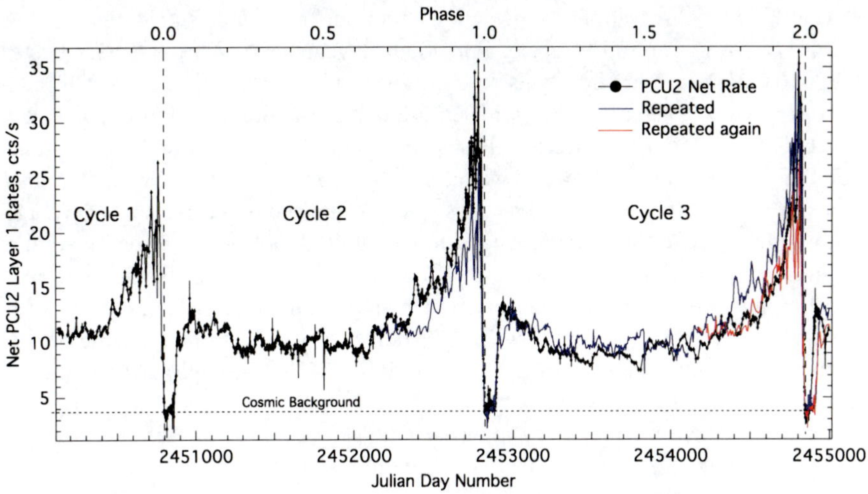

Fig. 9.2 The *RXTE* X-ray "lightcurve" of η Car at energies above 2 keV and below 10 keV, as measured by the PCA on *RXTE*. Three X-ray minima can be clearly seen near phases 0.0, 1.0 and 2.0. The "flat" bottom of the X-ray minimum seen by *RXTE* is produced when the X-ray flux from η Car falls below the cosmic background level (shown by the *horizontal dashed line*). The *blue* and *red lines* show the *RXTE* measures offset by 2,024 and 4,048 days, respectively (Adapted from Corcoran et al. [15])

X-rays were produced by collision of the wind with a circumstellar shell. However, variations in the temperature of the emission or changes in the intervening absorbing medium could not be ruled out.

Luck often plays a role in astronomical discovery; the mid-1992 *ROSAT* observations just happened to coincide with one of the spectroscopic events, and to coincide with the first observations of the star [33] by the radio telescopes of the Australia Telescope National Facility. At the same time η Car faded in the hard X-ray band, it also faded in the radio [33], but brightened in the near-IR [16, 96]. Thus, when the 5.5-year period was identified by Damineli shortly thereafter, an obvious question was whether X-ray variability was another aspect of the same cycle.

9.2.2 The X-ray Variability Cycle

Monitoring observations with the Proportional Counter Array (PCA) on the *Rossi X-ray Timing Explorer* (*RXTE*) [4] provided the next X-ray breakthrough. Unlike other X-ray observatories, *RXTE* can view η Car at any time of the year making it a near-ideal instrument to study the variability of the 2–10 keV flux. *RXTE* observations of η Car began in February 1996 and have continued through 2009 [7, 53]. Typically *RXTE* obtained 2–4 observations of η Car per month, with daily observations during important intervals. The *RXTE* X-ray count rates vs. time are shown in Fig. 9.2.

The *RXTE* observations show the important temporal signatures of the 2,024-day X-ray variability cycle. For about 4 years out of the 5.5 year cycle, the 2–10 keV X-ray luminosity is in a "quiescent" state, relatively constant at a level of $\sim 4 \times 10^{34}$ ergs s^{-1}, but with erratic small-amplitude variations. About a year before the fading of the He I 10830 line, both the intensity and the irregular variability progressively grow, and we see conspicuous X-ray "flares" with durations of weeks or days, typically 2 or 3 months apart. At the end of this phase, the average X-ray luminosity in the 2–10 keV band has roughly doubled to reach a maximum luminosity of $\approx 10^{35}$ ergs s^{-1}, while the interval between flare peaks drops to less than 1 month. About 40 days after reaching its maximum, the emission swiftly (with a few weeks) falls to a level below *RXTE*'s detectability. This rapid decline occurs along with similarly rapid fadings of the He I 10830 line strength [21] and the IR brightness [97]. The X-ray emission usually remains in this "low state" for about 3 months. At this time the X-ray intensity is lower than or near the cosmic X-ray background in the $\sim 1°$ PCA field of view, though η Car is still detectable throughout the minimum by other X-ray observatories with sufficient spatial sensitivity [14,40] at luminosities ranging from $\sim 6 \times 10^{32}$ ergs s^{-1} to 5×10^{33} ergs s^{-1} near the end of the minimum [40]. At the end of the low state the X-ray intensity rises rapidly to the "quiescent" level.

X-ray spectra of η Car show that the observable emission associated with the star is mostly in the 2–10 keV band, corresponding to a temperature range $T = E/k$ of roughly 20–110 million K. The PCA spectra show little flux above 10 keV at any phase. Near X-ray minimum, the observed emission declines more at lower energies showing the build up of circumstellar absorption, and the spectrum remains quite hard for an extended period even after the minimum ends.

9.3 Colliding Wind Emission as a Diagnostic of Massive Binary Systems

If the pan-chromatic periodicity is produced by the orbit of a companion star around η Car (as seems likely) then this offers the unique opportunity to dynamically measure the mass of a star near the Eddington Limit using conventional binary-star radial velocity analysis. Unfortunately attempts to solve the radial velocity curves of observable emission lines in the UV-to-IR [17, 19, 22, 30, 68] have not given trustworthy orbital elements, mostly because these lines are formed in a spatially complex mass flow and consequently are broad, asymmetric, variable, and highly model-dependent. For this reason, the X-ray variability, produced by the orbit-driven variation in the wind-wind collision zone, provides a very useful (though indirect) diagnostic of the orbit. We briefly review the physics of X-ray generation in colliding wind binaries to see why this is so.

Colliding wind binaries (CWBs) produce thermal X-rays in the strong shock where the stellar winds collide. The relevant geometry is crudely sketched in

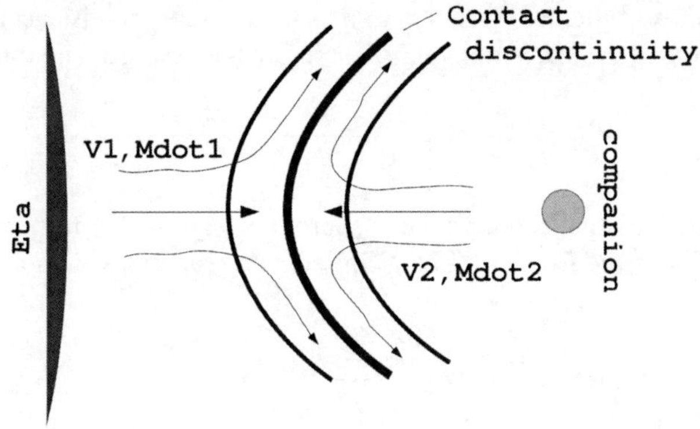

Fig. 9.3 A sketch of the η Car colliding wind binary system. Winds from two stars flowing radially from the stellar surface collide and form shocks separated by a contact discontinuity. In this simple model, the shocked gas forms two curved shock fronts between which the heated plasma flows out along the contact discontinuity. Temperatures on either side of the contact discontinuity are determined by the velocities of the stellar winds just before the collision

Fig. 9.3. The shock is concave toward the star with the "weaker" wind, i.e. the star with the lower wind momentum flux. The vertex of this shock, where the winds collide head-on close to the line of centers, is the "stagnation point" where the winds are fully decelerated. In the idealized case a contact discontinuity separates the shocked stellar winds from the primary star on one side and secondary star on the other. The temperature of the shocked gas on each side of the contact discontinuity is to first order determined by the pre-shock velocity of the stellar wind from the star on that side. This is important for η Car since the wind speed of η Car itself is only $\sim$500 km s^{-1}, which would produce only soft X-rays which would be unobservable due to heavy absorption by the cold gas in the Homunculus. The fact that hard X-rays are observed from η Car indicates the presence of a fast flow in the system. The main X-ray emission processes are thermal Bremsstrahlung from hot electrons, and inner-shell (K, L, M) transitions in nearly stripped ions. Standard examples of CWBs are HD 193793/WR 140, γ^2 Velorum, and HD 5980 in the SMC [67, 75, 98]. These massive binaries have observed X-ray luminosities in the range 10^{32}–10^{36} ergs s^{-1} $\approx$0.02–300 $L_\odot$, varying on orbital timescales. Though large, these luminosities are of course only a millionth or so of the immense stellar bolometric luminosity.

Theoretical descriptions of X-ray emission from wind-wind collisions in massive binaries go back to the 1970s, and the details can be quite complex [5, 76]. The essential behavior, however, can be roughly described by the following order-of-magnitude analysis by K. Davidson (private communication, 1999). Shocked gas cools by expansion and by emission of radiation as it flows outward away from the stagnation point. Characteristic flow speeds in that region should be comparable to the isothermal sound speed $w = (P/\rho)^{1/2} \propto T^{1/2}$. Based on Fig. 9.3, we also expect the lateral radius of the hottest region to be similar to d_2, the distance between the secondary star and the stagnation point. Therefore the *expansion* cooling time is

$$t_{\rm exp} \approx \frac{d_2}{w} \approx \frac{d_2}{(10^4\ {\rm cm\ s^{-1}\ K^{-1/2}})\, T^{1/2}}\,. \tag{9.1}$$

Meanwhile the radiative cooling time is approximately the thermal energy per unit volume divided by the Bremsstrahlung emission rate per unit volume:

$$t_{\rm rad} \approx \frac{3n_e kT}{AT^{1/2}n_e^2} \approx \frac{CT^{1/2}}{n_e}\,, \tag{9.2}$$

where n_e is electron density, $A \approx 10^{-27}$ erg cm^3 s^{-1} K$^{-1/2}$, and $C \approx 10^{11.5}$ cm^{-3} s K$^{-1/2}$. Thus

$$\frac{t_{\rm exp}}{t_{\rm rad}} \approx (10^{-15.5}\ {\rm cm^2\ K})\,\frac{n_e d_2}{T}\,. \tag{9.3}$$

We can relate T and n_e in the hot shocked gas to the speed and mass loss rate in the secondary wind. According to the Rankine-Hugoniot adiabatic jump conditions across a shock [58, 80], the maximum speed of sound and maximum temperature at the shock front are

$$w_{\rm max} = (3/16)^{1/2}\, v \quad \text{and} \quad T_{\rm max} = \frac{\mu m_H}{k}\, w_{\rm max}^2\,, \tag{9.4}$$

where v is the pre-shock wind velocity, m_H is the proton mass and μ is between 0.6 and 1.0 for the expected range of chemical composition. If the relevant average T in the above equations is half the maximum temperature, we find

$$T \approx (10^7\ {\rm K})\left(\frac{v}{1{,}000\ {\rm km\ s^{-1}}}\right)^2. \tag{9.5}$$

As noted above, this is the reason why η Car's primary wind does not produce observable X-rays: its typical speeds (which, it turns out, are latitudinally dependent [81]) of 200–1,000 km s^{-1} lead to average temperatures of only a few million K, and the resulting soft X-rays ($E < 0.7$ keV) are absorbed by the large column of gas ($N_H \sim 5 \times 10^{22}$ cm^{-2}) in the Homunculus nebula. The observed 2–10 keV X-ray spectrum indicates $T > 4 \times 10^7$ K, which requires $v > 2{,}000$ km s^{-1}. So far as observable X-rays are concerned, η Car's dense primary wind merely provides a wall for the secondary wind to collide into.

The jump conditions across the shock require that n_e in the shocked gas is about four times as large as the electron density of the pre-shock wind. The pre-shock density is related to the secondary star's mass-loss rate by mass conservation: $\dot{M} = 4\pi r^2 \rho v$. If the distance r from the secondary star to the shock is roughly equal to d_2, and the pre-shock velocity of the secondary's wind is v_2, then the density in the heated gas is

$$n_e \approx \frac{(10^{23.2}\ \mathrm{g}^{-1})\,\dot{M}_2}{{d_2}^2\, v_2}\,. \tag{9.6}$$

Substituting for n_e and v_2 in the earlier expression for $t_{\rm exp}/t_{\rm rad}$, we find

$$\frac{t_{\rm exp}}{t_{\rm rad}} \sim 10^7 \left(\frac{\dot{M}_2}{M_\odot\ \mathrm{year}^{-1}}\right)\left(\frac{1\ \mathrm{AU}}{d_2}\right)\left(\frac{10^6\ \mathrm{K}}{T}\right)^{3/2}. \tag{9.7}$$

Using $\dot{M}_2 \sim 10^{-5.5} M_\odot\ \mathrm{year}^{-1}$ [74], $d_2 \sim 1$ AU near periastron, and $T \sim 5\times10^7$ K (consistent with the observed 2–10 keV continuum), yields $t_{\rm exp}/t_{\rm rad} \sim 0.1$. This means that expansion dominates the total cooling rate but not by an enormous factor.

More formal and detailed analyses of colliding winds have been calculated [74, 88, 92], but the essential physics is captured by the simple analysis above. Numerical gas-dynamic simulations confirm our simplified reasoning in an order-of-magnitude sense, though incipient instabilities cause the shocked region to become spatially complex [74]. These detailed analyses suggest that the observed X-ray spectrum of η Car requires $v_2 \approx 3{,}000\ \mathrm{km\,s^{-1}}$ because of rapid cooling of the hot gas [74].

If $t_{\rm exp}/t_{\rm rad} \gtrsim 1$ radiative cooling dominates, and the shock structure becomes unstable [88]. For η Car this is expected to occur near periastron in the shocked primary-wind [74] on the left side of Fig. 9.3. In some proposed scenarios it occurs on both sides of the contact discontinuity, so that both shocks collapse near periastron [25, 61]. In this connection it is worth noting that here we have ignored likely inhomogeneities, which may exist in the wind or may be caused by shock instabilities. If the density is inhomogeneous, then the meaning of n_e in the expression for $t_{\rm exp}/t_{\rm rad}$ becomes less obvious. On average, inhomogeneities tend to increase the fraction of input energy that is radiated.

If the ratio $\zeta = t_{\rm exp}/t_{\rm rad}$ is small, then it approximately represents the efficiency for conversion of the secondary wind's kinetic energy to radiation. We noted above that the *observable* X-rays ($E > 1$ keV) in a CWB model for η Car are powered by the companion star's fast wind, whose energy outflow rate is $\dot{M}_2 v_2^2/2$. Consequently the observed X-ray luminosity should obey these approximate proportionalities:

$$L_x \propto \zeta \times \dot{M}_2 {v_2}^2 \propto \frac{\dot{M}_2}{d_2 T^{3/2}} \times \dot{M}_2 {v_2}^2 \propto \frac{\dot{M}_2^2}{d_2\, v_2}\,. \tag{9.8}$$

assuming that a constant fraction of the wind encounters the shock.

A key point is the inverse dependence of L_x on d_2; if the companion star follows an eccentric orbit, then the varying distance modulates the X-ray luminosity such that the intrinsic X-ray luminosity emitted by the shock is highest near periastron and lowest at apastron. Therefore the modulated intrinsic X-ray luminosity can in principle be used reconstruct the orbit by determining the stellar separation as a function of time. However it is often non-trivial to derive the intrinsic X-ray luminosity from the observed luminosity, which includes the effects of variable absorption from one (or both) of the highly structured stellar winds. If we exclude the complicated interval near periastron, the apparent flux variation shown in Fig. 9.2 is dominated by the combination of the $1/d_2$ variation of the intrinsic flux and the phase-dependent absorption along our line of sight. These two effects depend, in turn, on three classical binary-star parameters: (1) the inclination i of the orbit plane, usually assumed close to 45°, so that the orbital angular momentum is co-linear with the bipolar axis of symmetry of the Homunculus nebula [31]; (2) the orbit eccentricity e, which determines $r(t)$; (3) the azimuthal orientation of the orbit, specified by the longitude of periastron ω, customarily defined so that at periastron the secondary star is moving away from us at $\omega = 180°$ and is on the far side of the primary at $\omega = 270°$. As noted above, the intrinsic X-ray temperature can be used to constrain the speed of one (or both) of the stellar winds. The intrinsic X-ray temperature should vary little, because it depends chiefly on T and therefore on the wind speed v_2, which we expect to be more or less constant.

9.4 Modeling the *RXTE* X-ray Lightcurve of η Car

The first attempts [11, 52, 53, 86] to explain the *RXTE* X-ray lightcurve of η Car with colliding wind models were reasonably encouraging. Simple analytic models could describe the overall variation in X-ray flux and in N_H and the observed constancy of temperature T with plausible assumptions about the stellar winds, component masses and orbital inclination. The abrupt change from X-ray maximum to minimum is suggestive of an eclipse of the X-ray source region by the primary star's wind, so these first attempts assumed that the X-ray minimum was due entirely to an increase in absorbing column N_H, using simple, one-dimensional, spherically symmetric wind models. Since the duty cycle of the X-ray "eclipse" is so small, it is very likely that the X-ray minimum occurs near periastron, when the stellar separations and orbital geometry are changing most rapidly and column densities should naturally be large as the X-ray source becomes surrounded by the thick inner primary wind. These initial attempts required a high eccentricity ($e \approx 0.9$, similar to the eccentricity of WR 140) and were consistent with an orientation in which the companion star is in front of the primary near apastron and behind at periastron. However, because the orbital geometry changes so rapidly around periastron, the duration of the X-ray minimum in these simple "eclipse" models was always far shorter than the observed duration. There are many modifications possible to correct this, among them:

- The parameters i and ω can be altered [23, 25, 50, 70];
- Tidal and other effects may suddenly increase the primary wind's density near periastron [12, 22, 23, 61, 81];
- The large-scale structure of the primary wind is not spherical [81];

- The orientation of the shock structure may be significantly distorted by orbital motion [70, 74];
- Radiative cooling may destabilize the shock near periastron, possibly causing a general disruption or collapse of the X-ray source region [25, 61];
- And other effects [37, 82, 83, 87], including radiative inhibition of the winds near periastron [72].

Most of these possibilities are not mutually exclusive; some of them appear quite likely, several may occur together, and some can reinforce each other (Fig. 9.4).

Throughout the cycle the observed X-ray flux depends critically on absorption in the intervening part of the dense primary wind; N_H obviously varies with phase in the orbit. The wind geometry is complex, since the secondary wind produces a cavity on one side of the primary wind and orbital motion affects the cavity's orientation and shape. Analysis of the X-ray spectrum near the mid-point of the cycle [74] (i.e. near apastron in the orbit) showed that the velocity of the companion's wind is $\approx$3,000 km s^{-1}, while its wind mass-loss rate is $\approx 10^{-5}$ $M_\odot$ year^{-1}. Recent efforts to model η Car's colliding winds with gas-dynamic simulations [70, 72, 74] attempt to account for variations in the large-scale wind geometry. Dynamical wind models [71] and smoothed-particle simulations [70] show the distortion of the wind by the secondary star's orbital motion. Figure 9.5, for instance, shows a smoothed-particle model of the wind-wind interaction at a number of orbital phases [70]. A post-periastron buildup of material on the forward-facing shock surface causes extra absorption that extends the eclipse. Thus a pure eclipse model with suitable parameters can account for the duration of the X-ray minimum [70, 74], but apparently only if the X-ray emitting region is small. If the X-ray emitting region is extended, the eclipse model has difficulty reproducing the extended duration of the X-ray minimum [72]. This does not disprove the complications and alternative models listed above, however; observations in other wavelength regions, beyond the scope of this review, are harder to explain than the X-ray minimum [61]. In summary, at phases of the cycle *not* near periastron, i.e., not during a spectroscopic event, the models cited above are probably realistic in a semi-quantitative sense. They may be valid near periastron as well, but the complexity of the situation there makes this difficult to establish.

Altogether, then, one can attempt to fit most of the X-ray "light curve" with just two orbital parameters, e and ω, and the mass-loss rate and terminal velocity of the companion star's wind. In 2000 Ishibashi estimated $e \approx 0.8$ and $\omega \approx 200°$ by this method [50]. Implicitly using the same dependences, more elaborate calculations by Okazaki have led to $e \approx 0.9$ and $\omega \approx 243°$ [70]. Thus, if the reasoning is valid, periastron occurs 20– 70° before "superior conjunction" (the point where the secondary star is most nearly on the far side of the primary). Orientations in which the companion star is in front of η Car at periastron have also been suggested [34], but these require either a very large buildup of absorbing material behind the X-ray source or a dramatic change in the shock emissivity near periastron passage.

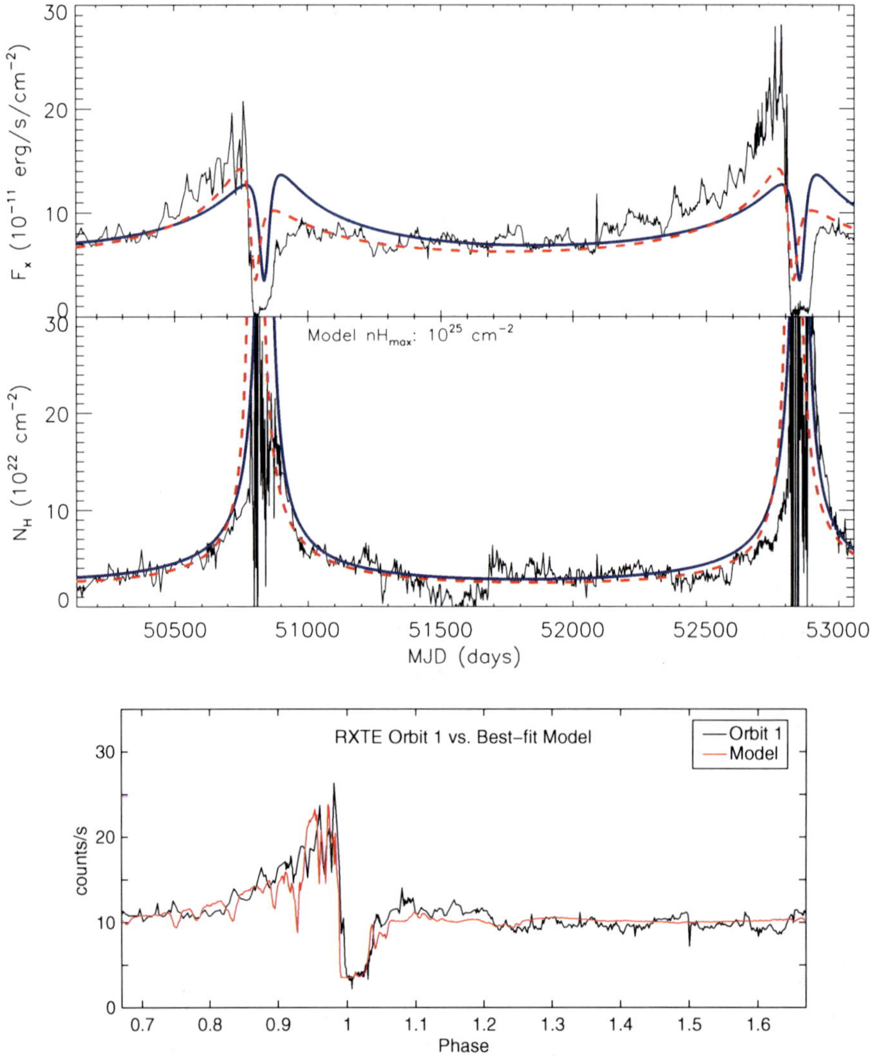

Fig. 9.4 Modeling the *RXTE* X-ray lightcurve of η Car. *Top*: The *solid blue line* shows an estimate [53] of the X-ray light curve using an analytical model [92] with published orbital parameters [17], assuming that the X-rays originate from a point near the apex of the shock cone. The *dotted red line shows* the results of the same analytic model based on more recently published orbital elements [70]. *Middle*: Variations in column depth N_H to the X-ray source along with the model variation [53]. *Line colors* are as above. *Bottom*: X-ray lightcurve (*black*) and model (*red*) based on a 3-D smoothed particle hydrodynamics model [70] which better accounts for the variation in column depth from the wind of the primary and successfully recovers the depth and duration of the X-ray minimum. None of the models account for observed cycle-to-cycle changes (The two upper panels are adapted from Ishibashi et al. [53], and the lower panel is from Okazaki et al. [70])

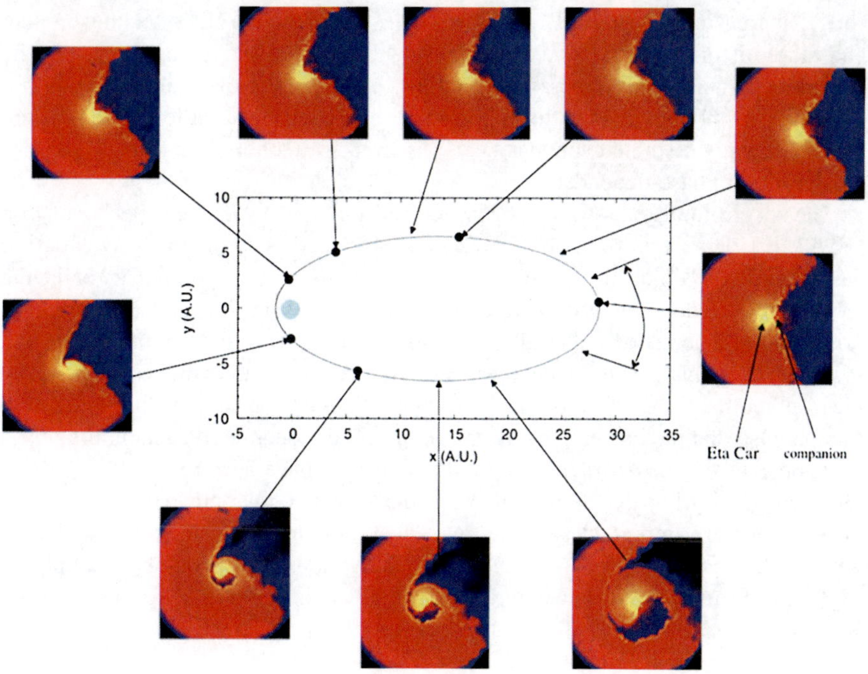

Fig. 9.5 Possible orbit of companion star. The orbit has an eccentricity $e = 0.9$ with semi-major axis $a \approx 15$ AU. Dots along the orbit show *CHANDRA* (CXO) observations, while images show the projected density in the orbital plane of the interacting winds of η Car (high density; *yellow-red*) and the companion (lower density, blue-black) in one type of 3-d model [70]. The curved arrowed line shows the best guess of the orientation of the line of sight projected onto the orbital plane

9.5 Dynamics of the X-ray Emitting Region

Thermal colliding-wind emission is rich in X-ray lines [91] that provide information on flow dynamics (via line widths and velocities), abundances (from line strengths) and temperatures and densities (from line ratios). The strongest lines are K-shell transitions in hydrogen-like and helium-like ions. In the case of η Car's wind, the observable X-ray features are broad lines of Si XIII, Si XIV, S XV, and S XVI, and Fe XXV at higher energies [41]. The implied elemental abundance ratios appear to be near solar [10, 41], which is not surprising since neither η Car nor its companion star is thought to be evolved enough to show enhanced silicon, sulfur or iron abundances at the surface. Unfortunately the carbon, nitrogen, and oxygen abundances, which help trace the evolutionary state of the system, are all at energies below 1 keV and so cannot be studied this way because of circumstellar absorption.

Observations in 2000–2003 with the transmission gratings on the *CHANDRA* X-ray telescope have resolved the broad lines and show that they apparently vary with phase in the 5.5-year cycle [41]. Near times of X-ray minimum these lines shift to increasingly negative Doppler velocities while their profiles become broader and more distorted [3, 41]. Interpretations of these changes are controversial. The observed variations may represent changes in the spatial distribution of the hot shocked gas [41], or perhaps the appearance of an additional outflow in the system [3]. So far only a single cycle has been observed so it's not clear if these changes are secular or phase-dependent.

The spectrum also shows a 6.4 keV K-shell line of less ionized species of iron [10, 13, 40]. This is presumably fluorescence produced as the main X-ray emission passes through "cold" iron rich material, either in the wind of η Car or in the Homunculus. The fluorescence line is broad, suggesting that at least part of it originates in fluoresced material in η Car's wind. During the minimum the line increases in equivalent width, but the measured increase is rather modest, which may mean either that the scattering zone is compact enough to be heavily obscured at that time, or else that the intrinsic X-ray emission actually declines during minimum.

Thomson-reflected X-rays from the Homunculus have been resolved in *CHANDRA* images during the minimum when the direct X-rays are weak [14]. Such observations are important since X-ray reflection could (in principle) provide us with a 3-D view of the cold absorbing wind material as a function of orbital phase. Light travel time delays are a complication, since light travel time for the reflected X-rays may be as much as 80 days towards the back-projected northwest lobe.

9.6 The X-ray Spectrum During the Minimum

The best measures of the X-ray spectrum during the 2003 minimum used the *XMM*-Newton and *CHANDRA* X-ray observatories [14, 40, 41]. *XMM*'s imaging detectors provide sufficient spatial resolution to exclude sources unrelated to η Car, while spectral imaging with *CHANDRA* resolved the source even from contamination by the Homunculus and outer ejecta. These observations [40] confirmed that the apparent X-ray flux decreased even at 10 keV, requiring either a column density $N_H \gtrsim 10^{25}\,\mathrm{cm}^{-2}$ lasting for 2 months or more or else a real decrease in the intrinsic hard X-ray emission. Meanwhile, analyses of the *XMM* spectra [40] suggested that the relevant average column density increased by only a factor of 5–10 during the minimum, reaching a maximum of roughly $4 \times 10^{23}\,\mathrm{cm}^{-2}$. This was not sufficient to explain the observed hard X-ray decrease. Either the analysis underestimated the maximum N_H, or else the intrinsic X-ray production considerably decreased, i.e. the event was not merely an eclipse [23, 29, 100].

During the minimum the hard X-ray flux varied strongly in timescales of days but not hours [40]; this seems reasonable if the lateral scale size of the emitting region was of the order of 1 AU at that time. The minimum apparent flux in the 2–10 keV band was only 0.7% of the maximum seen by *RXTE*. The slope of the X-ray continuum above 5 keV does not vary much, indicating that the electron temperature of the hottest plasma does not vary significantly at any phase. The X-ray emission seemed to show two states [40] during the minimum: first a "deep" state at the start of the minimum, in which the entire 2–10 keV flux was reduced, and then a "shallow" state beginning about midway through the minimum, when the $6-10$ keV emission increased and then remained constant for a brief period while the $E < 6$ keV emission continued to increase as the column density decreases.

The 2003 minimum also revealed a previously hidden "central constant emission" component of X-rays [40]. This dominated the system brightness at energies $E < 2$ keV during the minimum, with an intrinsic X-ray luminosity of $\sim 10^{34}$ ergs s^{-1} in the 0.1–10 keV band. Its origin is unknown; if thermal, the observed temperature implies a flow velocity of 1,000–2,000 km s^{-1}. It may represent the interaction of η Car's polar wind with nearby structures, perhaps the "Little Homunculus" [54] (though it's not clear that the Little Homunculus would survive any such collision for long); or it might be emission produced by outflowing shocked plasma interacting with cooler circumstellar material. It might also be intrinsic emission produced by shocks distributed in one of the stellar winds.

9.7 The Companion Star

To date, there has been no generally accepted direct detection of the secondary star,[1] but for reasons noted below this is not surprising. In the absence of direct UV-to-IR information, the X-ray data provide perhaps the most useful (though indirect) constraints on its physical parameters.

First, even if we were unaware of the minima and the 5.5-year cycle, the X-rays strongly suggest that η Car has a companion star. The average X-ray to total luminosity ratio, $L_X/L_{bol} \sim 5 \times 10^{-6}$, is very high for an evolved massive star, probably indicating that some sort of companion object is needed to help generate the X-rays. There's significant flux at high photon energies $E > 5$ keV, and such hard X-ray emission in evolved massive stars is almost unknown except in magnetized objects (whose X-rays vary on short rotational time scales) and binaries.

In a colliding-wind model, the X-ray temperature indicates the pre-shock speed of the wind; η Car's X-ray temperature requires a wind speed of about 3,000 km s^{-1} [23, 74]. This is far higher than speeds observed in the primary wind. Indeed it is near the upper limit of wind speeds observed for any massive stars. For a radiatively driven wind from a hot ($T_{phot} > 21{,}000$ K) massive star the wind terminal velocity is related to the photospheric escape speed V_{esc} as $V_\infty \approx 3 \times V_{esc}$ [1, 57], so that a terminal speed of 3000 km s^{-1} corresponds to a photospheric escape velocity of roughly 1000 km s^{-1}. For a massive, luminous star the escape velocity is modified from the usual relation to account for acceleration due to radiation

[1] Iping et al. asserted that certain variable emission features in the 905–1180 Å wavelength range were directly associated with the secondary wind [48]. This may be true but it is controversial [44,61,81]. The reported features were much narrower than one would expect for the fast secondary wind, and in some models they can originate in the primary wind. The wind velocities derived by Iping et al. were also about a factor of 2–3 too low to produce the colliding wind X-ray emission.

pressure on electrons: $V_{esc} = \sqrt{2GM(1-\Gamma)/R}$, where M and R are the mass and radius of the star, and Γ the ratio of radiative Thomson acceleration to gravitational acceleration [57].

In order to produce the observed X-ray luminosity, the same models show that the wind mass loss rate of the secondary star must be $\dot{M}_2 \sim 10^{-5}\,M_\odot\,\text{year}^{-1}$. This is exceptionally high for an O-type star, but seems marginally allowable for an object above $40\,M_\odot$. The determination of $\dot{M}_2$ is rather complex, depending on the poorly-constrained cooling ratio $t_{\rm exp}/t_{\rm rad}$ discussed in Sect. 9.3 above. For example, inhomogeneities in the wind might increase $t_{\rm exp}/t_{\rm rad}$. It's possible that improved models can attain the observed X-ray luminosity with a smaller value of $\dot{M}_2$.

In a "least surprising" evolutionary picture, one expects the companion object to be an early- or mid-O type star. Being less massive than η Car it should be less evolved and therefore near the main sequence. A mid-O spectral type with $T_{\rm eff} \sim 35{,}000$ K would account for the photoionization seen in the ejecta [23, 24, 29, 43, 47, 48]. Thus the most straightforward and most probable type of companion is a fairly "normal" star in the 30–60 $M_\odot$ mass range, though other possibilities can be devised. Other indirect arguments based on photoionization of circumstellar clouds by the UV flux from the companion star also point to a similar, though perhaps somewhat more evolved, object [62, 93]. The recent analysis by Mehner et al. [62] demonstrated that an O4–O6 giant with L $\sim 4 \times 10^5\,L_\odot$, $T_{eff} \approx$ 39,000 K, and M_{ZAMS} 40–50 $M_\odot$ would fit the observed range of parameters quite well. Such a star radiates mainly in the hard-to-observe far UV, should be less than 3% as bright as the primary at near-UV to IR wavelengths, and produces no spectral features that would be readily detected when superimposed on the complex primary wind spectrum [44]. The lack of a direct detection of the companion is therefore unsurprising. The only disquieting detail is the companion's high mass-loss rate required in some X-ray models, which might be viewed as evidence that the companion is a Wolf-Rayet star instead of an unevolved O star – making the system's evolutionary status problematic. However, we noted above that the deduced value of $\dot{M}_2$ is fairly uncertain.

9.8 Other Unresolved Issues

9.8.1 X-ray Flaring

The *RXTE*/PCA count rates (Fig. 9.2) show occasional brief maxima, which become dramatic in the year before the X-ray minimum. Early in 1997 Ishibashi noticed that the brightest X-ray "flares" tended to recur every 85 days [51]. This periodicity was statistically doubtful, but it correctly *predicted* the next two major flares in July and September 1997, and perhaps another in March 1998 after the X-ray minimum [52].

X-ray spectra show that they represent increased values of the volume emission measure [9, 10, 49], though some hardening of the spectrum during flares has been noted [7, 40]. The evidence so far favors a change in the hot colliding-wind plasma as the cause, rather than sporadic decreases in intervening absorption [9, 49]. The physical mechanism producing the X-ray flaring is not understood. In a simple qualitative model, localized density enhancements (clumps, or structured entities such as co-rotating streams [66]) in the wind eventually encounter the wind-wind shock and provide a brief increase in the amount of shocked X-ray emitting material. The flow timescale through the shock is $t_{\rm flow} \sim a/v_{\rm flow}$ where a is a characteristic size (thickness?) of the shocked region and $v_{\rm flow}$ is presumably a few hundred km s^{-1} based on the wind speeds, or roughly 0.1–0.4 AU per day. The observed 10–30 day flare durations are consistent with emission-region sizes of several AU [65].

The nature of the 85-day "quasi-period" is still not understood. A recurrence timescale of 85 days seems reasonable for either rotation or pulsation of the primary star; and in some models it might vary from cycle to cycle if each spectroscopic event alters the star's outermost structure [26]. But the interval between flare peaks varies with X-ray cycle phase, decreasing to as little as 20 days just before X-ray minimum, qualitatively consistent with expectations based on flow travel times [28] (Fig. 9.6).

9.8.2 *Cycle-to-Cycle Variations*

X-ray monitoring also revealed significant variations in the average X-ray flux level from cycle to cycle. Figures 9.2 and 9.7 shows *RXTE* count rates in the three 5.5-year cycles.The first and third cycles had very similar flux levels, but cycle 2 was significantly brighter. In a standard colliding-wind model the flux depends solely on orbital parameters, stellar mass-loss rates, and wind speeds (Sect. 9.3). Among these the mass-loss rate is probably most likely to change without perceptibly affecting the shape of the X-ray spectrum.

Cycle-to-cycle changes have also been observed in other wavelength bands. He I 10830Å emission showed a secular decrease for several cycles [16], centimeter-wavelength radio fluxes have been brighter in 2004–2008 than they were in 1998–2003 (S. White private communication, 2008), and the profiles of the extremely bright Hα and Hβ emission lines conspicuously differed between the 1997 and 2003 events [32]. Altogether, these developments likely indicate some undiagnosed change in the primary wind, although variations in other wavebands may be more complicated to interpret because they can be heavily influenced by secular changes in the Homunculus produced by expansion of the nebula and/or the creation or destruction of dust.

The most recent X-ray minimum which began on 16 January 2009, Fig. 9.7, was consistent with the phasings of the two previous minima. However, the X-ray recovery began on 18 February 2009, 1 month earlier than expected based on the duration of the two previous X-ray minima. During the 2009 recovery the X-ray hardness increased markedly, similar to the spectral change which occurred at the

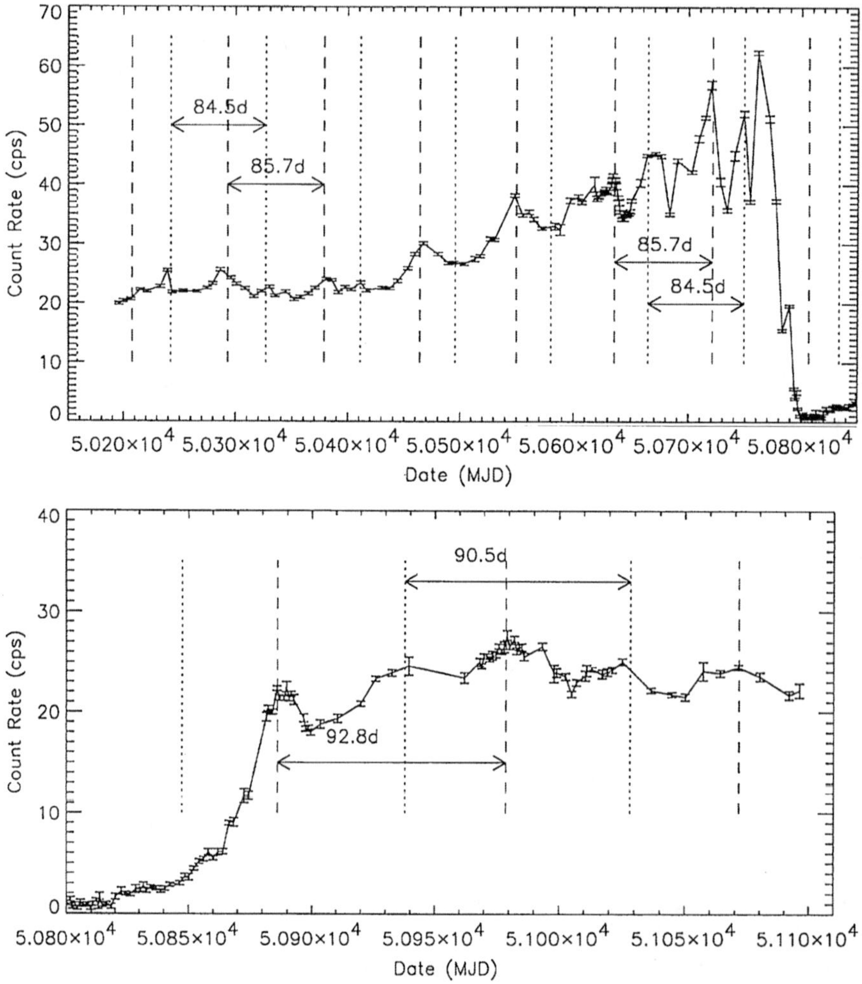

Fig. 9.6 Timing of major and minor X-ray flares in 1997–1998 according to Ishibashi [49]

end of the previous two minima, although the maximum hardness observed in 2009 was less than the maximum observed after the previous recoveries. This implies some fundamental change in the properties of the colliding wind emission. This may involve a decline in η Car's mass loss rate [15, 55, 63, 64],[2] though changes in the wind properties of the companion cannot be completely ruled out at this point.

[2]See Fig. 1.10 in the Chap. 1.

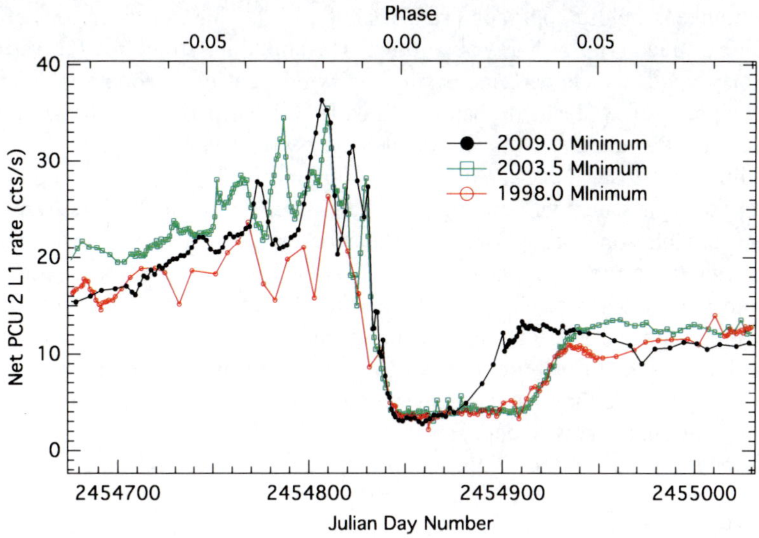

Fig. 9.7 Comparison of three X-ray minima observed by *RXTE*. The start of the 2009 minimum (16 January 2009) was consistent with the ephemeris derived from the previous two minima. The *bottom axis* shows the Julian Day Number corresponding to the 2009 minimum. The X-ray flux recovery occurred about 30 days early compared to recoveries from the previous two minima (From Corcoran et al. [15])

9.8.3 The Cause of the X-ray Minimum

The X-ray minimum may be caused by an increase in the intervening absorption, a decrease of the intrinsic emission, or both; in either case the maximum temperature is observed to be fairly constant [40, 53]. All models so far agree that the event happens near periastron and, if an eclipse plays a leading role, the eclipsing agent is the primary wind and not the primary star itself.

But several pieces of the puzzle do not easily fit into a simple eclipse model. The duration of the X-ray minimum is much longer than predicted in simple eclipse models, though more realistic wind models do better jobs in matching the eclipse duration [70,74]. If the X-ray minimum really occurs near periastron, simple models suggest that the intrinsic X-ray luminosity should be in excess of 10^{35} ergs s^{-1} at $E > 5$ keV at that time since $L_x \sim 1/d$. As noted above, hiding high-energy emission this bright requires an enormous wind column density which is about a factor of 100 higher than the observed column density. Another problem is that, during the minimum, a fully hidden X-ray source should produce a fluorescence iron line with an equivalent width about a factor of 6 larger than the observed value [40]. Also, the X-ray minimum shows two distinct phases (a deep stage and a shallow stage) [40], which is not expected in simple eclipse models.

Several non-eclipse ideas have been proposed. A spectroscopic event may be the sporadic ejection of extra material by the primary star, perhaps triggered by the companion star's close approach [12, 23, 61, 84, 100]. This causes various effects, including a sudden rise of X-ray absorption. The latitudinal structure of the primary wind may suddenly change, increasing the wind density at equatorial latitudes where the secondary star is probably located [81]. Near periastron, the wind-wind shocks may disintegrate, collapse, or at least become very complex because of instabilities that arise when radiative cooling dominates [25,61]. This would destroy most of the highest-temperature regions, thereby replacing the 2–10 keV emission by unobservable soft X-rays. Breaking the shock apart by accretion of part of the primary wind onto the companion star [83] could also reduce the colliding wind emission near periastron. Near periastron, the secondary star might move into a dense structure such as a disk around the primary. These possibilities are not all mutually exclusive; in fact the first, second, or fifth of them would make the third more likely. Nor are they inconsistent with an eclipse. Which, based on the likely orbit orientation, is likely to begin shortly thereafter [61].

The near-constant slope of the continuum above 5 keV is evidence that the highest temperatures are not reduced during periastron passage [40]. Finally, the strange He II λ4686 emission [61, 85], which peaked in a dramatic way during the 2003.5 event, seems directly pertinent to the X-rays for two reasons: it rose quickly in brightness during exactly the time when the 2–10 keV brightness was fading, and most likely it requires a remarkably increased flux of soft X-rays below 0.5 keV, which of course cannot be observed directly [61].

9.9 Future Prospects

X-ray observations provide unique direct information on the mass flows in the η Car system and indirect information about the system and stellar parameters. The recent X-ray minimum in January 2009 has been observed by a fleet of X-ray observatories including *RXTE*, *CHANDRA*, *XMM*, *Swift* and *Suzaku* with a primary goal of constraining the flow dynamics and determining the nature of the X-ray minimum. In the coming years high resolution X-ray spectra with X-ray calorimeters on observatories like the *Astro-H* satellite [90] should resolve the He-like iron K-line triplet at 6.7 keV for the first time and provide well-defined profiles of the the H-like iron line near 7 keV. These observations will be important since the high formation temperature of these lines mean that they must be formed near the stagnation point of the wind-wind collision; thus the line centroids can in principle be used to trace the orbit of the stagnation point, which can in turn be directly related to the radial velocity curve of the companion. This would enable us to dynamically constrain the system mass function and provide key answers about the evolution of the binary. Resolving the He-like iron K line triplet also provides important density and temperature diagnostics of the shocked gas near the stagnation point.

Even higher-energy images of hard X-ray and Gamma-ray emission from η Car have recently been reported [2, 59, 89], using modern observatories like the International Gamma-Ray Astrophysics Laboratory (*INTEGRAL*), the Italian-led

Gamma-ray satellite observatory *Astro-rivelatore Gamma a Immagini LEggero* (*Agile*), and the *Fermi* Gamma-Ray Space Telescope. As history shows, high-energy high-spatial resolution observations are extremely important, and revolutionary technology like the proposed space-based X-ray interferometer [38] which could reach resolutions of micro-arcseconds at X-ray energies, holds out the hope of directly measuring the geometry of the colliding wind shock as it moves around the orbit.

These observations will provide much of the the crucial information (system masses, stellar parameters) needed to fully understand the evolution of η Car. Looking even further down the road, X-ray observations will play an important role in understanding the eventual core-collapse of the star and the formation of a black hole and its associated supernova remnant.

Acknowledgements The authors would like to acknowledge many fruitful discussions with a large number of colleagues, among them Kris Davidson, Ted Gull, Roberta Humphreys, Augusto Damineli, Jean Swank, Stephen White, Julian Pittard, David Henley, Tony Moffat, Stan Owocki, John Hillier, Virpi Niemela, Nolan Walborn, and many others too numerous to fully mention. This work was supported in part by Chandra grant GO3-4008A and NASA grant NAS5-32490. This research made use of software and data available from the High Energy Astrophysics Science Archive Research Center, and NASA's Astrophysics Data System.

References

1. D.C. Abbott, The terminal velocities of stellar winds from early-type stars. ApJ. **225**, 893–901 (1978)
2. A.A. Abdo et al., Fermi/large area telescope bright gamma-ray source list. ApJ. Suppl. **183**, 46–66 (2009)
3. E. Behar, R. Nordon, N. Soker, A high-velocity transient outflow in η Carinae. ApJ. Lett. **666**, L97–L100 (2007)
4. H.V. Bradt, R.E. Rothschild, J.H. Swank, X-ray timing explorer mission. A&A. Suppl. **97**, 355–360 (1993)
5. A.M. Cherepashchuk, Detectability of Wolf-Rayet binaries from X rays. Sov. Astron. Lett. **2**, 138-+ (1976)
6. T. Chlebowski, F.D. Seward, J. Swank, A. Szymkowiak, X-rays from Eta Carinae. ApJ. **281**, 665–672 (1984)
7. M.F. Corcoran, X-ray monitoring of η Carinae: variations on a theme. AJ. **129**, 2018–2025 (2005)
8. M.F. Corcoran, G.L. Rawley, J.H. Swank, R. Petre, First detection of x-ray variability of eta carinae. ApJ. Lett. **445**, L121–L124 (1995)
9. M.F. Corcoran, K. Ishibashi, K. Davidson, J.H. Swank, R. Petre, J.H.M.M. Schmitt, Increasing X-ray emissions and periodic outbursts from the massive star Eta Carinae. Nature **390**, 587-+ (1997)
10. M.F. Corcoran, R. Petre, J.H. Swank, S.A. Drake, K. Koyama, Y. Tsuboi, R. Viotti, A. Damineli, K. Davidson, K. Ishibashi, S. White, D. Currie, The ASCA X-Ray spectrum of eta Carinae. ApJ. **494**, 381-+ (1998)
11. M.F. Corcoran, A.C. Fredericks, R. Petre, J.H. Swank, S.A. Drake, The X-Ray spectral evolution of η Carinae as seen by ASCA. ApJ. **545**, 420–428 (2000)

12. M.F. Corcoran, K. Ishibashi, J.H. Swank, R. Petre, The X-Ray light curve of η Carinae: refinement of the orbit and evidence for phase-dependent mass loss. ApJ. **547**, 1034–1039 (2001)
13. M.F. Corcoran, J.H. Swank, R. Petre, K. Ishibashi, K. Davidson, L. Townsley, R. Smith, S. White, R. Viotti, A. Damineli, The Chandra HETGS X-ray grating spectrum of η Carinae. ApJ. **562**, 1031–1037 (2001)
14. M.F. Corcoran et al., Waiting in the wings: reflected X-ray emission from the Homunculus nebula. ApJ. **613**, 381–386 (2004)
15. M.F. Corcoran, K. Hamaguchi, J.M. Pittard, C.M.P. Russell, S.P. Owocki, E.R. Parkin, A. Okazaki, Recent X-ray variability of η Carinae: the quick road to recovery. ApJ. **725**, 1528–1535 (2010)
16. A. Damineli, The 5.52 year cycle of Eta Carinae. ApJ. Lett. **460**, L49+ (1996)
17. A. Damineli, P.S. Conti, D.F. Lopes, Eta Carinae: a long period binary? New Astron. **2**, 107–117 (1997)
18. A. Damineli, D.F. Lopes, P.S. Conti, Eta Carinae Binarity after the 1998 event. ASP Conf. Ser. **179**, 288–294 (1999)
19. A. Damineli, A. Kaufer, B. Wolf, O. Stahl, D.F. Lopes, F.X. de Araújo, η Carinae: binarity confirmed. ApJ. Lett. **528**, L101–L104 (2000)
20. A. Damineli et al., The periodicity of the η Carinae events. MNRAS. **384**, 1649–1656 (2008)
21. A. Damineli et al., A multispectral view of the periodic events in η Carinae. MNRAS. **386**, 2330–2344 (2008)
22. K. Davidson, Is Eta Carinae a long-period Binary? New Astron. **2**, 387-+ (1997)
23. K. Davidson, Why the binary hypothesis isn't a Panacea. ASP Conf. Ser. **179**, 304–315 (1999)
24. K. Davidson, Unique spectroscopic problems related to Eta Carinae. ASP Conf. Ser. **242**, 3–12 (2001)
25. K. Davidson, Chandra meets Eta Carinae. ASP Conf. Ser. **262**, 267–274 (2002)
26. K. Davidson, The physical nature of η Carinae. ASP Conf. Ser. **332**, 101-+ (2005)
27. K. Davidson, R.M. Humphreys, Eta Carinae and its environment. Annu. Rev. Astron. Astrophys. **35**, 1–32 (1997)
28. K. Davidson, K. Ishibashi, M.F. Corcoran, The relationship between two periodicities observed in Eta Carinae. New Astron. **3**, 241–245 (1998)
29. K. Davidson, K., Ishibashi, T.R., Gull, R.M. Humphreys, HST/STIS observations of the star during its recent event. ASP Conf. Ser. **179**, 227–235 (1999)
30. K. Davidson, K. Ishibashi, T.R. Gull, R.M. Humphreys, N. Smith, η Carinae: testing a binary orbit model with the Hubble Space Telescope/Space Telescope Imaging Spectrograph. ApJ. Lett. **530**, L107–L110 (2000)
31. K. Davidson, N. Smith, T.R. Gull, K. Ishibashi, D.J. Hillier, The shape and orientation of the Homunculus Nebula based on spectroscopic velocities. AJ. **121**, 1569–1577 (2001)
32. K. Davidson, J. Martin, R.M. Humphreys, K. Ishibashi, T.R. Gull, O. Stahl, K. Weis, D.J. Hillier, A. Damineli, M. Corcoran, F. Hamann, A change in the physical state of η Carinae? AJ. **129**, 900–906 (2005)
33. R.A. Duncan, S.M. White, J. Lim, G.J. Nelson, S.A. Drake, M.R. Kundu, Intense radio outburst from the supermassive star Eta Carinae. ApJ. Lett. **441**, L73–L76 (1995)
34. D. Falceta-Gonalves, V. Jatenco-Pereira, Z. Abraham, Wind-wind collision in the η Carinae binary system: a shell-like event near periastron. MNRAS. **357**, 895–902 (2005)
35. A. Feinstein, H.G. Marraco, On a possible three-year cycle of eta Car. A&A. **30**, 271–273 (1974)
36. W. Forman, C. Jones, L. Cominsky, P. Julien, S. Murray, G. Peters, H. Tananbaum, R. Giacconi, The fourth UHURU catalog of X-ray sources. ApJ. Suppl. **38**, 357–412 (1978)
37. K.G. Gayley, S.P. Owocki, S.R. Cranmer, Sudden radiative braking in colliding hot-star winds. ApJ. **475**, 786-+ (1997)
38. K.C. Gendreau, W.C., Cash, P. Gorenstein, D.L. Windt, P. Kaaret, C. Reynolds, MAXIM: the black hole imager. Proc. SPIE Int. Soc. Opt. Eng. **5488**, 394–402 (2004)

39. R.E. Griffiths, A. Peacock, B.E.J. Pagel, Absence of soft X-rays from eta Car. Nature **250**, 714–715 (1974)
40. K. Hamaguchi, M.F. Corcoran, T. Gull, K. Ishibashi, J.M. Pittard, D.J. Hillier, A. Damineli, K. Davidson, K.E. Nielsen, G.V. Kober, X-ray spectral variation of η Carinae through the 2003 X-ray minimum. ApJ. **663**, 522–542 (2007)
41. D.B. Henley, M.F. Corcoran, J.M. Pittard, I.R. Stevens, K. Hamaguchi, T.R. Gull, Chandra X-ray grating spectrometry of η Carinae near X-ray minimum. I. Variability of the sulfur and silicon emission lines. ApJ. **680**, 705–727 (2008)
42. R.W. Hill, G. Burginyon, R.J. Grader, T.M. Palmieri, F.D. Seward, J.P. Stoering, A soft X-ray survey from the Galactic center to VELA. ApJ. **171**, 519-+ (1972)
43. D.J. Hillier, K. Davidson, K. Ishibashi, T. Gull, On the nature of the central source in η Carinae. ApJ. **553**, 837–860 (2001)
44. D.J. Hillier et al., The UV scattering Halo and central source associated with Eta Carinae. ApJ. **642**, 1098–1116 (2006)
45. D. Hoffleit, Elias and Francis Loomis on variable stars, 1866–69. JAAVSO. **13**, 67–77 (1984)
46. R.M. Humphreys, K. Davidson, The luminous blue variables: astrophysical geysers. PASP. **106**, 1025–1051 (1994)
47. R.M. Humphreys, K. Davidson, M. Koppelman, The early spectra of Eta Carinae 1892 to 1941 and the onset of its high excitation emission spectrum. AJ. **135**, 1249–1263 (2008)
48. R.C. Iping, G. Sonneborn, T.R. Gull, D.L. Massa, D.J. Hillier, Detection of a hot binary companion of η Carinae. ApJ. Lett. **633**, L37–L40 (2005)
49. K. Ishibashi, Multiwavelength investigation of the supermassive star Eta Carinae. Doctoral Dissertation, University of Minnesota, (1999)
50. K. Ishibashi, Searching for physically acceptable parameters for Eta Carinae's companion. ASP. Conf. Ser. **242**, 53–54 (2001)
51. K. Ishibashi, K. Davidson, M.F. Corcoran, J.H. Swank, R. Petre, K. Jahoda, Eta Carinae. IAU. Circ. **6668**, 1-+ (1997)
52. K. Ishibashi, M.F. Corcoran, K. Davidson, S.A. Drake, J.H. Swank, R. Petre, Resolving X-ray temporal variations in Eta Carinae. ASP. Conf. Ser. **179**, 266–274 (1999)
53. K. Ishibashi, M.F. Corcoran, K. Davidson, J.H. Swank, R. Petre, S.A. Drake, A. Damineli, S. White, Recurrent X-ray emission variations of η Carinae and the binary hypothesis. ApJ. **524**, 983–987 (1999)
54. K. Ishibashi et al., Discovery of a little homunculus within the Homunculus Nebula of η Carinae. AJ. **125**, 3222–3236 (2003)
55. A. Kashi, N. Soker, Explaining the early exit of Eta Carinae from its 2009 X-ray minimum with the accretion model. ApJ. Lett. **701**, L59–L62 (2009)
56. K. Koyama, I. Asaoka, N. Ushimaru, S. Yamauchi, R.H.D. Corbet, Hard X-ray emission from the Carina Nebula. ApJ. **362**, 215–218 (1990)
57. R.-P. Kudritzki, J. Puls, Winds from hot stars. Annu. Rev. Astron. Astrophys. **38**, 613–666 (2000)
58. L.D. Landau, E.M. Lifshitz, *Fluid Mechanics*, 2nd edn. (Pergamon, Oxford, 1987)
59. J.-C. Leyder, R. Walter, G. Rauw, Hard X-ray emission from η Carinae. A&A. **477**, L29–L32 (2008)
60. E. Loomis, On the period of η Argus. MNRAS. **29**, 298-+ (1869)
61. J.C. Martin, K. Davidson, R.M. Humphreys, D.J. Hillier, K. Ishibashi, On the He II emission in η Carinae and the origin of its spectroscopic events. ApJ. **640**, 474–490 (2006)
62. A. Mehner, K. Davidson, G.J. Ferland, R.M. Humphreys, High-excitation emission lines near Eta Carinae, and its likely companion star. ApJ. **710**, 729–742 (2010)
63. A. Mehner, K. Davidson, R.M. Humphreys, J.C. Martin, G.J. Ferland, N.R. Walborn, A sea change in Eta Carinae. ApJ. Lett. **717**, L22–L25 (2010)
64. A. Mehner, K. Davidson, J.C. Martin, R.M. Humphreys, K. Ishibashi, G.J. Ferland, Critical differences and clues in Eta Car's 2009 event. ApJ. **740**, 2 (2011)
65. A.F.J. Moffat, M.F. Corcoran, Understanding the X-ray flaring from η Carinae. ApJ. **707**, 603–704 (2009)

66. D.J. Mullan, Displaced narrow absorption components in the spectra of mass-losing OB stars – Indications of corotating interaction regions? A&A. **165**, 157–162 (1986)
67. Y. Nazé, M.F. Corcoran, G. Koenigsberger, A.F.J. Moffat, First detection of phase-dependent colliding wind X-ray emission outside the Milky Way. ApJ. Lett. **658**, L25–L28 (2007)
68. K.E. Nielsen, M.F. Corcoran, T.R. Gull, D.J., Hillier, K. Hamaguchi, S. Ivarsson, D.J. Lindler, Eta Carinae across the 2003.5 minimum: spectroscopic evidence for massive binary interactions. ApJ. **660**, 669–686 (2007)
69. J.J. Nugent, K.A. Jensen, J.A. Nousek, G.P. Garmire, K.O. Mason, F.M. Walter, C.S. Bowyer, R.A. Stern, G.R. Riegler, HEAO A-2 soft X-ray source catalog. ApJ. Lett. **51**, 1–28 (1983)
70. A.T. Okazaki, S.P. Owocki, C.M.P. Russell, M.F. Corcoran, Modelling the RXTE light curve of η Carinae from a 3D SPH simulation of its binary wind collision. MNRAS. **388**, L39–L43 (2008)
71. R. Parkin, J.M. Pittard, A 3D dynamical model of the colliding winds in binary systems. MNRAS. **388**, 1047–1061 (2008)
72. E.R. Parkin, J.M. Pittard, M.F. Corcoran, K. Hamaguchi, I.R. Stevens, 3D modelling of the colliding winds in η Carinae – evidence for radiative inhibition. MNRAS. **394**, 1758–1774 (2009)
73. C. Payne-Gaposchkin, The Galactic Novae. Dover Publications, (1957)
74. J.M. Pittard, M.F. Corcoran, In hot pursuit of the hidden companion of eta Carinae: an X-ray determination of the wind parameters. A&A. **383**, 636–647 (2002)
75. A.M.T. Pollock, M.F. Corcoran, I.R. Stevens, P.M. Williams, Bulk velocities, chemical composition, and ionization structure of the X-ray shocks in WR 140 near periastron as revealed by the Chandra gratings. ApJ. **629**, 482–498 (2005)
76. O.F. Prilutskii, V.V. Usov, X rays from Wolf-Rayet binaries. Sov. Astron. **20**, 2-+ (1976)
77. F.D. Seward, T. Chlebowski, X-ray emission from the Carina Nebula and the associated early stars. ApJ. **256**, 530–542 (1982)
78. F.D. Seward, C.G. Page, M.J.L. Turner, K.A. Pounds, X-ray sources in the southern Milky Way. MNRAS. **177**, 13P–20P (1976)
79. F.D. Seward, W.R. Forman, R. Giacconi, R.E. Griffiths, F.R., Harnden, C. Jones, J.P. Pye, X-rays from Eta Carinae and the surrounding nebula. ApJ. Lett. **234**, L55–L58 (1979)
80. S.N. Shore, An introduction to astrophysical hydrodynamics, chapter 4. (Academic, San Diego, 1992)
81. N. Smith, K. Davidson, T.R. Gull, K. Ishibashi, D.J. Hillier, Latitude-dependent effects in the stellar wind of η Carinae. ApJ. **586**, 432–450 (2003)
82. N. Soker, The binarity of Eta Carinae and its similarity to related astrophysical objects. ApJ. **619**, 1064–1071 (2005)
83. N. Soker, Accretion by the secondary in η Carinae during the spectroscopic event. I. Flow parameters. ApJ. **635**, 540–546 (2005)
84. O. Stahl, K. Weis, D.J. Bomans, K. Davidson, T.R. Gull, R.M. Humphreys, A spectroscopic event of η Car viewed from different directions: the data and first results. A&A. **435**, 303–312 (2005)
85. J.E. Steiner, A. Damineli, Detection of He II λ4686 in η Carinae. ApJ. Lett. **612**, L133–L136 (2004)
86. I.R. Stevens, J.M. Pittard, Eta Carinae: a colliding wind binary? ASP. Conf. Ser. **179**, 295–303 (1999)
87. I.R. Stevens, A.M.T. Pollock, Stagnation-point flow in colliding wind binary systems. MNRAS. **269**, 226-+ (1994)
88. I.R. Stevens, J.M. Blondin, A.M.T. Pollock, Colliding winds from early-type stars in binary systems. ApJ. **386**, 265–287 (1992)
89. M. Tavani et al.: Detection of gamma-ray emission from the Eta-Carinae region. ApJ. Lett. **698** L142–L146 (2009)
90. T. Takahashi, the NeXT team, The NeXT mission. **807** (2008) ArXiv e-prints
91. Y. Tsuboi, K. Koyama, M. Sakano, R. Petre, ASCA observations of Eta Carinae. Publ. Aston. Soc. Jpn. **49**, 85–92 (1997)

92. V.V. Usov, Stellar wind collision and X-ray generation in massive binaries. ApJ. **389**, 635–648 (1992)
93. E. Verner, F. Bruhweiler, T. Gull, The binarity of η Carinae revealed from photoionization modeling of the spectral variability of the weigelt blobs B and D. ApJ. **624**, 973–982 (2005)
94. S.M. White, R.A. Duncan, J. Lim, G.J. Nelson, S.A. Drake, M.R. Kundu, The radio source around Eta Carinae. ApJ. **429**, 380–384 (1994)
95. P.A. Whitelock, M.W. Feast, B.S. Carter, G. Roberts, I.S. Glass, The infrared spectrum and variability of Eta Carinae. MNRAS. **203**, 385–392 (1983)
96. P.A. Whitelock, M.W. Feast, C. Koen, G. Roberts, B.S. Carter, Variability of Eta-Carinae. MNRAS. **270**, 364-+ (1994)
97. P.A. Whitelock, M.W. Feast, F. Marang, E. Breedt, The 2003 shell event in Eta Carinae. MNRAS. **352**, 447–456 (2004)
98. A.J. Willis, H. Schild, I.R. Stevens, ROSAT observations of γ Velorum (WC8+O9I). I. The discovery of colliding-wind X-ray emission. A&A. **298**, 549-+ (1995)
99. R. Wolf, Schreiben des Herrn Prof. Wolf, Directors der Sternwarte in Zürich an den Herausgeber. Astron. Nachr. **60**, 59-+ (1863)
100. R. Zanella, B. Wolf, O. Stahl, Spectroscopy of the shell episode of Eta CAR (1981–1983). A&A. **137**, 79–84 (1984)

Chapter 10
Eta Carinae and the Luminous Blue Variables

Jorick S. Vink

Abstract We evaluate the place of Eta Carinae (η Car) among the class of luminous blue variables (LBVs) and show that the LBV phenomenon is not restricted to extremely luminous objects like η Car, but extends luminosities as low as log $(L/L_{\odot}) \sim 5.4$ – corresponding to initial masses $\sim 25\,M_{\odot}$, and final masses as low as $\sim$10–15 $M_{\odot}$. We present a census of S Doradus variability, and discuss basic LBV properties, their mass-loss behavior, and whether at maximum light they form pseudo-photospheres. We suggest that those objects that exhibit giant η Car-type eruptions are most likely related to the more common type of S Doradus variability. Alternative atmospheric models as well as sub-photospheric models for the instability are presented, but the true nature of the LBV phenomenon remains as yet elusive. We end with a discussion of the evolutionary status of LBVs – highlighting recent indications that some LBVs may be in a direct pre-supernova state, in contradiction to the standard paradigm for massive star evolution.

10.1 Introduction

Luminous Blue Variables (LBVs) are evolved, luminous hot stars that experience eruptions or outbursts and episodes of enhanced mass loss. During outburst they appear to make transits in the HR Diagram (HRD) from their normal hot quiescent state to lower temperatures. The LBVs include a number of famous stars such as P Cyg, S Dor, R127 and AG Car. Eta Car is often described as an LBV, although it is a more extreme example owing to its giant eruption.

During the late 1970s, it was recognized that the distribution of the most luminous hot stars on the HRD defines a locus of declining luminosity with decreasing

J.S. Vink (✉)
Armagh Observatory, College Hill, Armagh BT61 9DG, Northern Ireland
e-mail: jsv@arm.ac.uk

K. Davidson and R.M. Humphreys (eds.), *Eta Carinae and the Supernova Impostors*, Astrophysics and Space Science Library 384, DOI 10.1007/978-1-4614-2275-4_10,

temperature [49, 50, 57, 117]. Together with the fairly tight upper luminosity limit of the yellow and red supergiants at log $(L/L_{\odot}) \approx 5.8$ [50], this indicated that the most massive stars ($M > 60\, M_{\odot}$) do not evolve to cooler temperatures: the Humphreys-Davidson (HD) limit [50]. Humphreys and Davidson [50, 51] suggested that high mass-loss episodes, represented by stars like η Car, P Cyg, S Dor and the Hubble-Sandage variables in M31 and M33 [48], prevented the evolution of the most massive stars to cooler temperatures. With this addition of a post-main sequence period of high mass loss ($10^{-5} - 10^{-3}\, M_{\odot}\, \mathrm{year}^{-1}$) the evolutionary tracks of the most massive stars were shown to turn bluewards, towards the core He-burning Wolf-Rayet (WR) phase. During the WR phase, stars are anticipated to explode as supernovae (SNe) type Ib/c. However, at the end of this chapter, we challenge the canonical view that LBVs are always transitional between the O and WR stars, and suggest that some massive stars may already suffer their final explosion during or at the end of the LBV phase. One of the most relevant questions is therefore *do LBVs explode?*

The term LBV was introduced to describe the diverse group of unstable evolved hot stars in the upper HRD. Today we distinguish between more than one type of LBV [52]: (1) the normal LBV variability cycles with visual magnitudes changes of 1–2 magnitudes at essentially constant luminosity – on timescales of years to decades – represented by the prototype of the class, S Dor, in the Large Magellanic Cloud, and (2) the giant eruption LBVs represented by η Car and P Cyg with visual magnitudes changes of 3 magnitudes or more, during which the total bolometric luminosity increases [52, 54].

In this chapter, we focus on the S Dor-type variables and their transits in the HRD, the prime characteristics of the LBV class. Roughly 30 massive stars in the Galaxy ($\simeq$10) and Local Group ($\simeq$20) are known to be S Dor variables. By contrast, only two Galactic objects have been discovered to exhibit giant η Car-like eruptions: η Car itself and P Cygni, which suddenly appeared at naked-eye visibility in 1600. Due to their high luminosities at maximum, a significant number of LBV-like non-terminal eruptions have been discovered in external galaxies. Most are typical of giant eruptions whilst some appear to be more similar to S Dor type variables. These extragalactic LBVs are sometimes referred to as "SN imposters" (see the contribution in Chap. 11).

The circumstellar nebulae seen around many Galactic and MC LBVs [88, 135] may also have resulted from giant η Car eruptions, although stationary winds from prior evolutionary phases may constitute an alternative scenario for their creation [32]. Given the association of many LBVs with nebulae and ejecta and the close proximity of both the S Dor variables and the η Car-like variables to the Eddington limit, it is often suggested (though not proven) that their instabilities represent different manifestations of the same underlying evolutionary state. The fact that the giant eruptor P Cyg is also subject to small amplitude S Dor variability [24, 79] and that η Car's second outburst (during 1888–1895) was like that of a normal S Dor variable [54] lends support to this possibility.

We emphasize, however, that neither the η Car-type eruptions, nor the S Dor variations are understood. Worse still, we do not know whether all LBVs are subject to both types of variability, or in which order they may occur. Two pertinent questions are thus whether "*η Car is unique among the LBVs?*" and "*what is the root cause of the S Doradus variations?*"

10.2 Basic Properties of LBVs

Variability LBVs show significant spectroscopic and photometric variability on timescales of years (short S Dor phases) to decades (long S Dor phases, cf. van Genderen [126]). For completeness, we note that LBVs also show smaller amplitude "micro" variability, on shorter timescales (weeks, months), but this aspect of their variability is also a common feature of supergiants in general (see [71, 75] and references therein).

Luminosities The classical LBVs have log $(L/L_{\odot})$ larger than 5.8 with bolometric magnitudes in the range $M_{\rm bol}$ -9 to -11. There is an apparent "gap" in their luminosities just below log $(L/L_{\odot}) = 5.8$, and a separate group of less luminous LBVs with log $(L/L_{\odot}) = 5.4$–5.6, corresponding to bolometric magnitudes in the range $M_{\rm bol} \simeq -8$ to $\simeq -9$ [52]. We note that this separation in luminosity (see Fig. 10.1) may not be real, but due to small number statistics. As the less luminous LBVs are below the HD limit, they have presumably been red supergiants where they may already have shed a lot of mass. As a result, they may be equally close to their Eddington limit as the classical LBVs.

One of the most important properties of LBVs is that they appear to be close to the Eddington limit for stability against radiation pressure for their luminosities and current masses (see the subsection on stellar masses). The Eddington luminosity L_{Edd} is defined:

$$L_{\rm Edd} = \frac{4\pi cGM}{\kappa_{\rm F}} \tag{10.1}$$

where $\kappa_{\rm F}$ is the flux-mean opacity and M is the total mass of the star.

The dimensionless Eddington parameter Γ is defined as the ratio of radiative acceleration to gravitational acceleration (disregarding sign):

$$\Gamma = \frac{\kappa_{\rm F} L}{4\pi cGM} = (7.66 \times 10^{-5}\,{\rm g\,cm^{-2}})\,\kappa_{\rm F} \left(\frac{L}{L_{\odot}}\right)\left(\frac{M}{M_{\odot}}\right)^{-1} . \tag{10.2}$$

If opacity is entirely due to Thomson scattering by free electrons, as often assumed in this connection, then $\kappa_{\rm F} \approx 0.3\,{\rm cm^2\,g^{-1}}$; but this depends somewhat on chemical composition and ionization state.

Temperatures LBV temperatures are not only time-variable, but accurate $T_{\rm eff}$ determination requires sophisticated non-LTE spectral analyses for hot stars with extended envelopes and winds. Furthermore, it should be noted that for extreme

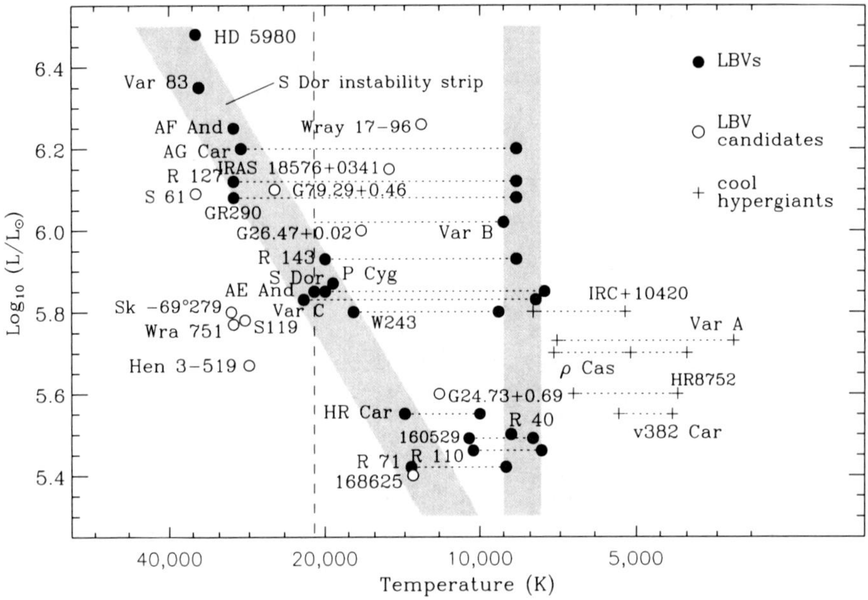

Fig. 10.1 The location of the LBVs (*black circles*) and candidates (*open circles*) in the Hertzsprung-Russell Diagram. The cool yellow hypergiants are indicated with *pluses*. The slanted and vertical *grey* bands represent visual minimum and maximum respectively. The *dashed vertical line* at 21,000 K indicates the location of the bi-stability jump (see Sect. 10.3) (The figure has been adapted from Smith et al. [109] and is similar to e.g. Fig. 9 in [52])

objects like η Car we cannot even estimate the radius based on $T_{\rm eff}$ as the wind is optically thick. In such a case, $T_{\rm eff}$ refers to the position in the wind where $R_{\rm ph} > R_{\rm sonic}$ (see below for a more extended discussion).

Figure 10.1 shows the confirmed and candidate LBVs on a schematic HRD with their transits (the dotted horizontal lines) between quiescence and their outburst state or visual maximum. At quiescence, or visual minimum, they appear to lie along a fairly narrow slanted band: the "S Dor instability strip" [138]. Their effective temperatures vary from over 30,000 K, corresponding to spectral type late O/early B, for the luminous classical LBVs to only $\simeq$15,000 K, corresponding to late B spectral types, for the less luminous LBVs. Objects from both subgroups transit redwards to $T_{\rm eff}$ not higher than $\simeq$8,000 K, corresponding to spectral types A to F, when they are "in outburst" or visual maximum. Wolf [138] noted that the amplitudes of the S Dor excursions become larger with increasing luminosity, and suggested that the most luminous LBVs could be cosmological distance indicators. Normal blue supergiants are also observed in and to the right of the S Dor instability strip, but they may not be close to their Eddington limit and may therefore not be subjected to the instability that characterizes LBVs. The cool or yellow hypergiants (YHGs) just below the HD limit [50, 52] are included as these might somehow represent the "missing" LBVs [109] – a possibility which will be discussed further in Sect. 10.3.

Table 10.1 The confirmed LBVs. See [81, 83, 124] for the latest additions

Galaxy:				
η Car	AG Car	HR Car	P Cygni	HD 160529
HD 168607	FMM 362	AFGl 2298	G24.73 + 0.69	W1-243
GCIRS 34W	Wra 751	G0.120−0.048		
LMC:				
S Dor	R71	R 110	R 116	R127
R 143	HD 269582	HD 269929		
SMC:				
R40	HD 5980			
M31:				
AE And	AF And	Var A-1	Var 15	
M33:				
Var B	Var C	Var 2	Var 83	GR 290
N 93351				
M81:				
I 1	I 2	I 3		
M101:				
V 1	V 2	V 10		
NGC 2403:				
V 12	V 22	V 35	V 37	V38
NGC 1058:				
SN 1961 V				

Membership Whether a particular blue hypergiant is a member of the select group of LBVs (see Table 10.1) is often a matter of debate (see [12, 52, 82]). In general, spectral resemblance to the P Cygni type spectrum of a known LBV is not a sufficient criteria for a star to be called an LBV. For this reason we include a group of LBV candidates (LBVc) listed in Table 10.2. To officially qualify as an LBV, an object should at least have shown the combination of spectral-type/photometric variations characteristic of S Dor-type variability due to changes in the star's *apparent* temperature. Spectral variability, e.g. due to mass-loss changes, on its own is not a criterion, as almost all massive stars would qualify. Changes in photometric color could be the result of $T_{\rm eff}$ changes, however obscuration by circumstellar dust may also give rise to color changes. Finally, the association of a supergiant with a circumstellar nebula is also not a criterion for an object to be added to the list of LBVs. The famous pistol star and LBV 1806−20 are thus strictly not LBVs.[1] These objects remain candidates until they have shown the characteristic spectral-type variations.

When LBVs are discussed in an evolutionary context, we should be aware that the LBV phenomenon is likely to be intermittent and that part of the population might be dormant. Disregarding this could lead to incorrect interpretations with respect to their relative numbers and their evolutionary state. When we discuss the LBV phenomenon, however, we should only include the confirmed LBVs.

[1] See Chap. 4.

Table 10.2 The LBV candidates. See [9, 40, 119] for the latest additions

Galaxy:				
Cyg OB2#12	Pistol star	HD 168625	HD 326823	HD 316285
He3-519	ζ^1 Sco	MWC 314	MWC 930	AS 314
G25.5 + 0.2	G79.29 + 0.46	G26.47 + 0.02	Wra17-96	WR102ka
LBV1806−20	Sher 25	W51 LS1	GCIRS 16NE	GCIRS 16C
GCIRS 16SW	GCIRS 16NW	GCIRS 33SE		
MN112				
LMC:				
R 4	R 66	R 74	R 78	R 81
R 84	R 85	R 99	R 123	R 128
R 149	S 18	S 22	S 61	S 119
S 134				

Abundances As massive stars evolve on the main sequence their atmospheric abundances are expected to undergo a transition in chemical abundances from solar He and CNO abundances to He-enriched and nuclear-equilibrium CNO abundances (with N enhanced, C/O depleted) [77]. For massive stars with $\simeq 60\,M_\odot$, this transition occurs rather rapidly after about 2 Myr of evolution [140]. LBVs show a wide span of CNO ratios. Davidson et al. [19] showed that the ejecta around η Car are enhanced in both nitrogen and helium, and Pasquali et al. [93] found that the shell ejected by the LBVc HD 168625 is also N-enriched compared to the interstellar medium by a factor of several. Smith et al. [107] studied a sample of LBV nebulae and found the ejecta to be generally N-enhanced, although most nebulae have not reached abundances characteristic of CNO equilibrium.

The origin of the circumstellar nebulae associated with some LBVs however is uncertain. Photospheric abundance measurements might therefore provide a more direct means to constrain the LBV evolutionary state. These abundances however depend not only on the details of the atomic physics but also on the complexities of non-LTE spectral analyses. For example, Hillier et al. [47] showed that the H/He ratio in the wind of η Car is ambiguous due to a strong coupling with the mass-loss rate. Eta Car however, could be an exception with its huge luminosity and high mass-loss rate. Najarro et al. [86] studied the atmospheric He abundance of P Cygni and quoted a best value of n(He)/n(H) of 0.3, corresponding to a mass fraction $Y = 0.63$. However, they noted a huge uncertainty due to the trade-off in ionization and He abundance and provided a He abundance range of n(He)/n(H) = 0.25–0.55. Hillier et al. [46] also ruled out a solar He abundance for the extreme P Cygni star HD 361285, but admitted that the uncertainty in n(He)/n(H) could be as large as a factor 20! Evidence for advanced CNO processing was found by Lennon et al. [74] for the LBV R71, but this is not a well-established result for LBVs in general. Although most LBVs have He and N enhanced atmospheres, it seems unlikely that all of them have reached equilibrium CNO values in their outer atmospheres.

Stellar masses Masses are most accurately determined using detached binary systems. However, almost all LBVs are single. Eta Car is the best-known exception to this rule, with its 5.5 years periodicity [16, 137] attributed to a companion. The current mass of η Car is thought to be at least $90\,M_{\odot}$, to avoid violating the Eddington Limit at its high luminosity (see Chaps. 3 and 12).

LBVs with log $(L/L_{\odot}) > 5.8$, are usually assumed to be evolved from the most massive stars with $M > 50\,M_{\odot}$ (e.g. [103]). Mass measurements for these high-luminosity objects are scarce and uncertain. Pauldrach and Puls [96] quote a mass of $23\,M_{\odot}$ for P Cyg. Vink and de Koter [130] estimated LBV masses from time-variable mass-loss rates and found a mass of $35\,M_{\odot}$ for AG Car. These results suggest that classical LBVs have already lost a significant fraction of their initial mass probably through the combined effects of line-driven mass loss during the OB and LBV phase, and/or through prior major eruptions. The quoted mass estimates are highly model dependent. For instance, wind clumping was not considered for the mass estimate of AG Car. However, if these objects have lost half their initial mass, their L/M ratio is quite large, with $\Gamma_e \simeq 0.5$, and they are close to the Eddington Limit for their luminosities.

The initial masses of the less luminous LBVs, log $(L/L_{\odot}) \simeq 5.4$, is of the order $M \sim 25\,M_{\odot}$; e.g. [103]). Comparing this mass with current-day mass estimates of $M \sim 12\,M_{\odot}$ for R71 [73], $M \sim 10\,M_{\odot}$ for R110 [115], and $M \sim 13\,M_{\odot}$ for HD 160529 [118], suggests that these low-luminosity LBVs may already have lost more that half of their initial mass (e.g. during a prior RSG phase). The low-luminosity LBVs are equally close to their Eddington limit with $\Gamma_e \simeq 0.6$. All of these empirical mass estimates are uncertain by at least a factor 2.

Interestingly, Martins et al. [80] recently reported that the LBVc GCIRS 16SW may be an eclipsing binary, with both components weighing $\simeq 50\,M_{\odot}$. If it is a genuine LBV, it may be a key object for constraining massive star evolution models, because current models with rotation [84] suggest such massive objects will not pass through the LBV phase at all during their evolution.

10.3 Mass-Loss Properties – Do LBVs form Pseudo-Photospheres?

With its current mass-loss rate of order $10^{-3}\,M_{\odot}\,\mathrm{year}^{-1}$ it is clear that η Car has formed an optically-thick wind and a pseudo-photosphere, but whether the S Dor-type variables have optically thick winds is less well established. It has been suggested that the temperature changes during S Dor cycles do not represent true stellar temperature changes, but are due to the formation of the dense, optically-thick wind. As the mass-loss rate increases, the effective photosphere of the moves out into the wind, and the apparent effective temperature of the star drops, whilst the apparent stellar radius increases – without an actual expansion of the star [3, 18, 62]. Davidson [18] also showed that the minimum temperature the wind can achieve as the mass loss increases, in the LBV eruption, is $\sim$7000 K, in rough agreement with the apparent temperatures of the LBVs at visual maximum.

Normal OB stars have winds that are optically thin in the continuum and we see through the entire wind – down to the photosphere. In other words, the wind is formed outside the photosphere. However, when the mass-loss rates increases, the wind becomes less transparent, and the photosphere from which the optical light originates is now at larger radii. When the wind has become optically thick, the photosphere is formed above the sonic velocity. In other words, the wind is accelerated "inside" the photosphere – forming an opaque wind with $R_{ph} > R_{sonic}$.

Leitherer et al. [73] and de Koter et al. [27] performed detailed NLTE modelling of this process, showing that the extent of a pseudo-photosphere that results from increased mass loss is relatively modest in comparison to that of a WR star. The conclusion was that the underlying LBV radius itself must become larger due to an as yet unidentified sub-photospheric mechanism. Furthermore, it became clear that empirical mass-loss rate during S Dor redward excursion do not always increase during outburst [72, 130]. More recently, Smith et al. [109] showed that pseudo-photospheres might be feasible under certain special circumstances discussed later in Sect. 10.3.3. Before we provide a more detailed account of the possibility of pseudo-photosphere formation, we give an overview of our current knowledge of stationary LBV mass loss. The radiative forces that may be responsible for giant eruptions are discussed elsewhere (Chap. 12).

10.3.1 Observed Mass-Loss Rates

Analysis of the blue-shifted absorption components in the P-Cygni profiles of strong emission lines such as Hα is commonly used to determine mass-loss rates from massive stars.

Terminal velocities The terminal velocities of LBV winds measured from the blue edge of the P Cygni absorption component are in the range 100–250 km s^{-1}, with η Car having a v_∞ of ~500 km s^{-1} (cf. [72]). These wind velocities are significantly lower than those of normal OB supergiants, which have $v_\infty \sim$ 1,000–3,000 km s^{-1}. The mass-loss rates of LBVs are also a factor of 10–100 larger than those of normal supergiants, so their wind densities, $\rho(r) = \dot{M}/4\pi r^2 v(r)$, are much higher, giving the line profiles their characteristic P Cygni shapes. As the LBV mass-loss rate is variable, some LBVs exhibit profile shape changes and variability in the absorption profile, such as the split blue-shifted absorption components seen in the Hα line of AG Car (see Fig. 10.2) and other LBVs such as R 127 [113, 114]. P Cyg, R 66, and HD 160529 also exhibit shell components in their Fe II lines (cf. [66] for P Cyg). The split Hα absorption components in LBVs have recently been proposed to be the result of an abrupt bi-stability jump [35].

Mass-loss rates Mass-loss rates of most LBVs have been determined using non-LTE models such as CMFGEN [45]. The very high mass-loss rate for η Car ($3 \times 10^{-3} M_\odot$ year^{-1}) has been determined by several methods using radio,

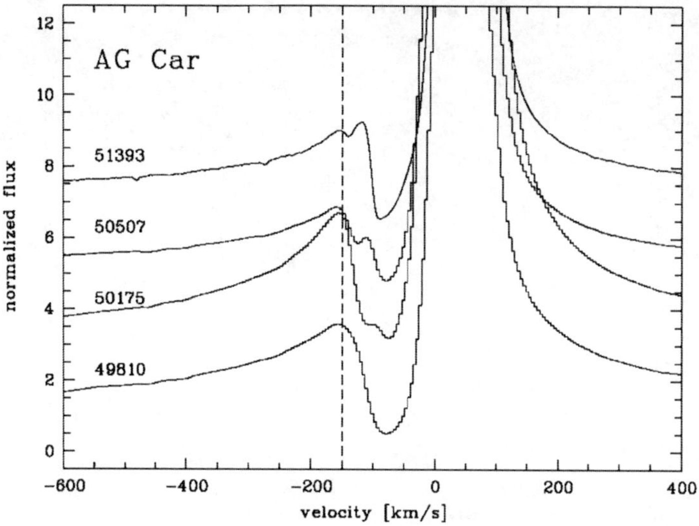

Fig. 10.2 The appearance of double-split absorption components in Hα in the spectrum of AG Car during the years 1996–1999. The spectra are marked in Julian days (The figure has been taken from Stahl et al. [116])

mm-wavelength, and Hubble Space Telescope data [14, 20, 136], and the non-LTE model results for η Car [47] yield a similar answer. Sophisticated NLTE mass-loss determinations for other LBVs include a value of $3 \times 10^{-5} M_{\odot}$ year^{-1} for the extreme twin of P Cyg: HD 316285 [46], while Najarro et al. [86] derived $3 \times 10^{-5} M_{\odot}$ year^{-1} for P Cyg itself, using similar analysis tools. These results are all much higher than for normal OB supergiants of comparable temperatures.

The non-LTE model based mass-loss determinations however are based on assumed spherical winds, but the wind of η Car, is bipolar [108, 125]. Whether non-spherical winds have a large effect on the mass-loss determination, remains to be shown. A more serious assumption may be homogeneity. Davies et al. [21] performed a linear spectropolarimetry survey of Galactic and MC LBVs and found large line polarizations in over half of their survey targets. This is a higher incidence of polarization line effects than in O and WR stars where asphericities of resp. $\sim$25% [43, 133] and $\sim$15% [42, 129] have been reported. Rather than attributing the polarization to large-scale axi-symmetry (e.g. [104]), Davies et al. [21] attribute the linear polarization of LBVs like AG Car to wind clumping, because the position angle in the polarization was shown to vary significantly, as was the case for P Cyg [87]. A similar conclusion regarding overall sphericity of LBV winds was drawn by Guo and Li [39] on the basis of modelling LBV continuum energy distributions.

The LBV polarization variability implies that the clumps must arise close to the photosphere, and many small clumps predominate over a few larger ones [22]. The quantitative implications of wind clumping for the absolute mass-loss rate of LBVs and in early-type stars, in general, have yet to be established [41].

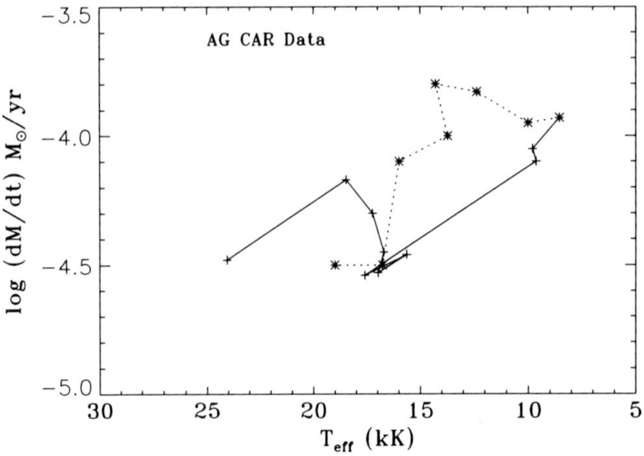

Fig. 10.3 Time-variable empirical mass-loss rates for AG Car as a function of $T_{\rm eff}$, as analyzed by Stahl et al. [116] (Figure taken from Vink and de Koter [130])

Although most LBVs have been monitored photometrically, only a handful have been subject to quantitative spectroscopic analysis at various epochs, and mass-loss rates have rarely been reported for different S Dor phases. In this respect, AG Car is the best studied LBV. Stahl et al. [116] investigated AG Car's mass-loss behavior over the period December 1990 – August 1999 and modelled the Hα profiles in detail. Their empirical mass-loss rates for the cycle from visual minimum to maximum – and back to minimum – are plotted against apparent temperature in Fig. 10.3. The mass-loss rate rises, drops, and rises again towards visual maximum (solid line) due to ionization changes of the Fe lines that drive the wind [130]. We note that there is a difference in mass-loss behavior from visual minimum to maximum and in the opposite direction (dotted line). We suspect that this is due to the breakdown of the assumption of stationarity from outburst to quiescence. Due to the larger radii, the dynamical flow times are much longer at maximum than they are at minimum light, which implies that material that was lost in this phase may still be near the photosphere, which may significantly affect the mass-loss determinations, resulting in erroneously large mass-loss rates for the route back to minimum. A second reason for the difference may be related to the release of gravitational energy when the star returns to quiescence. If this plays a role, the assumption of constant bolometric luminosity may no longer hold (see also [13,37]. Due to the above-mentioned complexities, we focus the comparison of mass-loss predictions to empirical mass-loss rates for the outburst phase (solid line) only.

10.3.2 Theoretical Mass-Loss Rates

Mass loss from a star with a stationary stellar wind is assumed to be due to an outward acceleration larger than the inward directed gravitational acceleration.

For early-type stars, this acceleration has been identified with the radiation force, which depends on both the available photospheric flux and the cross section of the particles that can intercept this radiation.

In hot-star winds, nearly all H is ionized by the strong radiation field, which implies that there is an enormous number of free electrons present which are the main contributors to the continuum opacity. The radiative acceleration due to photon scattering off free electrons is subject to the same $1/r^2$ radius dependence as is the gravitational acceleration, and for this very reason cannot drive a stellar wind by itself. Lucy and Solomon [76] showed that a stationary wind would occur when scattering by optically thick spectral lines was included. The interested reader is referred to the introductory book on stellar winds by Lamers and Cassinelli [63] for an overview of the line acceleration of optically thin and thick lines.

The line acceleration due to *all* spectral lines is often expressed in terms of the radiative acceleration due to electron scattering times a certain multiplication factor: the force multiplier $M(t)$. Using this method, one can parametrize the line force, and solve the equation of motion in a rather straightforward manner [11]. In this approach, the radiation is assumed to emerge directly from the star. The effects of diffuse radiation and multiple scatterings are not taken into account.

Abbott and Lucy [1] showed that calculated mass-loss rates can also be obtained using Monte Carlo simulations, counting the cumulative radiative accelerations due to photon interactions with gas particles of different chemical species (mostly Fe). However, the main challenge in radiation-driven wind dynamics is that the line acceleration $g_{\rm line}$ depends on the velocity gradient (dV/dr), but the velocity $V(r)$, hence dV/dr, in turn depends on $g_{\rm line}$. Due to its non-linear character, the dynamics of line-driven winds is quite complex. Fortunately, observational analyses provide accurate information on wind velocities, which can be used to constrain the wind dynamics.

Vink and de Koter [130] adopted an empirical velocity stratification,[2] $V(r)$, and predicted stationary mass-loss rates for LBVs as a function of $T_{\rm eff}$ in a similar vein to their mass-loss prescriptions for OB supergiants [132]. They studied the effects of lower masses and modified He/H and CNO abundances in comparison to normal OB supergiants and found that the main difference in mass-loss rate is attributable to the lower masses of LBVs compared to OB supergiants, resulting in a larger Eddington parameter $\Gamma_{\rm e}$. The increase in He abundance changed the mass-loss properties by only very small amounts (up to about 0.2 dex in log $\dot{M}$). CNO processing also had only a minor effect on the mass-loss rate, because Fe was found to be the dominant contributor to the line force in the inner wind. CNO lines contribute mostly to the line force in the outer wind where the terminal velocity is set [100, 131].

They [130] also compared their LBV mass-loss predictions with observational analyses and showed that the mass-loss variability during the S Dor cycles may arise from changes in the ionization balance of Fe: the bi-stability mechanism, first noticed in model calculations of the wind of P Cygni [96]. The wind either

[2] Note that our treatment of the winds dynamics has now been updated [85].

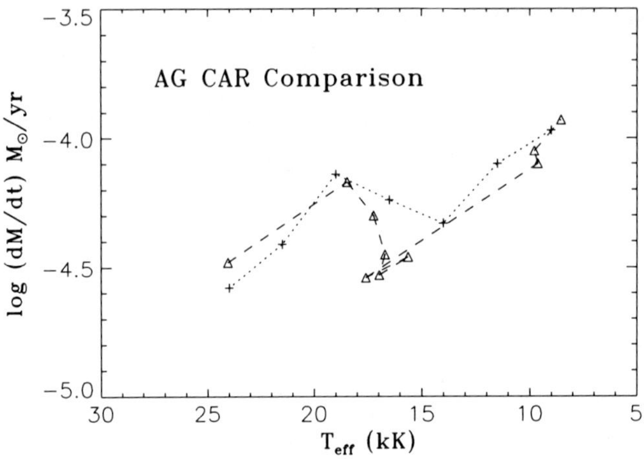

Fig. 10.4 Predicted (*dotted line*) and empirical (*dashed line*) mass-loss rates versus $T_{\rm eff}$ for the LBV AG Car. Note that both the qualitative behaviour and the amplitude of the mass-loss variations are well reproduced, provided the predictions are shifted by $\Delta T_{\rm eff} = -6{,}000$ K. See [130] for details

had a high v_∞ and low $\dot{M}$, or visa versa. The location of the jump near spectral type B1 (21,000 K) was established by Lamers et al. [67] from a v_∞ study of OB supergiants and improved by Crowther et al. [15]. The nature of the jump was originally attributed to the optical thickness of the Lyman continuum [65], but Vink et al. [131] showed that most of the line driving for both the hot and cool side of the jump was due to Fe in the Balmer continuum, with the jump in mass loss being the result of the recombination of Fe IV, as Fe III has more lines available to drive the wind. The first empirical evidence for a jump in $\dot{M}$ may have been found by Benaglia et al. [5], even though the rates at later spectral types appear to drop below those predicted [78, 121, 132].

The intriguing case of AG Car is depicted in Fig. 10.4, where we compare the predictions with the Stahl et al. [116] rates when the apparent temperature decreased from 24,000 to 9,000 K. In these computations, it was assumed that log $(L/L_\odot) = 6.0$; $M = 35\,M_\odot$; the He mass fraction Y $= 0.60$, and the ratio of the terminal over escape velocity was 1.3. The luminosity and He abundance are similar to those assumed by Stahl et al. We used the mass-loss behavior to constrain the stellar mass of the LBV. Unfortunately, the terminal velocity is poorly constrained by observations, as the v_∞ determination from Hα only allowed for a lower limit [116]. We note that the adopted ratio of the terminal over the effective escape velocity may impact the mass determination. Furthermore, the Fe recombination temperatures show an offset compared to empirical constraints from the drop in terminal velocities at spectral type B1 in OB supergiants.

Figure 10.4 shows that after accounting for a corrective shift $\Delta\ T_{\rm eff}$, the observed and predicted mass loss agree within $\simeq$0.1 dex. As $\dot{M}(T_{\rm eff})$ shows a complex behavior, with fluctuations of over 0.5 dex, this is a satisfactory result, confirming that AG Car's mass-loss variability is the result of changes in the ionization of the dominant line-driving element Fe.

10.3.3 Do S Dor Variables Form Pseudo-photospheres?

Now that we have gathered information on the empirical and theoretical mass-loss rates of LBVs, we can start addressing the question of whether these rates are large enough to be capable of forming a pseudo-photosphere. In most modern non-LTE atmosphere codes, the core radius follows from the relation $L = 4\pi R_{\rm in}^2 \sigma T_{\rm in}{}^4$. As the inner boundary is chosen to be deep in the stellar photosphere the input temperature does not necessarily equal the output effective temperature. The effective temperature $T_{\rm eff}$ is defined at the position where the thermalization optical depth at 5,555 Å equals $1/\sqrt{3}$. We intentionally choose the thermalization optical depth over purely thermal optical depth, as we wish to include the effects of dilution by scattering (see [27] and references therein for more extensive discussions). For stars with modest mass fluxes, such as normal O stars, the winds are optically thin and $T_{\rm eff}$ is only slightly lower than $T_{\rm in}$. For LBVs, with $\dot{M} \sim 10^{-4} M_{\odot}\,{\rm year}^{-1}$, there may be a significant difference between these temperatures. If the wind is so strong that the optical light originates from the depth of rapid acceleration, the object is considered to be forming a pseudo-photosphere.

The formation of pseudo-photospheres in LBVs may be favored by their lower masses, providing an increased mass-loss rate and an increase of the photospheric scale-height. Leitherer et al. [73] and de Koter et al. [27] assessed whether variable wind properties might explain $\Delta V \simeq$ 1–2 mags during S Dor cycles, and concluded that pseudo-photospheres are unlikely to form in LBVs. However, they did not investigate the effect of an order of magnitude change in the wind density of a star that is close to the bi-stability and Eddington limit.

Smith et al. [109] investigated the thermalization optical depth in the inner wind and showed that models on the cool side of the bi-stability jump may start to form optically thick winds which could lead to the formation of a pseudo-photosphere – *if* the objects are close to the Eddington limit. Figure 10.5 shows that for stars with log $(L/L_{\odot}) = 5.6$–5.8, if the current mass is below $\sim 12\,M_{\odot}$, corresponding to $\Gamma_{\rm e} = 0.8$, the star would start to form an extended optically thick wind when it crosses the bi-stability jump, and for stars with masses below $10.5\,M_{\odot}$ (at log $(L/L_{\odot}) = 5.7$) the pseudo-photosphere starts to form at $\rm T_{eff}$ 13,000 K. Applying the $\Delta T_{\rm eff}$ shift of $-6{,}000$ K to match the empirical temperature of the bi-stability jump would bring the effective temperature on the cool side of the bi-stability jump down to ~7,000 K, which agrees with the location of LBVs in eruption and the YHGs.

The scenario described above hinges critically on the large value of $\Gamma_{\rm e}$ of >0.8, but it remains to be seen if these high $\Gamma_{\rm e}$ values are realistic for LBVs. We quoted earlier $\Gamma_{\rm e}$ values in the range $\Gamma_{\rm e} \simeq 0.5$ for classical, and $\Gamma_{\rm e} \simeq 0.6$ for low-luminosity LBVs. It is indeed possible that the missing LBVs had larger $\Gamma_{\rm e}$ values, but clearly, more work is needed to determine LBV masses and luminosities to reliably establish the proximity of LBVs to the Eddington limit.

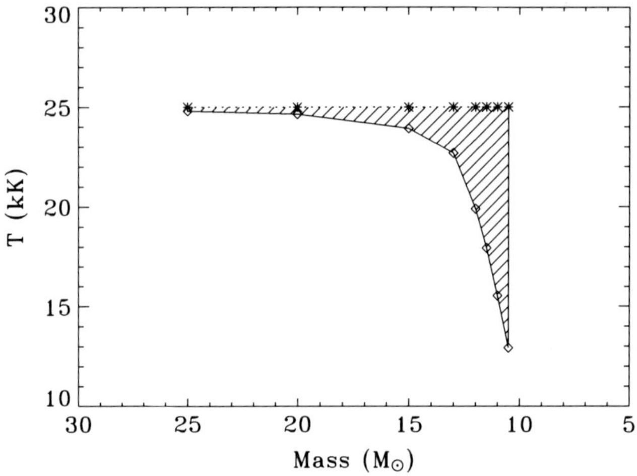

Fig. 10.5 The possible formation of a pseudo-photosphere. The figure shows the drop in apparent temperature that results from crossing the bi-stability jump which causes an order of magnitude increase in wind density. The apparent temperature drop from $T_{\rm in} = 25{,}000$ K is shown as a function of the stellar mass – for objects with $\log(L/L_\odot) = 5.7$. The computed effective output temperatures are denoted with $\diamond$. When the stellar mass drops below $M \sim 10.5 M_\odot$, the objects starts to form an extended pseudo-photosphere, which results in an effective temperature of ~13,000 K. Applying a corrective shift of 6,000 K [130] this comes down to $T_{\rm eff} \sim 7{,}000$ K – corresponding to the location of the YHGs. The figure indicates that supergiants that have lost significant mass during prior (e.g. RSG) phases may be susceptible to pseudo-photosphere formation (The figure has been adapted from [109])

10.4 Theoretical Models for S Dor Variability

In the previous section we addressed the issue of whether the S Dor variability is the result of a sub-photospheric effect that actually increases the radius resulting in a decrease in $T_{\rm eff}$, or whether the apparent decrease in temperature is due to an increase in the mass-loss rate and the formation of an the optically thick wind. It appears that the jury is still out on this, and we should thus consider both atmospheric as well as sub-photospheric mechanisms (see also [52]).

In those cases where a pseudo-photosphere would form as a result of increased mass loss, the root cause of such a sudden $\dot{M}$ increase still needs to be established. The bi-stability mechanism is a good candidate because it can account for observed LBV mass-loss behavior. In the following discussion we include bi-stability under the more general topic of "radiation pressure instabilities".

To investigate which mechanism might be responsible for the large visual brightness and spectroscopic variations in S Dor variables, we must also consider the momentum balance defined as:

$$-g_{\rm eff} = v\frac{dv}{dr} + \frac{1}{\rho}\frac{dP}{dr}\,. \qquad (10.3)$$

Here $g_{\rm eff}$ is the net result of gravity, $g_{\rm G}$, minus the outward-directed radiative and turbulent accelerations $g_{\rm rad}$ and $g_{\rm turb}$.

Radiation pressure instability Luminous evolved stars have reduced stability with respect to radiation pressure due to their reduced mass for their luminosity and are thus close to the Eddington limit. However massive stars are also rotating. The critical velocity is defined as $v_{\rm crit}^2 = (1 - \Gamma)GM/R$. When rotation is included via the $v_{\rm rot}^2$ term in the equation of motion, objects may become unstable when $\Omega = v_{\rm rot}/v_{\rm crit} > 1$, before arriving at the classical or opacity-modified Eddington limit [69]. For example, the projected rotational velocity for AG Car in the hot phase, of $190 \pm 30\,{\rm km\,s^{-1}}$ [36] is close to its critical velocity, and a similar result was recently found for HR Car as well [38]. Thus LBVs are considered to be in close proximity to both the Eddington and the Omega limits. In the sense usually intended, the Ω limit is really the Eddington limit modified by rotation.[3] Thus, in the following we include the Omega (Ω) limit under the general topic of the "Eddington limit".

Radiation-pressure driven instabilities occur because as the temperature drops, the opacity rises (e.g. due to bi-stability), and the radiative acceleration $g_{\rm rad}$ increases. The opacity–modified Eddington limit was initially introduced to explain the great eruption of η Car [17], and subsequently for the temperature dependence of the HD limit [51,62], and the instability of LBVs [3,4]. Lamers and Fitzpatrick [64] computed the location of the opacity-modified Eddington limit, including metal-line opacities from model atmospheres in addition to electron scattering. They suggested that S Dor variations could result from a conflict between a star's tendency to expand (following core H-burning) and strong mass loss close to the Eddington limit, requiring the star to shrink as the mass decreases. However there are issues with this simplified approach. When the ratio of radiative to gravitational force approaches unity, an instability could be expected, but as the atmosphere expands and density decreases, the ionization increases thereby reducing the absorptive opacity, and instead it approaches the classical Eddington limit due to electron scattering (which is not temperature dependent). However, if the instability would occur at Γ somewhat less than unity, the density decrease would not eliminate absorption, and the concept of the modified Eddington limit might nonetheless work.

The attraction of scenarios based on the Eddington limit are clear; they naturally explain the temperature dependent luminosity limit in the HRD, the S Dor variability, and the two states of LBVs, their high $T_{\rm eff}$ (just on the hot side of the bi-stability jump), and their low $T_{\rm eff}$ limits. They could also lead to enhanced mass loss, increased density in the winds and the formation of a pseudo-photosphere. However, there are no self-consistent models that provide a sound theoretical basis for for scenarios involving pseudo-photospheres [27].

---

[3]In principle this is a simplification of the real physics, because rotation tends to make the equator cooler than the poles (von Zeipel effect). The resulting temperature gradient affects opacity and possible instabilities, potentially making the combination of radiation pressure and rotation synergistic as noted in [65,97] as well as Sect. 5 of [141].

Turbulent pressure instability As a star approaches the Eddington limit, the outermost layers of the envelope become convective (e.g. [10]) and turbulent pressure gradients may provide an additional acceleration, $g_{\rm turb}$ to the momentum equation. De Jager [25] showed how supersonic turbulence may destabilize the atmosphere, and as the mechanism becomes more efficient at higher luminosity, the mass-loss rate increases. This is also true for radiation pressure forces and it may be difficult to distinguish between these two atmospheric instabilities.

Vibrations and dynamical instability Together with the radiation pressure-based instabilities, sub-photospheric dynamical mechanisms are the most promising explanation for the LBV/S Dor variability (see [34] and references therein). In these models, "strange modes" and dynamical instabilities are caused by the bump or increase in the opacity due to iron at the base of the photosphere leading to a strong ionization-induced instability in the outer envelope as stars transit the HRD after the end of core H-burning. In the models of Stothers and Chin [120] the star keeps re-adjusting itself on thermal timescales after periods of strong mass loss, whilst shrinking in radius. These models provide the correct S Dor timescales and also appear to "behave" properly at constant bolometric luminosity. The strange mode calculations by Glatzel and Kiriakidis [33] reproduce the S Dor instability strip and the upper luminosity boundary in the HRD quite well. Dynamical instability thus remains one of the more promising candidates to explain LBV variability.

Vibrational or pulsational instability was once thought to be one of the main contenders for instability and mass loss in the most massive stars (e.g. [2]), but the ε-mechanism is energized in the core, appears to grow too slowly (e.g. [91]) and is therefore no longer considered valid for LBVs. Another sub-photospheric instability, the κ mechanism responsible for pulsation in massive stars such as the Cepheids may also cause pulsations in the outer envelope. It may be responsible for some of the micro-variability seen in LBVs and other supergiants ([71] and references therein) which occur on timescales of weeks to months. The timescale of the S Dor variations however is much longer and therefore unlikely to be due to pulsations.

Binarity Most LBVs are apparently single, and although. η Car may have a companion, it seems clear that binarity can neither be the root cause of the S Dor variations, or for the giant eruptions, as the only other local example, P Cyg, is single (but see [60], or unless it formed through merging, cf. [98]).

10.5 Evolutionary State

There is no doubt that LBVs are evolved, unstable massive hot stars. The more massive classical LBVs have apparently evolved off the main sequence, while the less luminous LBVs may be post-red supergiants. In the generally accepted view of massive star evolution, the classical LBVs are considered "transitional" objects in a phase before entering the He-burning WR stage [70], by the end of which the star is anticipated to explode as a type Ib/c supernova. Many LBVs are known to be N and He rich compared to O stars, but H rich compared to the more evolved WR stars. This situation is somewhat more complex as there is also a group of high-luminosity late-type H-rich WR stars, which appear closely related to many classical LBVs in quiescence the Ofpe/WN stars [134]. These and other luminous stars, the B[e] supergiants [68], and and the cool or yellow Hypergiants (YHGs) [26], that show evidence for high mass loss and instabilities which may be related to the LBV state.

10.5.1 The Evolutionary Neighbours

Ofpe/late-WN stars The "slash" stars are a group of luminous hot stars with very strong emission lines due to their strong mass loss. They have He and N enhanced atmospheres indicative of an evolved state [92], but are generally not believed to be highly variable. These stars however may be closely related to the LBVs in quiescence. The S Dor-type variable, R127, was a late-WN star prior to its long-term outburst beginning in the early 1980s [114]. Either the late-WN stars evolve *into* LBVs, or they may represent a dormant phase of LBV evolution. Either way, Ofpe/late-WN stars are thought to be evolved massive stars in a transitional stage for objects with initial masses $M > 50–60\,M_\odot$. Since no evolved stars are observed redwards of the Humphreys-Davidson (HD) limit at these high luminosities and masses , high mass loss during the Ofpe/late-WN and LBV phases may reverse the evolutionary track back to the hotter part of the HRD, where they should appear as He-burning WR stars.

B[e] supergiants The spectra of the B[e] stars [68] show an abundance of high-excitation permitted and forbidden emission lines that are thought to arise from an equatorially enhanced outflowing disk. Zickgraf et al. [142] proposed a 2-component wind with a normal fast polar wind and a dense slow outflowing equatorial "disk". A popular mechanism to explain this 2-component wind is the rotationally induced bi-stability mechanism [65, 97]. The pole is hotter than the equator, due to the Von Zeipel gravity darkening effect which could lead to a fast, low $\dot{M}$, polar wind driven by Fe IV lines, and a slow, high $\dot{M}$, equatorial wind driven by the more effective Fe III lines [131]. This mechanism is expected to occur predominantly at spectral type B. However the star is expected to rotate rapidly but $v_{\rm rot}$ measurements are difficult for B[e] stars, with most of the lines in emission.

The B[e] supergiants may represent a subset of massive stars with high rotational velocities. The B[e] supergiants were originally not thought to be variable, but there is now evidence for large amplitude variability for some B[e] supergiants [127], suggesting a closer evolutionary link between LBVs and B[e] supergiants than previously acknowledged.

Yellow or cool hypergiants YHGs are found just below the HD-limit [50, 52] at intermediate temperatures with A to G spectral types. Many of these stars show spectroscopic and photometric variability, high mass-loss rates, large infrared excesses and visible circumstellar ejecta, all evidence for instability. The YHGs are often assumed to be post-RSGs [26, 90], although the evolutionary state is not established for all of them. The intriguing object IRC +10420 has been shown to be a post-RSG [59, 89] and numerous studies have revealed its complex ejecta [53, 55] and large-scale asymmetry [23, 28, 95]. Given their variability and high mass loss, the YHGs are likely close to their Eddington limit, with a large Γ_e, and it is thus probable that many of them are post-RSGs Nevertheless, it is not clear whether objects like IRC +10420 are on an evolutionary blueward journey towards the WR phase [55, 89] bouncing against the yellow void [26], or are on the cool side of the bi-stability jump [109].

10.5.2 Do LBVs Explode?

The picture of the LBVs as a high mass loss, relatively short-lived (some 10^4 years) and presumably core H-burning phase prior to a much longer (a few times 10^5 years) core He-burning WR phase seemed well established – until recently.

There is increasing observational evidence that LBVs could be *direct* progenitors of some SNe. Kotak and Vink [61] proposed that the quasi-periodic modulations seen in the radio light curves of transitional SNe such as SN 2001ig and SN 2003bg are the manifestation of variable mass loss during S Dor excursions. Although several other possibilities have been put forward to explain these modulations [102, 112], none has been entirely satisfactory. The recurrence timescale of the variability, as well as the amplitude of the radio modulations are consistent with those of S Dor variables and their scenario [61] provides a rather natural explanation for a behaviour that is expected on theoretical grounds [130].

The same wind bi-stability mechanism may be able to account for wind-velocity variations seen spectroscopically in SN 2005gj [122] in which the variable winds are inferred from double P Cygni components (see Fig. 10.6) which appear almost identical to those seen in the Hα profiles of S Dor variables like AG Car and HD 160529 (see also [35]). Both the timescales and the spectroscopically measured wind velocities of SN 2005gj, with $v_\infty \simeq 100$–$200\,\mathrm{km\,s^{-1}}$, are consistent with those of LBVs, but not with those of the much slower RSG winds ($\sim 10\,\mathrm{km\,s^{-1}}$), or the much faster WR winds ($\simeq 1{,}000$–$5{,}000\,\mathrm{km\,s^{-1}}$). See Van Marle et al. [128] and [122] for more information.

The progenitor star of the recent supernova SN 2006jc had a giant eruption just 2 years before its terminal explosion. Foley et al. [30] and Pastorello et al. [94] suggested that the progenitor star was either a WR star that exhibited an "LBV or η Car-like" eruption, or that the progenitor was part of a binary system including both an LBV and a WR star, with the WR star exploding and the giant eruption attributable to the LBV, as WR stars have never been observed to have an η Car type eruptions. A more direct application of Occam's razor would be to accept that the progenitor object exploded during or at the end of the LBV phase or an η Car-like giant eruption [61].

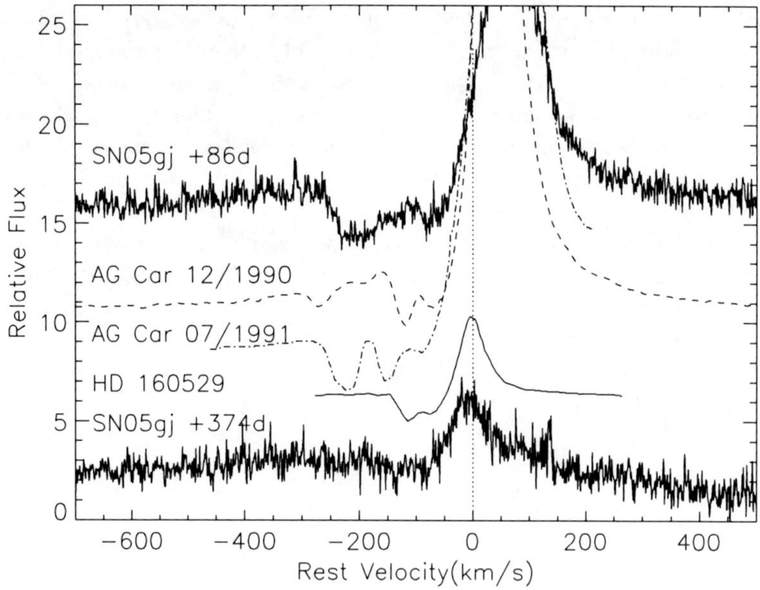

Fig. 10.6 Multiple absorption components seen in the P Cygni Hα profile of SN 2005gj (*top*) in comparison to the LBVs AG Car and HD 160529 (The figure has been taken from Trundle et al. [122])

Gal-Yam et al. [31] discovered a luminous source (with $M_{\rm bol} = -10.3$) in the pre-explosion image of SN 2005gl. Although, the properties of the progenitor are consistent with those of LBVs, they are equally consistent with a luminous blue supergiant that has not exhibited spectral type variations (see the membership discussion in Sect. 10.2), and the progenitor of SN 2005gl may potentially be classified as an LBVc. The resulting SN explosion was of type IIn, indicating the presence of a dense circumstellar medium, additional evidence for prior η Car-type eruptions.

Other hints that LBVs may explode come from similarities in the morphology of LBV nebula and the circumstellar medium of SN 1987A [105], while the very luminous SN 2006gy – hypothesized to be an exotic pair-instability SN – may also have undergone an η Car-type eruption before exploding [110]. An alternative scenario for a giant η Car-type outburst was suggested by Woosley et al. [139] who attributed the dense shells around this luminous SN to pulsational pair instability.[4]

We emphasize that the evolutionary status of LBVs remains uncertain (see the discussion in [29]). Evolutionary models have been constructed to allow LBVs in a transitional phase between a core H-burning main sequence and a core He-burning WR phase. This model naturally accounts for the the chemical abundances (He, N)

[4]See Chap. 13.

of LBVs which are intermediate between those of O and WR stars. The concept of an LBV exploding is certainly at odds with current stellar evolution models and an exploding LBV scenario was until recently considered “wildly speculative” [106]. Nevertheless, the evolutionary models do not provide a straightforward explanation for the wide range of phenomena described above and a simple explanation for this could be that at least some massive stars in an LBV state could precede the terminal explosion. This would allow some massive stars to skip the WR phase – in contradiction to the basic framework of massive star evolution. This also suggests that LBVs are already in the core He-burning phase of evolution. Given the intriguing variations seen in radio light curves and the double absorption features in the P Cygni line profiles, I suggest that the changing winds of LBVs may help address the issue of whether the LBV phase may indeed represent the evolutionary endpoint for some of the most massive stars.

10.6 Outlook

We have examined observational, atmospheric modelling, and theoretical aspects of the current status of our knowledge of the LBV instability and their role in massive star evolution. Although radiation pressure as well as a dynamical instability are strong candidates for explaining S Dor variability, we conclude that the mechanism at the origin of the LBV phenomenon remains elusive. One of the most relevant issues to be addressed relates to the nature of the S Dor variability itself. With respect to massive star evolution, and in particular whether LBVs are in a transitional or final phase of evolution, the last few years have seen a flurry of activity, but we should not yet draw any definitive conclusions regarding their evolutionary state.

Progress is expected to be made on a number of fronts. First of all, the number of known LBVs is small and for a proper understanding of any individual member, photometric, polarimetric, and spectroscopic monitoring on the timescale of the S Dor variations is required. Atmospheric modelling is necessary to determine the stellar parameters and place the LBVs properly onto the HRD. There are still enormous uncertainties in the values shown in Fig. 10.1. Furthermore, LBV masses, nor their proximity to the Eddington limit, are known with any level of certainty, with profound consequences for theoretical interpretation. Therefore, binary searches, orbit determinations, and spectroscopic modelling are strongly encouraged to determine this most basic parameter.

As is the case for the normal OB supergiants, the LBV mass-loss rates are uncertain due to wind clumping. Progress will undoubtedly be made regarding the role of of wind clumping and its impact on mass-loss rates in OB stars and LBVs. However, the LBVs themselves provide an ideal laboratory for studies of wind clumping, because the polarization variability is most extreme, due to the combined effect of low outflow velocities and high mass-loss rates [21]. Mass-loss variability – in conjunction with $T_{\rm eff}$ determinations – could also be utilized to constrain the stellar masses, as was exemplified for the object AG Car [130].

A number of developments provide new opportunities for LBV research. The infrared era has properly come of age with satellites such as *Spitzer*. Recent activities involve both in-depth studies of known LBVs [123], as well as the discoveries of new ones, even beyond the Local Group. The so-called intermediate luminosity "red" transients, where "intermediate" means that their luminosities are in between classical novae and supernovae are examples. The nature of these transients, incl. SN 2008S and NGC 300 OT2008-1, remains as yet elusive. Some authors have argued for a SN origin [8, 99], whilst others [6, 111] suggest they are due to the non-terminal outburst of an LBV-type star or an evolved intermediate mass, post-AGB star [7, 56]. Another intriguing direction for LBV research is provided by the detection of LBVs and candidates in very low metallicity environments [44, 58, 101]. The mere presence of LBVs at such low metallicities presents a challenge for theory, because of the reduced opacity available to drive an LBV wind.

Further theoretical work on the formation of pseudo-photospheres and the more general question regarding the origin of the LBV variations are badly needed. Such studies may result in a better understanding of the origin of the variability which is necessary to place the LBV phase correctly in the evolution of massive stars.

The final message to emerge from this chapter is that η Car may be one of the most extreme LBVs but it is not unique among the LBVs. Given that η Car's second outburst (during 1888–1895) was like that of a normal S Dor variable, the key to our understanding of η Car's great eruption may thus not exclusively lie in the study of η Car itself, but pivotal clues may be obtained through a better understanding of the more typical S Dor variations that define the LBVs as a class among the most massive stars.

The most outstanding question for all these stars, S Dor-type LBVs and the giant eruptions like η Car, is still, what is the underlying origin of their instabilities?

Acknowledgements I am grateful for the many discussions I have had with colleagues from the massive star community and in particular to Alex de Koter, Henny Lamers, and Rubina Kotak.

References

1. D.C. Abbott, L.B. Lucy, Multiline transfer and the dynamics of stellar winds. ApJ. **288**, 679–693 (1985)
2. I. Appenzeller, Mass loss rates for vibrationally unstable very massive main-sequence stars. A&A. **9**, 216–220 (1970)
3. I. Appenzeller, Instability in massive stars – an overview. in *Proceedings of the IAUS*, vol. 116 (D. Reidel, Dordrecht, 1986), pp. 139–148
4. I. Appenzeller, The role of radiation pressure in LBV atmospheres, in *Proceedings of the IAUC*, ed. by K. Davidson, A.F.P. Moffat, H. Lamersvol, vol. 113 (Kluwer, Dordrecht, 1989), pp. 195–202
5. P. Benaglia, J.S. Vink, J. Marti, J. Maiz Apellaniz, B. Koribalski, P.A. Crowther, Testing the predicted mass-loss bi-stability jump at radio wavelengths. A&A. **467**, 1265–1274 (2007)
6. E. Berger, A.M. Soderberg, R.A. Chevalier et al., An intermediate luminosity transient in NGC 300: the eruption of a dust-enshrouded massive star. ApJ. **699**, 1850 (2009)

7. H.E. Bond, L.R. Bedin, A.Z. Bonanos et al., The 2008 luminous optical transient in the nearby Galaxy NGC 300. ApJ. **695**, 154 (2009)
8. M.T. Botticella, A. Pastorello, S.J. Smartt et al., SN 2008S: an electron-capture SN from a super-AGB progenitor? MNRAS. **398**, 1041 (2009)
9. C. Buemi, E. Distefano, P. Leto et al., Photometric monitoring of Luminous Blue Variables. BSRSL 80, 335 (2011)
10. M. Cantiello, N. Langer, I. Brott et al., Sub-surface convection zones in hot massive stars and their observable consequences. A&A. **499**, 279–290 (2009)
11. J.I. Castor, D.C. Abbott, R.I. Klein, Radiation-driven winds in of stars. ApJ. **195**, 157–174 (1975)
12. J.S. Clark, V.M. Larionov, A. Arkharov, On the population of galactic Luminous Blue Variables. A&A. **435**, 239–246 (2005)
13. J.S. Clark, P.A. Crowther, V.M. Larionov et al., Bolometric luminosity variations in the luminous blue variable AFGL2298. A&A. **507**, 1555–1565 (2009)
14. P. Cox, P.G. Mezger, A. Sievers et al., Millimeter emission of eta Carinae and its surroundings. A&A. **297**, 168–174 (1995)
15. P.A. Crowther, D.J. Lennon, N.R. Walborn, Physical parameters and wind properties of galactic early B supergiants. A&A. **446**, 279–293 (2006)
16. A. Damineli, The 5.52 year cycle of Eta Carinae. ApJ. **460**, L49–L52 (1996)
17. K. Davidson, On the nature of Eta Carinae. MNRAS. **154**, 415–427 (1971)
18. K. Davidson, The relation between apparent temperature and mass-loss rate in hypergiant eruptions. ApJ. **317**, 760–764 (1987)
19. K. Davidson, N.R. Walborn, T.R. Gull, The remarkable spectrum of some material ejected by Eta Carinae. ApJ. **254**, L47–L51 (1982)
20. K. Davidson, D. Ebbets, G. Weigelt, HST/FOS spectroscopy of ETA Carinae: the star itself, and ejecta within 0.3 arcsec. AJ. **109**, 1784–1796 (1995)
21. B. Davies, R.D. Oudmaijer, J.S. Vink, Asphericity and clumpiness in the winds of Luminous Blue Variables. A&A. **439**, 1107–1125 (2005)
22. B. Davies, J.S. Vink, R.D. Oudmaijer, Modelling the clumping-induced polarimetric variability of hot star winds. A&A. **469**, 1045–1056 (2007)
23. B. Davies, R.D. Oudmaijer, K.C. Sahu, Integral-field spectroscopy of the post-red supergiant IRC + 10420: evidence for an axisymmetric wind. ApJ. **671**, 2059–2067 (2007)
24. M. de Groot, C. Sterken, A.M. van Genderen, Cyclicities in the light variations of S Doradus stars III. P Cygni. A&A. **376**, 224–231 (2001)
25. C. de Jager, The stability limit of hypergiant photospheres. A&A. **138**, 246–252 (1984)
26. C. de Jager, The yellow hypergiants. A&ARv. **8**, 145–180 (1998)
27. A. de Koter, H.J.G.L.M. Lamers, W. Schmutz, Variability of Luminous Blue Variables. II. Parameter study of the typical LBV variations. A&A. **306**, 501–518 (1996)
28. W.J. de Wit, R.D. Oudmaijer, M.A.T. Groenewegen, M.G. Hoare, F. Malbet, Resolving the ionized wind of the post-red supergiant IRC +10 420 with VLTI/AMBER. A&A. **480**, 149–155 (2008)
29. V.V. Dwarkadas, On Luminous Blue Variables as the progenitors of core-collapse supernovae, especially type IIn supernovae. MNRAS. **412**, 1639–1649 (2011)
30. R.J. Foley, N. Smith, M. Ganeshalingam et al., SN 2006jc: a wolf-rayet star exploding in a dense he-rich circumstellar medium. ApJ. **657**, L105–L108 (2007)
31. A. Gal-Yam, D.C. Leonard, D.B. Fox et al., On the progenitor of SN 2005gl and the nature of type IIn supernovae. ApJ. **656**, 372–381 (2007)
32. G. Garcia-Segura, N. Langer, M.-M. Mac Low, The dynamical evolution of circumstellar gas around massive stars. I. The impact of the time sequence Ostar – LBV – WR star. A&A. **305**, 229–244 (1996)
33. W. Glatzel, M. Kiriakidis, Stability of massive stars and the Humphreys-Davidson limit. MNRAS **263**, 375–384 (1993)
34. G. Grafener, S.P. Owocki, J.S. Vink, Stellar envelope inflation near the Eddington limit. Implications for the radii of Wolf-Rayet stars and luminous blue variables. A&A. 538, idA40 (2012)

35. J.H. Groh, J.S. Vink, The bi-stability jump as the origin for multiple P-Cygni absorption components in Luminous Blue Variables. A&AL. **513**, 10–13 (2011)
36. J.H. Groh, D.J. Hillier, A. Damineli, AG Carinae: a luminous blue variable with a high rotational velocity. ApJ. **638**, L33–L36 (2006)
37. J.H. Groh, D.J. Hillier, A. Damineli, P.A. Whitelock, F. Marang, C. Rossi, On the nature of the prototype luminous blue variable AG Carinae. I. Fundamental parameters during visual minimum phases and changes in the bolometric luminosity during the S-Dor cycle. ApJ. **698**, 1698–1720 (2009)
38. J.H. Groh, D.J. Hillier, A. Damineli et al., Bona fide, strong-variable galactic luminous blue variable stars are fast rotators: detection of a high rotational velocity in HR Carinae. ApJL. **705**, L25–L30 (2009)
39. J.H. Guo, Y. Li, Estimating the structure and geometry of winds from Luminous Blue Variables via fitting the continuum energy distributions. ApJ. **659**, 1563–1575 (2007)
40. V.V. Gvaramadze, A.Y. Kniazev, S. Fabrika et al., MN112: a new Galactic candidate Luminous Blue Variable. MNRAS. **405**, 520–524 (2010)
41. W.-R. Hamann, A. Feldmeier, L.M. Oskinova, *Clumping in Hot-Star Winds: Proceedings of an International Workshop held in Potsdam*, Germany, 18–22 June 2007, Potsdam Universitätsverlag, Potsdam, 2008
42. T.J. Harries, D.J. Hillier, I.D. Howarth, A spectropolarimetric survey of northern hemisphere Wolf-Rayet stars. MNRAS. **296**, 1072–1088 (1998)
43. T.J. Harries, I.D. Howarth, C.J. Evans, Spectropolarimetry of O supergiants. MNRAS. **337**, 341–355 (2002)
44. A. Herrero, M. Garcia, K. Uytterhoeven et al., The nature of V39: an LBV candidate or LBV impostor in the very low metallicity galaxy IC 1613? A&A. **513**, 70 (2010)
45. D.J. Hillier, D.L. Miller, The treatment of non-LTE line blanketing in spherically expanding outflows. ApJ. **496**, 407 (1998)
46. D.J. Hillier, P.A. Crowther, F. Najarro, A.W. Fullerton, An optical and near-IR spectroscopic study of the extreme P Cygni-type supergiant HDE 316285. A&A. **340**, 483–496 (1998)
47. D.J. Hillier, K. Davidson, K. Ishibashi, T. Gull, On the nature of the central source in η Carinae. ApJ. **553**, 837–860 (2001)
48. E. Hubble, A. Sandage, The brightest variable stars in extragalactic nebulae. I. M31 and M33. ApJ. **118**, 353–361 (1953)
49. R.M. Humphreys, Studies of luminous stars in nearby galaxies I supergiants and O stars in the Milky Way. ApJ. Suppl. **38**, 309 (1978)
50. R.M. Humphreys, K. Davidson, Studies of luminous stars in nearby galaxies. III – comments on the evolution of the most massive stars in the Milky Way and the Large Magellanic Cloud. ApJ. **232**, 409–420 (1979)
51. R.M. Humphreys, K. Davidson, The most luminous stars. Science **223**, 243–249 (1984)
52. R.M. Humphreys, K. Davidson, The Luminous Blue Variables: astrophysical geysers. PASP. **106**, 1025–1051 (1994)
53. R.M. Humphreys, N. Smith, K. Davidson, et al., HST and infrared images of the circumstellar environment of the cool hypergiant IRC + 10420. AJ. **114**, 2778 (1997)
54. R.M. Humphreys, K. Davidson, N. Smith, Eta Carinae's second eruption and the light curves of the eta Carinae variables. PASP. **111**, 1124–1131 (1999)
55. R.M. Humphreys, K. Davidson, N. Smith, Crossing the yellow void: spatially resolved spectroscopy of the post-red supergiant IRC + 10420 and its circumstellar ejecta. AJ. **124**, 1026–1044 (2002)
56. R.M. Humphreys, H. E. Bond, L. Bedin, A.Z. Bonanos, K. Davidson, B. L. A. G. Monard, J. Prieto, F. M. Walter, The photometric and spectral evolution of the 2008 luminous optical transient in NGC 300. ApJ. **743**, id 118 (2011)
57. J.B. Hutchings, Stellar winds from hot supergiants. ApJ. **203**, 438 (1976)
58. Y.I. Izotov, T.X. Thuan, Luminous blue variable stars in the two extremely metal-deficient blue compact dwarf galaxies DDO 68 and PHL 293B. ApJ. **690**, 1797–1806 (2009)

59. T.J. Jones, R.M. Humphreys, R.D. Gehrz et al., IRC + 10420 – A cool hypergiant near the top of the H-R diagram. ApJ. **411**, 323–335 (1993)
60. A. Kashi, An indication for the binarity of P Cygni from its seventeenth century eruption, MNRAS. in press, http://arxiv.org/abs/0912.3998 (2010)
61. R. Kotak, J.S. Vink, Luminous Blue Variables as the progenitors of supernovae with quasi-periodic radio modulations. A&A. **460**, L5–L8 (2006)
62. H.J.G.L.M. Lamers, P Cygni type stars – evolution and physical processes. in *Proceedings of the IAUS*, vol. 116 (D. Reidel, Dordrecht, 1986), pp. 157–178
63. H.J.G.L.M. Lamers, J.P. Cassinelli, *Introduction to Stellar Winds* (Cambridge University Press, Cambridge, 1999)
64. H.J.G.L.M. Lamers, E.L. Fitzpatrick, The relationship between the Eddington limit, the observed upper luminosity limit for massive stars, and the Luminous Blue Variables. ApJ. **324**, 279–287 (1988)
65. H.J.G.L.M. Lamers, A.W.A. Pauldrach, The formation of outflowing disks around early-type stars by bi-stable radiation-driven winds. A&A. **244**, L5–L8 (1991)
66. H.J.G.L.M. Lamers, P. Korevaar, A. Cassatella, The ejection of shells in the stellar wind of the hypergiant P Cygni (B1 Ia/+/). A&A. **149**, 29–40 (1985)
67. H.J.G.L.M. Lamers, Th.P.Snow, D.M. Lindholm, Terminal velocities and the bistability of stellar winds. ApJ. **455**, 269 (1995)
68. H.J.G.L.M. Lamers, F.-J. Zickgraf, D. de Winter, L. Houziaux, J. Zorec, An improved classification of B[e]-type stars. A&A. **340**, 117–128 (1998)
69. N. Langer, The Eddington limit in rotating massive stars. ASPC. **120**, 83 (1997)
70. N. Langer, W-R. Hamann, M. Lennon et al., Towards an understanding of very massive stars. A new evolutionary scenario relating O stars, LBVs and Wolf-Rayet stars. A&A. **290**, 819–833 (1994)
71. K. Lefever, J. Puls, C. Aerts, Statistical properties of a sample of periodically variable B-type supergiants. Evidence for opacity-driven gravity-mode oscillations. A&A. **463**, 1093–1109 (2007)
72. C. Leitherer, Mass loss from LBVs: observational constraints. ASPC. **120**, 58–65 (1997)
73. C. Leitherer, W. Schmutz, D.C. Abbott, W.-R Hamann, U. Wessolowski, Atmospheric models for Luminous Blue Variables. ApJ. **346**, 919–931 (1989)
74. D.J. Lennon, D. Wobig, R.-P. Kudritzki, O. Stahl, The atmospheric composition, extinction and luminosity of the LBV star R71. SSRv. **66**, 207 (1993)
75. L.B. Lucy, An analysis of the variable radial velocity of alpha Cygni. ApJ. **206**, 499–508 (1976)
76. L.B. Lucy, P.M. Solomon, Mass loss by hot stars. ApJ. **159**, 879–893 (1970)
77. A. Maeder, Evolution of chemical abundances in massive stars. I – OB stars, Hubble-Sandage variables and Wolf-Rayet stars – Changes at stellar surfaces and galactic enrichment by stellar winds. II – Abundance anomalies in Wolf-Rayet stars in relation with cosmic rays and 22/Ne in meteorites. A&A. **120**, 113–135 (1983)
78. N. Markova, J. Puls, Bright OB stars in the galaxy. IV. Stellar and wind parameters of early to late B supergiants. A&A. **478**, 823–842 (2008)
79. N. Markova, N. Morrison, I. Kolka, H. Markov, P Cygni in a short S Doradus phase. Spectroscopic and photometric evidences. A&A. **376**, 898–906 (2001)
80. F. Martins, S. Trippe, T. Paumard et al., GCIRS 16SW: a massive eclipsing binary in the galactic center. ApJ. **649**, L103–L106 (2006)
81. O. Maryeva, P. Abolmasov, Modeling the optical spectrum of Romano's star. MNRAS. submitted (arXiv:1109.0443), (2011)
82. P. Massey, R.T. McNeill, K.A.G. Olsen et al., A survey of local group galaxies currently forming stars. III. A search for Luminous Blue Variables and other H emission-line stars. AJ. **134**, 2474–2503 (2007)
83. J.C. Mauerhan, M.R. Morris, A. Cotera et al., Discovery of a luminous blue variable with an ejection nebula near the quintuplet cluster. ApJ. **713**, 33 (2010)

84. G. Meynet, A. Maeder, Stellar evolution with rotation. XI. Wolf-Rayet star populations at different metallicities. A&A. **429**, 581–598 (2005)
85. P.E. Müller, J.S. Vink, A consistent solution for the velocity field and mass-loss rate of massive stars. A&A. **492**, 493–509 (2008)
86. F. Najarro, D.J. Hillier, O. Stahl, A spectroscopic investigation of P Cygni. I. H and HeI lines. A&A. **326**, 1117–1134 (1997)
87. K.H. Nordsieck, J. Wisniewski, B.L. Babler et al., Ultraviolet and visible spectropolarimetric variability in P Cygni. ASPC. **233**, 261–274 (2001)
88. A. Nota, M. Livio, M. Clampin, R. Schulte-Ladbeck, Nebulae around Luminous Blue Variables: a unified picture. ApJ. **448**, 788–796 (1995)
89. R.D. Oudmaijer, M.A.T. Groenewegen, H.E. Matthews, J.A.D.L. Blommaert, K.C. Sahu, The spectral energy distribution and mass-loss history of IRC + 10420. MNRAS. **280**, 1062–1070 (1996)
90. R.D. Oudmaijer, B. Davies, W.-J. de Wit, M. Patel, Post-red supergiants, in *ASP Conf. Series*, vol. 412, ed. by D.G. Luttermoser, B.J. Smith, R.E. Stencel (Astronomical Society of the Pacific, San Francisco, 2009), p. 17
91. J.C.B. Papaloizou, Non-linear pulsations of upper main sequence stars-II.Direct numerical integrations. MNRAS. **162**, 169 (1973)
92. A. Pasquali, N. Langer, W. Schmutz, O stars in transition. II. Fundamental properties and evolutionary status of Ofpe/WN9 stars from Hubble space telescope ultraviolet observations. ApJ. **478**, 340–357 (1997)
93. A. Pasquali, A. Nota, L.J. Smith et al., Multiwavelength study of the nebula associated with the galactic LBV candidate HD 168625. AJ. **124**, 1625–1635 (2002)
94. A. Pastorello, S.J. Smartt, S. Mattila et al., A giant outburst two years before the core-collapse of a massive star. Nature **447**(7146), 829–832 (2007)
95. M. Patel, R.D. Oudmaijer, J.S. Vink et al., Spectropolarimetry of the massive post-red supergiants IRC + 10420 and HD 179821. MNRAS. **385**, 967–978 (2008)
96. A.W.A. Pauldrach, J. Puls, Radiation-driven winds of hot stars. VIII – The bistable wind of the luminous blue variable P Cygni (B1 Ia/+/). A&A. **237**, 409–424 (1990)
97. I. Pelupessy, H.J.G.L.M. Lamers, J.S. Vink, The radiation driven winds of rotating B[e] supergiants. A&A. **359**, 695–706 (2000)
98. P. Podsiadlowski, P.C. Joss, S. Rappaport, A merger model for SN 1987 A. A&A. **227**, L9–L12 (1990)
99. J.L. Prieto, M.D. Kistler, T.A. Thompson, Discovery of the dust-enshrouded progenitor of SN 2008S with spitzer. ApJ. **681**, 9 (2008)
100. J. Puls, U. Springmann, M. Lennon, Radiation driven winds of hot luminous stars. XIV. Line statistics and radiative driving. A&AS. **141**, 23–64 (2000)
101. S.A. Pustilnik, A.L. Tepliakova, A.Y. Kniazev, A.N. Burenkov, Discovery of a massive variable star with Z = Zsolar/36 in the galaxy DDO 68. MNRAS. **388**, L24–L28 (2008)
102. S.D. Ryder, E.M. Sadler, R. Subrahmanyan et al., Modulations in the radio light curve of the Type IIb supernova 2001ig: evidence for a Wolf-Rayet binary progenitor? MNRAS. **349**, 1093–1100 (2004)
103. G. Schaller, D. Schaerer, G. Meynet, A. Maeder, New grids of stellar models from 0.8 to 120 solar masses at Z = 0.020 and Z = 0.001. A&AS. **96**, 269–331 (1992)
104. R.E. Schulte-Ladbeck, G.C. Clayton, D.J. Hillier, T.J. Harries, I.D. Howarth, The axisymmetric stellar wind of AG Carinae. ApJ. **429**, 846–856 (1994)
105. N. Smith, Discovery of a nearby twin of SN 1987A's nebula around the luminous blue variable HD 168625: was Sk -69 202 an LBV? AJ. **133**, 1034–1040 (2007)
106. N. Smith, S.P. Owocki, On the role of continuum-driven eruptions in the evolution of very massive stars and population III stars. ApJ. **645**, L45–L48 (2006)
107. L.J. Smith, A. Nota, A. Pasquali et al., Ejected nebulae as probes of the evolution of massive stars in the Large Magellanic Cloud. ApJ. **503**, 278 (1998)
108. N. Smith, K. Davidson, Th.R. Gull, K. Ishibashi, D.J. Hillier, Latitude-dependent effects in the stellar wind of eta Carinae. ApJ. **586**, 432–450 (2003)

109. N. Smith, J.S. Vink, A. de Koter, The missing Luminous Blue Variables and the bistability jump. ApJ. **615**, 475–484 (2004)
110. N. Smith, W. Li, R.J. Foley et al., SN 2006gy: discovery of the most luminous supernova ever recorded, powered by the death of an extremely massive star like Eta Carinae. ApJ. **666**, 1116 (2006)
111. N. Smith, M. Ganeshalingam, R. Chornock et al., ApJ. **697**, 49 (2009)
112. A.M. Soderberg, R.A. Chevalier, S.R. Kulkarni, D.A. Frail, The radio and X-Ray luminous SN 2003bg and the circumstellar density variations around radio supernovae. ApJ. **651**, 1005–1018 (2006)
113. O. Stahl, B. Wolf, Circumstellar shells around luminous emission-line stars in the Large Magellanic Cloud. A&A. **158**, 371–381 (1986)
114. O. Stahl, B. Wolf, G. Klare et al., R 127 – an S DOR type variable intermediate between Of and WN. A&A. **127**, 49–62 (1983)
115. O. Stahl, B. Wolf, G. Klare, A. Juettner, A. Cassatella, Observations of the new luminous blue variable R 110 of the Large Magellanic Cloud during an F star-phase. A&A. **228**, 379–386 (1990)
116. O. Stahl, I. Jankovics, J. Kovács et al., Long-term spectroscopic monitoring of the luminous blue variable AG Carinae. A&A. **375**, 54–69 (2001)
117. C. Sterken, B. Wolf, Mass loss of B1 Ia-O supergiants. A&A. **70**, 641–651 (1978)
118. C. Sterken, E. Gosset, A. Juttner et al., HD160529 – a new galactic luminous blue variable. A&A. **247**, 383–392 (1991)
119. C. Sterken, A.M. van Genderen, A. Plummer, A.F. Jones, Wra 751, a luminous blue variable developing an S Doradus cycle. A&A. **484**, 463–467 (2008)
120. R.B. Stothers, C.-W. Chin, Dynamical instability as the cause of the massive outbursts in Eta Carinae and other Luminous Blue Variables. ApJ. **408**, L85–L88 (1993)
121. C. Trundle, D.J. Lennon, Understanding B-type supergiants in the low metallicity environment of the SMC II. A&A. **434**, 677–689 (2005)
122. C. Trundle, R. Kotak, J.S. Vink, W.P.S. Meikle, SN 2005 gj: evidence for LBV supernovae progenitors? A&A. **483**, L47–L50 (2008)
123. G. Umana, C.S. Buemi, C. Trigilio, J.L. Hora, G.G. Fazio, P. Leto, The dusty nebula surrounding HR car: a spitzer view. ApJ. **694**, 697–703 (2009)
124. A.F. Valeev, O. Sholukhova, S. Fabrika, A new luminous variable in M33. MNRAS. **396**, 21L–25L (2009)
125. R. van Boekel, P. Kervella, M. Schöller, Direct measurement of the size and shape of the present-day stellar wind of eta Carinae. A&A. **410**, L37–L40 (2003)
126. A.M. van Genderen, S Doradus variables in the galaxy and the Magellanic Clouds. A&A. **366**, 508–531 (2001)
127. A.M. van Genderen, C. Sterken, Light variations of massive stars (alpha Cyg variables). XVIII. The B[e] supergiants S 18 in the SMC and R 66 = HDE 268835 and R 126 = HD 37974 in the LMC. A&A. **386**, 926–935 (2002)
128. A. J. van Marle, N. Langer, G. Garcia-Segura, Constraints on gamma-ray burst and supernova progenitors through circumstellar absorption lines. II. Post-LBV Wolf-Rayet stars. A&A. **469**, 948–948 (2007)
129. J.S. Vink, Constraining GRB progenitor models by probing Wolf-Rayet wind geometries in the Large Magellanic Cloud. A&A. **469**, 707–711 (2007)
130. J.S. Vink, A. de Koter, Predictions of variable mass loss for Luminous Blue Variables. A&A. **393**, 543–553 (2002)
131. J.S. Vink, A. de Koter, H.J.G.L.M. Lamers, On the nature of the bi-stability jump in the winds of early-type supergiants. A&A. **350**, 181–196 (1999)
132. J.S. Vink, A. de Koter, H.J.G.L.M. Lamers, New theoretical mass-loss rates of O and B stars. A&A. **362**, 295–309 (2000)
133. J.S. Vink, B. Davies, T.J. Harries, On the presence and absence of disks around O-type stars. A&A. **505**, 743 (2009)

134. N.R. Walborn, Ofpe/WN9 circumstellar shells in the Large Magellanic Cloud. ApJ. **256**, 452-459 (1982)
135. K. Weis, On the structure and kinematics of nebulae around LBVs and LBV candidates in the LMC. A&A. **408**, 205–229 (2003)
136. S.M. White, R.A. Duncan, J. Lim et al., The radio source around Eta Carinae. ApJ. **429**, 380–384 (1994)
137. P.A. Whitelock, M.W. Feast, C. Koen, G. Roberts, B.S. Carter, Variability of Eta-Carinae. MNRAS. **270**, 364–372 (1994)
138. B. Wolf, Empirical amplitude-luminosity relation of S Doradus variables and extragalactic distances. A&A. **217**, 87–91 (1989)
139. S.E. Woosley, S. Blinnikov, A. Heger, Pulsational pair instability as an explanation for the most luminous supernovae. Nature **450**, 390–392 (2007)
140. L.R. Yungelson, E.P.J. van den Heuvel, J.S. Vink, S.F. Portegies Zwart, A. de Koter, On the evolution and fate of super-massive stars. A&A. **477**, 223–237 (2008)
141. T. Zethson, S. Johansson, K. Davidson, R.M. Humphreys, K. Ishibashi, D. Ebbets, Strange velocities in the equatorial ejecta of Eta Carinae. A&A. **344**, 211–220 (1999)
142. F.-J. Zickgraf, B. Wolf, C. Leitherer, I. Appenzeller, O. Stahl, B(e)-supergiants of the Magellanic Clouds. A&A. **163**, 119–134 (1986)

Chapter 11
The Supernova Impostors

Schuyler D. Van Dyk and Thomas Matheson

Abstract The "supernova impostors" in nearby galaxies mimic the appearance of a true supernova, but rather than a terminal explosion, the impostors appear to be massive stars that have undergone a powerful eruption. Several of these have energetics comparable to true supernovae, and may, in fact, be analogous to the Great Eruption of η Carinae in the 1800s. We distinguish the observed characteristics of impostors from true supernovae. We discuss in detail the prototypical impostor, SN 1961V. We also discuss other known examples where available data imply that these are also impostors. Finally, we connect the impostors to true supernova events—in other words, several recent supernovae appear to have progenitors which experienced high mass-loss events immediately prior to explosion, with one case clearly having an impostor-like eruption only 2 years before discovery as a supernova. The supernova impostors represent rare instances of analogs to η Car and other known luminous blue variables from which we can acquire important insight into the final stages in the evolution of very massive stars.

11.1 What are Supernova Impostors?

The post-main-sequence evolution for stars with $M_{\rm ZAMS} \gtrsim 30$–$40\,M_{\odot}$ is thought to progress (depending on $M_{\rm ZAMS}$) through a blue supergiant phase, possibly to a red supergiant phase, or directly to a short-lived luminous blue variable (LBV) phase, to become H-poor or H-free Wolf-Rayet (WR) stars preceding termination

S.D. Van Dyk (✉)
Spitzer Science Center, Caltech 220-6, Pasadena, CA 91125, USA
e-mail: vandyk@ipac.caltech.edu

T. Matheson
National Optical Astronomy Observatory, 950 N. Cherry Avenue, Tucson, AZ 85719-4933, USA
e-mail: matheson@noao.edu

K. Davidson and R.M. Humphreys (eds.), *Eta Carinae and the Supernova Impostors*, Astrophysics and Space Science Library 384, DOI 10.1007/978-1-4614-2275-4_11,

as supernovae (SNe), e.g., [61, 79, 115]. The most massive stars, with $M_{ZAMS} \gtrsim 60\,M_\odot$, may proceed directly to the pre-SN, pre-LBV, H-rich WNL, or "WNH," phase without becoming red supergiants [48, 104]. Since WR stars have $M \lesssim 20\,M_\odot$ [23], their high-mass progenitors must lose most of their mass through eruptive mass ejections [107], probably as LBVs. Powerful eruptions of massive stars, such as η Carinae (e.g., [25] and Humphreys and Martin this volume), are often referred to as "supernova impostors," e.g., [124], because some observational aspects can mimic the appearance of a true supernova (SN). During the Great Eruption from 1837 to 1858 of η Car ($M_{ini} > 150\,M_\odot$), the star greatly exceeded the Eddington limit, with the bolometric luminosity increasing by $\sim$2 mag. The total luminous output of such an eruption ($\sim 10^{49.7}$ erg) can rival that of a SN, to such a degree that some impostors initially are assigned designations, historically, as SNe. A SN is an explosive event that ends the life of a star, although a compact object may form in the process. This is the one key distinction from the impostors. If, after a bright optical transient is observed, a star remains present, then it is an impostor (see, e.g., the discussion of SN 1961V, Sect. 11.2). Not all impostors are as powerful as η Car; some cases are more like the "classical" LBVs (e.g., S Dor, AG Car, R 127), where the bolometric luminosity remains nearly constant during an eruption, as the star's envelope expands or its wind becomes optically thick and the apparent temperature cools to $\sim$7000–8000 K, see Chap. 10.

Spectroscopically, LBVs at quiescence have the spectra of hot (12000–30000 K) supergiants, with H and He I lines in emission, often with P Cygni profiles, or in some cases, such as R 127, they are Of/WNL stars. At visual maximum, the optically thick, slowly expanding ($\sim$100–200 km s^{-1}), dense ($\sim 10^{11}$ cm^{-3}) wind, or "pseudo-photosphere," causes the spectrum of the otherwise-hot star to appear as a much cooler, F-type ($\sim$7,000–8,000 K) supergiant. Fe II and [Fe II] emission lines are often present in quiescence or eruption, or both. Many LBVs show evidence for circumstellar matter (CSM); see Weis, this volume. The dusty "Homunculus" around η Car, which obscures the light from the star, is the extreme example.

Knowledge of how a SN appears spectroscopically and photometrically is necessary to distinguish the impostor from the "real McCoy." SNe have traditionally been classified by the features present in their spectra near maximum brightness [80]. The original distinction between types I and II was based on the appearance of the hydrogen lines. The characteristics of SNe are summarized in Table 11.1, and the various types are illustrated in Fig. 11.1 with representative light curves in Fig. 11.2. For more complete reviews of SN classification, see, e.g., [35, 121].

The impostors are sometimes mistaken for a subset of SNe II referred to as "IIn," where "n" is for "narrow" [97]. The coincidence of a hydrogen envelope with similar kinetic energies makes the spectrum of the strong Balmer lines, typically a narrow emission profile atop an "intermediate width" base (with velocities of $\sim$1,500–2,000 km s^{-1}, [110]) look similar for both the SNe IIn and the impostors.[1]

[1]The impostor spectra do appear unusual, though, relative to the prototypical SNe IIn, e.g., SN 1988Z [122], SN 1995N [42], and SN 1998S [43].

Table 11.1 Summary of SN types

Type	Defining spectral characteristics	Progenitor[a]	Circumstellar interaction[a]
Ia	No Hydrogen, strong Silicon II absorption, IME[b]	White dwarf[c] + ?	No?[d]
Ib	No Hydrogen, strong Helium I absorption	WR? mass-transfer binary?	Yes[e]
Ic	No (or weak[f]) Hydrogen, no (or weak[f]) Helium	WR? mass-transfer binary?	Yes[e]
II-P[g]	Hydrogen, P Cygni emission profiles	Red supergiants ($\lesssim 20\,M_\odot$)[h]	Yes[i]
II-L[j]	Hydrogen, only broad emission	Red hypergiants?[k]	Yes[l]
IIb[m]	Hydrogen, then Helium (then Hydrogen)	Mass-transfer binary?[n]	Yes[o]
IIn	Hydrogen, narrow/intermediate-width emission	LBVs?[p]	Yes?[q]

[a]Question marks indicate the current uncertainty in our knowledge. [b]IME = intermediate mass elements. [c]Thermonuclear disruption near Chandrasekhar limit [10, 82]. [d]E.g., [1, 19]. [e]E.g., [13, 14]. [f]E.g., [32]. [g]Light curve type "Plateau" [3]. [h]E.g., [100]. [i]The interaction is weak; e.g., [14, 16]. [j]Light curve type "Linear" [3]. [k][31]. [l]E.g., [14, 15, 81]. [m]E.g., SN 1993J [4, 36, 55, 68, 73, 74, 77, 78, 138]. [n]E.g., [75]. [o]E.g., [14, 134]. [p][44, 45]. [q]E.g., [14, 18, 42, 122, 136]; however, see [125]

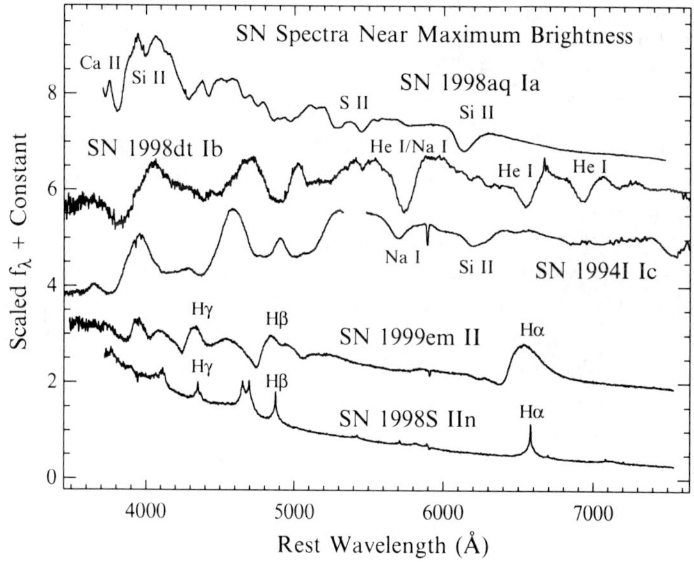

Fig. 11.1 Optical spectra of the main types of SNe, observed near maximum brightness

The lines are narrow in the SNe IIn, presumably because of shock/circumstellar matter interactions, e.g., [122], resulting in luminous and long-lived radio, X-ray, and optical emission, e.g., [14, 15] (although see [125] regarding the radio emission).

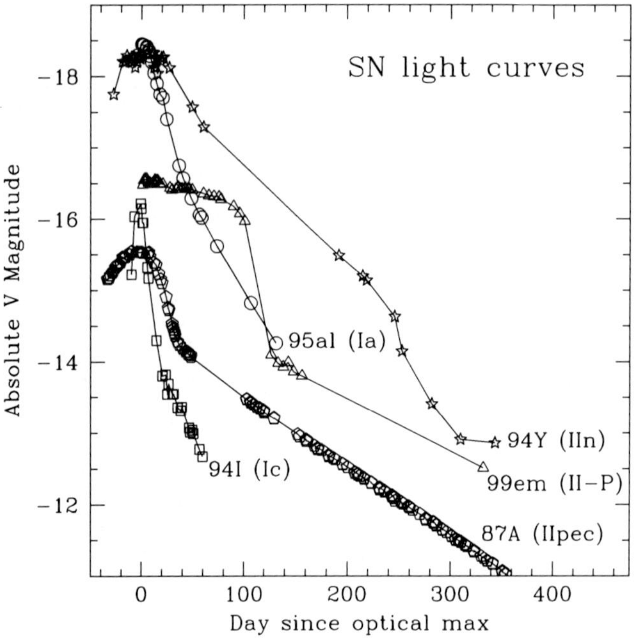

Fig. 11.2 V-band absolute light curves for the main types of SNe: SN 1995al (Ia, *circles*; [96]), SN 1999em (II-P, *triangles*; [64]), SN 1994I (Ic, *squares*; [95]), SN 1994Y (IIn, *stars*; [47]), and SN 1987A (II pec, *pentagons*; [12, 117])

SNe IIn constitute ~5% of all SNe and so, are relatively rare [11,66]. In the cases of SNe 1988Z [2, 136] and 1995N [42], various indicators suggest that the progenitor was a very massive star—in fact, based on the circumstellar interaction of the latter, the implication is that all SNe IIn may originate from high-mass red supergiants, or hypergiants, experiencing a superwind phase, such as VY CMa and IRC + 10420 [42]. The latter star, which is more yellow than red, e.g., may be making a transition to the LBV phase [51]. For the progenitor of SN IIn 2005gl, Gal-Yam et al. [44,45] have shown that SNe IIn may be associated directly with hypergiant stars, which may have undergone LBV-like eruptions.

The absolute luminosity is, therefore, the best means to distinguish between a true SN and an impostor, if the distance of the host galaxy is known. The impostors peak near an absolute magnitude $M_V \approx -14$, e.g., [48], while SNe IIn have peak absolute magnitudes around $M_V \approx -19$, e.g., [2, 33, 35]. Even so, mistakes are still made, as is clear from the discussion, below, of individual SN impostors.

Two of the best-known, best-studied impostors are SN 1954J in NGC 2403 [59] and SN 1961V in NGC 1058 [139], identified as members of Zwicky's now-obsolete "Type V" SN classification [140]. Until the early 1970s, η Car itself was considered to be a slow, peculiar SN [84] of Type V [139]! SN 1961V is the "classical prototype." With caveats that this is not necessarily a complete list, more recent examples of impostors include SNe 1997bs (often looked upon as the "modern prototype"), 1999bw, 2000ch, 2001ac, 2002kg, 2003gm, 2006fp, 2006qq, 2007sv, 2009ip, 2010C, 2010da, 2010dn, NGC 1511-OT1, SNHunt23 and SNHunt32, both in NGC 3016, PSN J10523453+2256052, PSN J17592296+0617267, PSN J12304185+4137498, and PSN J12355230+2755559. It has been argued that SNe 2002bu and 2006bv may also be impostors [119]. Other related objects, not strictly mistaken for SNe, include NGC 2363-V1, UGC 2773-OT20091, and the "possible SN in NGC 4656" from 2005. The natures of SN 2008S, M85 OT2006-1, and NGC 300 OT2008-1 are still, as of this writing, being hotly debated, and we do not discuss these here. SN 2010U has been shown to be a luminous nova [54]. We also argue, below, that SN 2003hy is probably not an impostor, but an actual SN.

Like η Car, the precursor star for each impostor is expected to survive the eruption and return to relative quiescence. Therefore, a primary goal of studying the impostors is to recover the survivor. We have also had the incredibly good fortune in a few cases of detecting the precursor star prior to the outburst or eruption. The *Hubble Space Telescope* (*HST*) has been particularly invaluable in our efforts so far in "unmasking" the impostors, since they are occurring well outside the Local Group. Information about the environments of these events can also provide clues about these stars: η Car itself sits in the cluster Trumpler 16, with some of the most massive stars known, in one of the most stunning regions of our Galaxy, the Carina Nebula [25]; see Walborn this volume.

Since SN impostors are massive stars, we expect them at some point to explode. Recent examples exist of actual SNe (mostly, IIn) whose progenitors were likely SN impostors. As such, the comparison of SNe and SN impostors comes full circle.

11.2 SN 1961V: The Prototypical Impostor

The spectrum (Fig. 11.3) and light curve (Fig. 11.4) of SN 1961V in NGC 1058 were highly unusual. The early-time optical spectrum of SN 1961V was dominated by narrow emission lines of H, He I, and Fe II (indicating a relatively low maximum expansion velocity of $\sim$2000 km s^{-1} [139]), similar to the SNe IIn [35, 97].

Bertola [5] and Zwicky [139] inspected Lick, Mt. Wilson, and Palomar pre-SN photographic plates of the host galaxy, dating from 1937 to 1954, and located a $m_{pg} \approx 18$ mag star at the site of SN 1961V. At the host galaxy distance (adopting 9.3 Mpc, the Cepheid distance to NGC 925, in the same "NGC 1023" group as NGC 1058 [99]), the precursor star is an incredible $M_{pg} \approx -12$ mag, implying it was the most luminous known star in *any galaxy*. Utrobin [123] modelled the SN's unusual light curve as the explosion of a (almost certainly improbable) 2000 $M_\odot$ star! Bertola's [6] long-term photometric monitoring, however, showed that the SN's colors remained relatively constant, resembling those of an F supergiant.

The SN 1961V environment is among one of the most fascinating, with several luminous supergiants and much nebulosity, possibly in the form of an extended H II region, filaments, and bubbles, not unlike that of η Car itself (the Carina Nebula). This is a conspicuous region of recent star formation toward the outskirts of the host

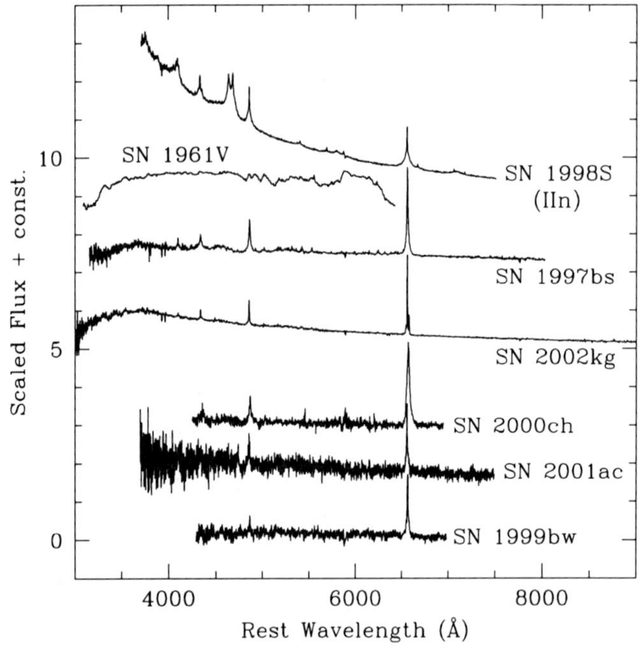

Fig. 11.3 Optical spectra of the SN impostors, compared to the spectrum of the SN IIn 1998S. The SN 1998S spectrum was obtained near maximum. The epochs for the impostors spectra are as follows: SN 1961V, 1961 Nov 12 [7]; SN 1997bs, 1997 Apr 16 [127]; SN 1999bw, 1999 Apr 24; SN 2000ch, 2000 May 31; SN 2001ac, 2001 Mar 14; and, SN 2002kg, 2003 Jan 6 [130]. No spectral data exist for SN 1954J

galaxy's bright disk. No other H II regions in the host galaxy exhibit detectable 20 cm radio emission [9]. The high [O III]/Hβ ratios indicate a high-excitation region, and the metallicity is subsolar [46].

It has been contentious whether this star has survived and is now detectable. Using groundbased images from 1983 with 1$''$ resolution, Fesen [34] identified a faint knot of Hα emission very near the accurate SN position [57], in the easternmost of two H II regions (each separated by $\sim$4$''$, with $\log L_{\mathrm{H}\alpha} \approx 37.2$ [46]). In 1984 Branch and Cowan [9] detected a source of nonthermal radio emission with the Very Large Array (VLA) very near the optical SN position. The conclusion they reached was that this emission was from an aging (radio) SN, or very young remnant, corresponding to SN 1961V. However, Lick spectra of the SN environment from 1986 show an intermediate-width ($\approx$2100 km s^{-1} FWHM) base to otherwise-narrow, unresolved Hα emission-line profile at the SN position [46]. This indicated that SN 1961V should *not* be considered a true SN, but a possible η Car analog, behind high circumstellar extinction.

In 1991 Filippenko et al. [37] obtained *VRI* images of the field of SN 1961V with the pre-refurbishment *HST*/WFPC, detecting an apparent cluster of ten stellar-like objects, all within $\sim$5$''$. From those relatively low-quality data, they isolated

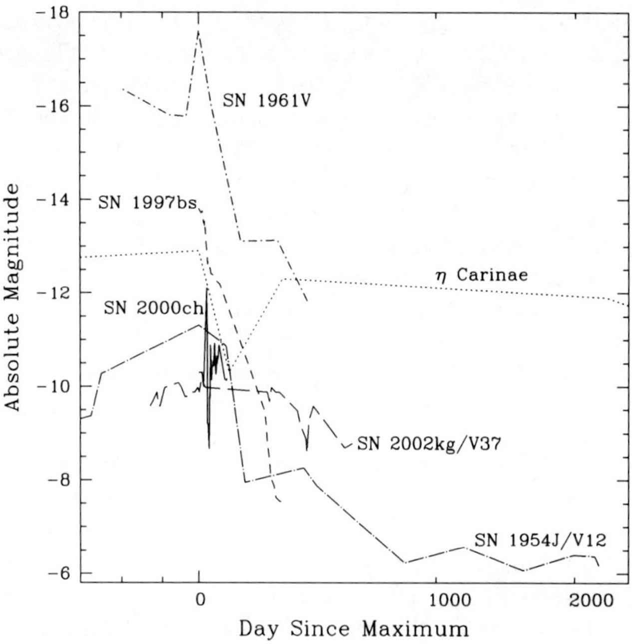

Fig. 11.4 Absolute light curves for the SN impostors, compared to the light curve of η Carinae during its Great Eruption

"Object 6" in that field as the most likely contender for the survivor of the 1961 super-outburst, although other stars in the environment could not be completely excluded based on astrometry alone. This presumably blue star (with $V \approx 25.6$ mag) now appears to have the colors of a red supergiant, possibly as a result of substantial dust ($A_V \approx 4$ mag).

If SN 1961V were a giant LBV outburst, this still could not explain the nonthermal radio emission (η Car is a thermal bremsstrahlung radio source; e.g., [22]). Stockdale et al. [114] confirmed the nonthermal radio emission with more recent (1999/2000) VLA observations and, furthermore, it appeared that the source had declined in flux since the previous radio observations: The inference was that SN 1961V is a fading, core-collapse radio SN. If the radio source were the SN, then a fainter, red object in the environment, "Object 11," could be the optical counterpart, based on shallow, archival *HST*/WFPC2 images containing the SN site [128].

Chu et al. [17] subsequently obtained *HST*/STIS G430M and G750M grating spectra (along with an unfiltered 50CCD image) of the environment in 2002, to search for the old SN. What was detected, instead, was the spectrum of exactly *one* point source within the slit, exhibiting exactly *one* bright emission line—that of Hα, with a broad line profile of width up to 550 km s^{-1}. The spectrum most closely resembles that of η Car [17]! The spectrum of a decades-old SN would more likely be dominated by emission lines, such as [O III] λ5007, with much higher velocities, >2,000 km s^{-1} (e.g., SN 1957D [67]). Although the STIS astrometry is inconclusive, it appears that the spectrum can be attributed to Filippenko et al.'s "Object 7" and is a possible LBV in the field, but not necessarily SN 1961V [17]. Nonetheless, if there were an old SN in the field, detectable optical emission should accompany the nonthermal radio emission for SNe (e.g., [15]).

Further analysis of the existing data [124] showed that the accurate optical position of SN 1961V [57] is most coincident with Object 7. The nonthermal radio emission does not appear that well correlated with the stellar objects in the environment. It may be synchrotron radiation generated within what appears to be a complex, turbulent ISM. The existing STIS 50CCD image [17] hints at the spectacular interstellar environment. The fact that the radio source was at 0.091 ± 0.005 mJy at 6 cm in 2000, based on a remeasurement of those data [124], and 0.11 ± 0.03 mJy in 1986 [21] raises the question whether the radio flux had actually declined.

Although an η Car-like star has been detected in the SN environment, it has been uncertain if it can be directly linked to a survivor of the 1961 outburst. Our knowledge hinged on multi-band photometry of stars from Lucy-Richardson-deconvolved WFPC images, from disparate WFPC2 archival images of relatively low signal-to-noise ratio (S/N), the low S/N STIS spectrum, and the unfiltered STIS 50CCD image (with no color information). If Object 7 [37], the one closest to the accurate optical position measured for the SN, is responsible for the broad Hα emission line seen in the STIS spectrum, such emission could easily account for the observed excesses in the WFPC/F702W and WFPC2/F606W bands (Fig. 11.5). The star appears most consistent with a luminous, blue star behind circumstellar extinction of $A_V \approx 1.5$–2 mag, with a more realistic $M_V^0 \approx -7$ mag (Fig. 11.6). If it experienced only moderate extinction ($A_V \approx 0.5$ mag), the star would have the intrinsic colors and brightness ($M_V^0 \approx -6$ mag) of an A-type supergiant, however, then it would be difficult to account for the broad Hα emission if this star is the source.

A source was detected in *HST*/WFPC2 images from 2008 at the position of Object 7 at $m_{\mathrm{F555W}} = 24.70$ and $m_{\mathrm{F658N}} = 20.45$ mag. Comparison of the F555W magnitude with that from Filippenko et al. [37] and with the F606W magnitude from Van Dyk et al. [128] shows that this object may have declined somewhat in brightness in that band. If we consider a model for the Hα emission, consisting of a line with a Lorentzian profile of FWHM ~310 km s^{-1} (a model which provides an approximate fit to a STIS spectrum of η Car A), then the integrated flux in this hypothetical line that could account for the observed F658N brightness is $\sim 7.5 \times 10^{-16}$ erg cm^{-2} s^{-1}. Assuming a distance to η Car of 2.3 kpc and $A_V \sim 1.7$ mag to the central star [25], and $A_{\mathrm{H}\alpha} = 0.81 A_V$ [132], the extinction-corrected observed Hα flux of η Car [69] agrees astonishingly well with that from Object 7, when scaled to the assumed distance to SN 1961V. We consider it, based on these data, to be highly suggestive that Object 7 is, indeed, the η Car-like survivor of SN 1961V.

Much of the light we see from η Car is reflected by dust in the Homunculus. Since the formation of dust occurred in the Homunculus soon after the cessation

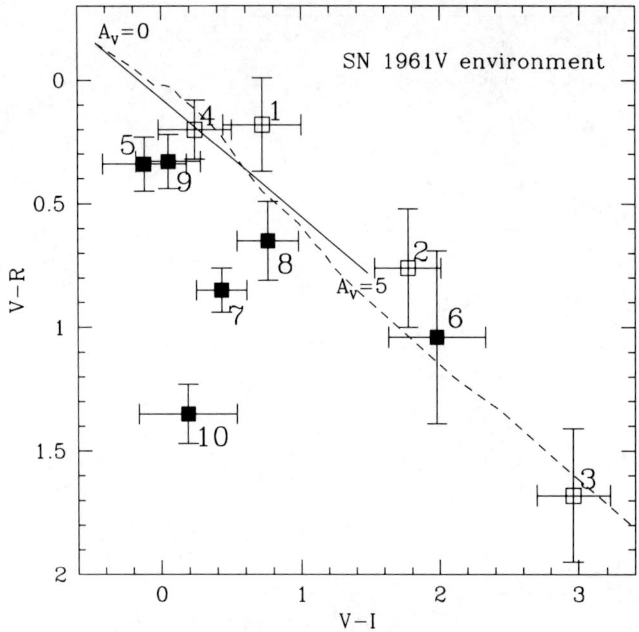

Fig. 11.5 Color-color diagram for the SN 1961V environment based on photometry derived from Lucy-Richardson-deconvolved WFPC images [37]. Stars nearest the SN position are shown as *filled squares*; the *open squares* are the other stars in the field. The *solid line* is a reddening vector to $A_V = 5$ mag. The *dashed line* is the locus for the supergiants

of the Great Eruption [53, 101, 105], we might expect the impostors to also show evidence for dust emission. This should be particularly true for SNe 1961V and 1954J (see Sect. 11.3.1), which are older events and, at least in the case of SN 1954J, show evidence for a dusty Homunculus-like nebula. We have examined the available archival image data from 2004 to 2007 obtained with the *Spitzer Space Telescope* in the mid-infrared for NGC 1058, and find that a source is detected in the SN 1961V environment in the IRAC wavebands at fluxes $<3\,\mu$Jy at 3.6 μm, $<5\,\mu$Jy at 4.5 μm, 51 μJy at 5.8 μm, and 150 μJy at 8.0 μm (*Spitzer* images in the longer wavelength bands do not have sufficient spatial resolution to distinquish sources in the impostor environments). Such a source is consistent with a blackbody of temperature $T \approx 350$ K. However, the source is not spatially coincident with either Objects 6 or 7, but, instead, with Object 8 (cf. [37])[2].

[2]Editor's note: After this chapter was finished Kochanek et al. [58] also noticed the apparent lack of dust emission and suggested that SN 1961V was therefore a true SN and not an impostor. Although, the question of whether all of these objects should necessarily produce dust, depending on the nature of the eruption, is not known. Van Dyk and Matheson [131] continue to argue for impostor status for SN 1961V.

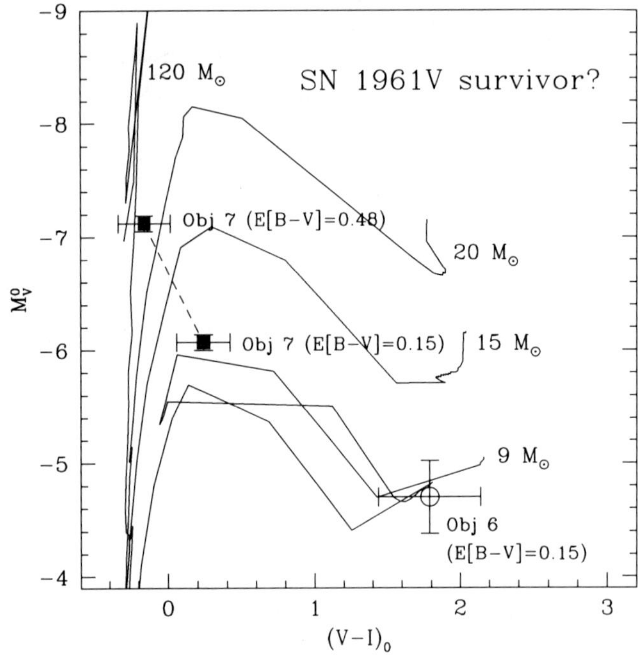

Fig. 11.6 Color-magnitude diagram, based on photometry derived from Lucy-Richardson-deconvolved WFPC images [37], showing Objects 6 and 7. Object 7 is shown with one estimate of reddening ($E(B-V) = 0.15$ mag [37]) and another derived from Fig. 11.5 ($E(B-V) = 0.48$ mag). Also shown are theoretical evolutionary tracks for subsolar metallicity and enhanced mass loss [63]. Note that if Object 7 is SN 1961V *and* the star corresponding to the broad Hα emission in the *HST* STIS spectrum [17], i.e., an η Car analog, it could be very massive ($M_{ZAMS} \gtrsim 20\,M_{\odot}$)

11.3 Other Impostors

11.3.1 SN 1954J/Variable 12

The identification of SN 1954J in NGC 2403 as a genuine SN has always been debatable, since no spectra of it exist. Its peculiar light curve (Fig. 11.4) showed similarities to the one for SN 1961V, leading it to be classified [60] as a SN of Type V [140]. Based on its unusual light curve, it had been identified as an irregular blue variable, Variable 12 (V12) [118]. In quiescence the star had $B \approx 21$ mag, and between 1949 and 1954, the star fluctuated in brightness, at one point becoming the brightest star in the host galaxy, at $B \lesssim 16.5$ mag. It subsequently faded below its pre-outburst state by late 1954. SN 1954J/V12 had behaved quite unlike any genuine SN [50].

Smith et al. [108] identified an object through ground-based, multi-band optical and near-infrared imaging, which they considered the post-outburst survivor star. This object was later confirmed by Van Dyk et al. [129] via further ground-based imaging with the Keck-I 10-m telescope and the 2.56-m Nordic Optical Telescope, as well as a Keck spectrum of the source. The line shape and equivalent width of the broad Hα emission line, the only line in the spectrum, appear very similar to those of η Car. Van Dyk et al. [129] further imaged the source with *HST* in *BVI* and Hα, and found that the ground-based image is really a composite of four stars.

One of the stars showed a clear excess of emission in the *HST* Hα image, relative to the other three stars, and is therefore logically the source of the Hα line in the spectrum, based on the offset from a fiducial star in the field, which also was on the slit during the Keck spectroscopy [129]. The broadband colors of the Hα-excess object are consistent with G-type star with little reddening, and its observed brightness indicates it has to be a supergiant. It is difficult, however, to account for the presence of the broad Hα in a G-type supergiant spectrum, although yellow hypergiants with emission line spectra do exist. This "Star 4," however, could also be a reddened, early-type supergiant, i.e., a very luminous O-type or early-B star hidden behind a dusty ($A_V \approx 4$ mag) shell or LBV-like nebula (with $R(\mathrm{H}\alpha) \approx 0.1$ pc), much like η Car and its Homunculus [129]. The shell could have been ejected during the 1954 event. After correcting for this assumed reddening, the intrinsic colors ($(B-V)_0 = -0.22$ and $(V-I)_0 = -0.21$ mag) and brightness ($M_V{}^0 \approx -8.0$ mag) are consistent with the evolutionary tracks of stars with initial mass $M_{\mathrm{ini}} \gtrsim 20\,M_\odot$. This star is, thus, the likely luminous, massive survivor of the SN 1954J/V12 eruption [129]. The other stars in the SN 1954J/V12 environment appears to be a sparse cluster of roughly coeval, massive supergiants.

A mid-infrared source is detected in the SN 1954J environment with *Spitzer* /IRAC in 2004, at fluxes 63 μJy at 3.6 μm, 40 μJy at 4.5 μm, $\lesssim$80 μJy at 5.8 μm, and $\lesssim$320 μJy at 8.0 μm. This can be fit by a $T \simeq 4{,}000$ K blackbody with a luminosity $L \simeq 2.0 \times 10^{39}$ erg s^{-1} and radius $R \simeq 1.0 \times 10^{14}$ cm. It is, however, not certain whether this source is associated with SN 1954J; it could well be emission from one of the other supergiants in the environment.

11.3.2 SN 1997bs

SN 1997bs in NGC 3627 (M66) has been discussed elsewhere in detail [126, 127]. The spectrum resembled that of a SN IIn, with relatively narrow (FWHM $\simeq 1000$ km s^{-1}) Balmer emission lines on a featureless continuum, and many weaker Fe II emission lines (Fig. 11.3). However, both the luminosity and the *BVRI* light curves, based on monitoring with the 0.75-m Katzman Automatic Imaging Telescope (KAIT [39]) and, at later times, from *HST* data, were unusual (Fig. 11.4). At maximum SN 1997bs was only at $M_V \simeq -13.8$ mag.

One of the notable discoveries about SN 1997bs was Van Dyk et al.'s [126] identification of a star at the position of the SN in a *HST* F606W image from 1994 December. This star had $m_{\mathrm{F606W}} = 22.86$ mag. Assuming the estimated $E(B-V) = 0.21$ mag, and that $m_{\mathrm{F606W}} \approx V$, the star had $M_V \simeq -8.1$ mag, qualifying it as an extremely luminous supergiant star [49]. Unfortunately, without an additional band, no color information exists for the star, and we therefore do not know the original spectral type of the outburst precursor. However, the outburst spectrum at maximum is consistent with a hot star. Based on the V-band alone and considering the range of possible spectral types for supergiants, the star had $M_{\rm ini} > 20\,M_{\odot}$ and $T_{\rm eff} > 6,300$ K, i.e., hotter than spectral type F6 [124]. If it were an O- or B-type supergiant, as expected for LBVs, the optically thick cooler wind, or pseudo-photosphere, formed during the eruption, would approach a minimum temperature of $\sim$6,500–7,000 K, i.e., an A- or F-type supergiant, as the optical depth increases [24, 48], consistent with the color of SN 1997bs near maximum.

The evolution of the SN 1997bs light curves was fairly smooth. The color became progressively redder as well, from $V-I \simeq 0.62$ mag at V maximum, to $V-I = 1.43$ mag at the end of KAIT monitoring, to the reddest measured, $V-I \approx 3.4$ mag, in 1997 December, based on the *HST* data. SN 1997bs subsequently became bluer, at $V-I \approx 3$ mag in early 1998 [127]. In additional *HST*/WFPC2 imaging in 2001, SN 1997bs was still detected, albeit 2.9 mag fainter than the precursor [127], and had apparently become much bluer with $V-I < 0.8$ mag (V [F555W] $\sim$25.8 mag and undetected at I; [65]).

The fact that the observed color became as red as a M star may indicate the formation of dust in or around the pseudo-photosphere. This dust could have been destroyed, as the star's observed color subsequently became bluer while fading to minimum between 1998 and 2001.[3] If dust formed during the SN 1997bs outburst, we might be able to detect evidence for it in available infrared image data. Mid-infrared emission is detected near the position of SN 1997bs in *Spitzer* IRAC images from 2004 May of the host galaxy, with 32 μJy at 3.6 μm and 40 μJy at 4.5 μm, and undetected at 5.8 μm and 8.0 μm. This can be fit by a $T \simeq 970$ K blackbody with a radius $R \simeq 1.4 \times 10^{15}$ cm and luminosity $L \simeq 1.2 \times 10^{39}$ erg s^{-1}. It is not absolutely certain that this emission was arising from SN 1997bs, but the positional coincidence is fairly high. In a near-infrared image obtained at the Palomar Observatory Hale 5-m in 2005 January, SN 1997bs is not detected to $K_s \gtrsim 19.7$ mag ($\lesssim$9 μJy), consistent with this assumed blackbody emission.

The evidence that the star survived what appears to have been a powerful eruption is quite strong. In *HST*/ACS F555W and F814W images obtained in 2009 of the host galaxy, the star was recovered, at $m_{\rm F555W} = 26.08$ and $m_{\rm F814W} = 25.08$ mag. The brightness and color are both consistent with the previous detection in 2001. It is interesting that when the star was detected prior to outburst, it was much brighter than at the observed minimum. It is possible that by 1994 the star was already in some pre-eruptive phase.

The precursor star is not in an H II region or obvious association of blue stars.

---

[3]Editor's note: Given this star's fading by nearly 3 magnitudes below the apparent brightness of the precursor, together with its color evolution from red to a bluer color as it got much fainter, suggests that the dust grains may be large, producing obscuration without significant reddening, similar to η Car [25].

11.3.3 SN 2000ch

SN 2000ch in NGC 3432 had been studied in some detail by Wagner et al. [133]. The object was discovered using KAIT in 2000 May [85]. The light curve, monitored primarily by KAIT and United States Naval Observatory telescopes, was highly erratic, varying between $R = 17.4$ and $R = 20.8$ mag (Fig. 11.4). At maximum, SN 2000ch reached $M_R^0 \approx -12.7$ mag. The deep minimum in its light curve after maximum may have been the result of dust formation, an occultation, or from the growth of an optically thick, cooling, envelope.

The spectra of SN 2000ch all showed strong Balmer emission on a relatively flat continuum, with a Hα FWHM of $\approx$1,550 km s^{-1} (Fig. 11.3). However, the spectra lacked Fe II and [Fe II] emission lines, as well as P Cygni-like profiles, which are generally characteristic of LBVs in both outburst and quiescence. The general absence of forbidden lines, particularly [O I], indicated very high electron densities of 10^9–10^{10} cm^{-3} in the ejecta . From this density, and the Hα luminosity, a radius for the emission region could be calculated, at $\sim$0.2 pc, comparable to the size of η Car's Homunculus [133].

The *UBV* colors for SN 2000ch were relatively constant. The observed blue color excess is possibly attributable to Balmer continuum emission. Color variability was observed in $V - R$ and $R - I$, likely reflecting changes in the Hα emission strength. The locus of SN 2000ch in the ($U - B$, $B - V$) color-color diagram throughout its outburst was close to the blackbody line, which is true for LBVs, including η Car. Archival imaging prior to the discovery (1996–2000) show that the star had $R \approx 19.4$ mag, implying that, pre-outburst, the star had an astounding $M_R^0 = -10.7$ mag. Together with the maximum brightness, if the star were constrained by the Eddington limit, its mass $M \gtrsim 40\ M_\odot$ [133].

We recovered SN 2000ch in images obtained at the Palomar 5-m, with $R = 19.39$, $B - V = 0.29$ and $V - R = 0.95$ mag in 2004 May and $K_s = 18.21$ mag in 2005 January (although the source is somewhat blended; see below). The *R* brightness is consistent with the pre-outburst value; the star was in relative quiescence, but was still extraordinarily luminous. The $B - V$ color is similar to that during outburst [133] and is consistent with a late F-type star (without correcting for reddening). The $V - R$ color is significantly redder than during outburst [133], possibly indicating that the emission line strengths, particularly at Hα, had increased further. A continuum-subtracted Hα image, also obtained at Palomar (along with the broadband images), shows SN 2000ch to be a very luminous source at Hα at late times. The excess at *R* is consistent (to within a factor of 2) with the net Hα (+[N II]) flux in this image, plus a late-F star continuum. The K_s brightness for the star is $\sim$1 mag brighter than what one would expect for a late-F star, implying an infrared excess, possibly from dust in the nebula around the star.

In *Spitzer* IRAC images of the host galaxy from 2007 December we detected a mid-IR source with 63 μJy at 3.6 μm, 50 μJy at 4.5 μm, 41 μJy at 5.8 μm, and 18 μJy at 8.0 μm. It is somewhat unclear whether or not this emission is associated with SN 2000ch. The emission at K_s, mentioned above, e.g., is not strictly coincident with the mid-infrared emission. The mid-infrared source is consistent with a blackbody at ~2,500 K, which corresponds to luminosity $L = 1.3 \times 10^{39}$ erg s^{-1} and radius $R = 2.1 \times 10^{14}$ cm. We note that these are quite similar to the properties of the dusty cocoons around the luminous supergiants NML Cyg and VY CMa [52].

SN 2000ch has recently exhibited subsequent outbursts [30, 89]. High-resolution images of SN 2000ch and its immediate environment have also been obtained with *HST* during its revived/ongoing outburst. SN 2000ch is in an obvious cluster of stars. Analysis of its stellar environment in these images is clearly needed to better understand this object's star's nature and possible evolutionary state.

11.3.4 SN 1999bw and SN 2001ac

Comparatively far less is known about SN 1999bw in NGC 3198 and SN 2001ac in NGC 3504, discussed briefly by Matheson [71]. SN 1999bw had spectral characteristics similar to SN 1997bs and was also quite subluminous for a true SN [38]. SN 2001ac showed a blue continuum and narrow Balmer emission lines, without the broad component, and also resembled SN 1997bs more than a true SN IIn [72]. We show spectra for both of these in Fig. 11.3, but early-time photometry is available only for SN 1999bw [65].

Li et al. [65] identified a $V \simeq 24.1$ mag source with the late-time (2001 January) detection of SN 1999bw, based on *HST*/WFPC2 imaging. However, we have reanalyzed these WFPC2 data, imposing a more accurate astrometric grid on the F555W image and remeasuring the accurate SN position (which differs from the published one by 0.″6), and find that the actual SN position is 1.″0 to the southwest of this identified source. No source is therefore detected in the WFPC2 image at the SN position at that epoch to $V \gtrsim 26.4$ mag. For a distance modulus $\mu_0 = 30.80$ mag (distance 14.5 Mpc) [56] and extinction $A_V = 0.04$ mag [98] toward the host galaxy, this magnitude limit constrains the luminosity of any survivor to $M_V^0 \gtrsim -4.4$ mag. In a *HST*/ACS F606W image from 2006 October, there may be one source with $V \simeq 27.6$ mag, and two with $V \simeq 27.9$ mag, within the positional uncertainty. No source is detected within the positional uncertainty of the SN from that epoch in the F658N (Hα) band as well. If none of these is SN 1999bw, then the limit to any survivor may be as faint as $M_V^0 \gtrsim -3$ mag. We stacked the WFPC2 F606W and F814W images from 2008, but no source is detected exactly at the SN's position; we have not yet estimated a luminosity limit for any survivor, based on these data.

Curiously, despite the lack of a visible detection of a survivor, a source near the position of SN 1999bw is clearly detected at late times in archival mid-infrared images obtained of the host galaxy in 2004 May with *Spitzer* [116]. The published photometry is $0.02 \pm 0.01, 0.04 \pm 0.01, 0.11 \pm 0.02$, and 0.19 ± 0.04 mJy at 3.6, 4.5, 5.8, and 8.0 μm, respectively. Sugerman et al. [116] fit the resulting spectral energy distribution (SED) with a 450 K blackbody, of luminosity $L = 6.2 \times 10^{38}$ erg s^{-1}

and a blackbody radius $R = 1.6 \times 10^{16}$ cm in relative agreement with an assumed $\sim$1,000 km s^{-1} ejecta expansion. The inference from their modelling is that this mid-IR excess arises from dust condensed within the SN ejecta. We have remeasured the photometry from these images and obtain 12, 41, 126, and 295 μJy, respectively. This implied SED can be fit reasonably well by a slightly cooler blackbody at 400 K, with a somewhat higher luminosity, 9.2×10^{38} erg s^{-1}, and larger blackbody radius, 2.5×10^{16} cm. (We have not included any model for dust emissivity here.) What is most striking, however, is that in subsequent *Spitzer* imaging in these same wavebands from 2005 November to 2006 May, this source has significantly faded. Additionally, from ground-based near-infrared imaging at the Palomar 5-m in 2005 January, SN 1999bw is not detected to $K_s > 17.0$ mag ($<$11 μJy). The transient nature of this source strongly implies that it is SN 1999bw.

Such behavior of a fading transient, both in the optical and the infrared, may be analogous to what has been recently found for SN 2008S [93]. (Again, it has not yet been established whether or not SN 2008S is itself a SN impostor; e.g., [8, 111]). In both cases, it is possible that a surviving precursor has become (once again) self-obscured by dust. In the case of SN 2008S, its precursor was only detected at *Spitzer* wavelengths [92]. Of course, no precursor for SN 1999bw (or, SN 2001ac, for that matter) was detected at any wavelength.

Similar behavior may have been observed for SN 2001ac. In multi-band *HST* /WFPC2 images from 2008 nothing is convincingly detected near the nominal position of SN 2001ac; the closest object has $m_{\rm F814W} = 24.6$ mag at 3σ ($M_I \sim -6.1$ mag, for Galactic foreground reddening and an assumed distance to the host galaxy of 13.6 Mpc). However, a source is clearly detected very near the SN position with *Spitzer*/IRAC at 3.6, 4.5, and 5.8 μm (detection at 8.0 μm is less clear), also in 2008. The mid-infrared emission from the source can be modeled as a blackbody with $T \simeq 675$ K, a radius 4.5×10^{15} cm, and a luminosity 3.3×10^{39} erg s^{-1}.

11.3.5 SN 2003gm

SN 2003gm in NGC 5334, discovered at $M_R^0 \approx -14.4$ mag, $\sim$1 mag brighter than SN 2000ch at maximum (see Sect. 11.3.3). The spectrum of SN 2003gm showed a featureless continuum dominated by narrow and broad Balmer emission lines [90], without P Cygni profiles and without [N II] lines [76], similar to what was seen for SN 2000ch. The strengths and widths of the Balmer lines decreased with time. A late-time Keck spectrum of SN 2003gm obtained by Maund et al. [76] showed a greatly reduced continuum and unresolved Balmer and forbidden emission lines of comparable strengths which may not have been from the star, but from an H II region in the star's environment. There are very few photometric observations of SN 2003gm. However, the assessment was that it is photometrically and spectroscopically similar to SN 1997bs.

Maund et al. [76] identified the precursor of SN 2003gm in archival pre-outburst *HST* WFPC2 images. The site of the precursor was precisely located using *BVI*

HST ACS/HRC images. At the distance of the host galaxy and accounting for the reddening toward SN 2003gm ($E(B-V) = 0.05$ mag), the star had $M_V^0 = -7.48$, comparable to the luminosity of the SN 1997bs precursor, and $(V-I)_0 = 0.8$ mag, which is consistent with a luminous F- or G-type supergiant. At discovery during the outburst, the star thus increased by $\Delta m_V \approx -7$ and $\Delta m_I \approx -5$ mag, much more than is expected for normal, or "classical," LBVs. The initial mass of the supergiant star was estimated to be $M_{\rm ini} = 20\,M_\odot$ [76].

11.3.6 SN 2002kg/Variable 37

SN 2002kg in NGC 2403 was identified as a previously known irregular blue variable (V37 [118]) in the host galaxy [124, 135]. A more detailed discussion is to be found elsewhere [76, 130]. A spectrum is shown in Fig. 11.3 and is characterized by broad and normal Balmer emssion lines, without P Cygni profiles. The broad component of the Balmer lines implies an expansion velocity $v_{\rm exp} \sim 390\,{\rm km\,s^{-1}}$, consistent with the wind velocities of LBVs. The spectrum showed strong [N II] emission, with weaker Fe II and [S II] emission, but no lines attributable to oxygen. Uncharacteristic of SN spectra were the appearance of various absorption lines, e.g., Ca II, Si II, Na I, which faded in strength with time.

The light curve is highly erratic and unusual, relative to normal behavior for a SN, reaching only $M_R^0 \approx -10.2$ mag. Similar to SN 2000ch, it showed a broad, pronounced dip, possibly as a result of changes in opacity in the ejecta. Interestingly, the spectrum near maximum resembled that of S Dor *at minimum* (see [70]). SN 2002kg/V37 is thus likely an S Dor-type LBV [130, 135]. At outburst it developed an optically thick, expanding pseudo-photoshere, with the spectrum and colors of a late A- to early F-type supergiant.

Van Dyk et al. [130] measured the properties of the $V = 20.6$ mag V37 prior to the SN 2002kg event from high-quality ground-based, multiband imaging from the Nordic Optical Telescope [62]. Assuming that the reddening to V37 is $E(B-V) = 0.19$ mag, the progenitor star was very bright, ${M_V}^0 = -7.4$ mag, and blue, $(U-B)_0 = -1.1$ mag. Since for normal LBVs the bolometric luminosity does not change during outburst, then, near maximum, when the apparent temperature[4] $T_{\rm app} \simeq 8{,}000$ K for the star and the bolometric correction is essentially zero, $M_{\rm bol} \approx M_V \approx -9.9$ mag.[5] The star's luminosity and temperature at quiescence put it within the "S Dor instability strip" on the HR diagram [48, 137]. V37's initial mass is estimated at $M_{\rm ini} \gtrsim 40\,M_\odot$ [130].

[4]The effective temperature, $T_{\rm eff}$, of a LBV during outburst cannot be defined, since the radius, R, of the star cannot really be measured; hence, the concept of an "apparent temperature," $T_{\rm app}$, for the star has been introduced [48].

[5]This assumes that the measurement of $V = 18.3$ mag in 2003 March is near maximum and the color of the star, $V - R \approx 0.3$ mag during outburst remained relatively constant.

Net Hα (+[N II]) emission was visible at the location of V37 prior to the SN 2002kg outburst [130, 135] suggesting that V37 had an associated nebula possibly from some prior eruptive event. No obvious cluster of stars is seen at the site of SN 2002kg/V37. SN 2002kg/V37 is not detected in a deep coaddition of *Spitzer* images of NGC 2403, obtained between 2004 and 2009, to very low limits ($\approx$few μJy) in all IRAC bands. This indicates that the dust emission at the site of SN 2002kg/V37 must be quite faint. It is possible that the reddening to the precursor has been overestimated, or, alternatively, that any pre-existing dust prior to outburst has been significantly destroyed.

11.3.7 NGC 2363-V1

Although strictly speaking NGC 2363-V1 is not a SN impostor, since it was never identified as a SN, mistakenly or otherwise, it has all of the characteristics of an impostor. It was initially discovered in 1996 in *HST*/WFPC2 images of the giant extragalactic H II region NGC 2363 in the nearby galaxy NGC 2366 [28]. NGC 2363 is similar to 30 Doradus in the LMC, in terms of size and luminosity, and the H II region and the host galaxy are also subsolar in metallicity. V1's spectrum is very blue and is dominated by strong Balmer emission lines. The strong Hα, in particular, is interpreted as indicative of a very high mass-loss rate. Drissen and collaborators [28,29,91] have monitored V1 photometrically and spectroscopically with *HST* and ground-based observations from 1997 to 2004. The star was fainter than $V = 22$ mag prior to 1994, although it may have been present in 1995 in ground-based data from the Canada-France-Hawaii Telescope. V1 peaked at $V = 17.4$ mag in late 1997, which, at a distance of 3.44 Mpc and with low extinction, $E(B-V) = 0.06$ mag, corresponds to $M_V^0 = -10.4$ mag. The star has shown little visible variability, gradually fading 0.3 mag from 1997 to 2004, but with a steady increase in the UV flux over the same period, indicating an increase in temperature [29, 91].

Changes in the emission lines in the UV spectra from *HST*/STIS from 1997 to 2003 correspond to a significant increase in temperature and UV flux and the appearance of a pronounced Balmer jump in emission [91]. There was no corresponding decrease in the visual flux when the UV increased between 1997 and 2000, indicating an actual increase in the star's total luminosity. The spectrum transitioned from an apparent spectral type of ~B6 in 1997 to ~B0 in 2003. In the UV, it resembles an early B-type supergiant, although without the expected P Cygni profiles. Further indication of an increasing apparent temperature is the appearance of strong He I lines in 2001. The Balmer lines decreased in strength between 1997 and 2004; the Hα line width reached a minimum ($v_{\rm wind} \approx 230\,{\rm km\,s^{-1}}$) in 2001, but then subsequently increased, most likely due to changes in the wind and/or the wind velocity.

V1 is a hot, very luminous star with a high mass-loss rate and a slow wind [91]. In the UV and optical, it most closely resembles P Cygni, i.e., an extreme B-type supergiant, rather than η Car. However its outburst is not like that of a normal

LBV/S Dor-type excursion in the HR diagram. Since 1994, V1's total luminosity, mass-loss rate and radius have increased, and since 1997 both the temperature and luminosity have increased. It is thus more like a giant eruption, although the total amount of energy involved is much less than η Car's great eruption.

V1's relative faintness prior to eruption suggests a rather low initial mass for the progenitor compared to most LBVs and impostors. It is possible that the star may have been obscured by dust. Any infrared emission from NGC 2363-V1 detectable by *Spitzer*, though, is confused with the brightness of dust in the surrounding H II region. The mass of ejected gas is small ($\lesssim 0.1\,M_\odot$), so the amount of dust formed during the eruption is also likely to be correspondingly small [91].

11.3.8 SN 2009ip and UGC 2773-OT20091

SN 2009ip in NGC 7259 and UGC 2773-OT20091 (or, UGC 2773-OT) have been linked together, as a result of their discoveries within $\sim$ a week of each other. However, the two are different in their observed properties and phases of variability. They have been both discussed in detail elsewhere in two quite similar studies [41,112]. What distinguishes these two objects the most is that their precursors were both identified in pre-outburst *HST* optical images, and that data exist spanning over years prior to each object's outburst. All indications are that these are both impostors. These extensive pre-outburst datasets are unprecedented for the study of impostors in general. The spectrum of SN 2009ip near maximum showed a hot ($\sim$10,000 K) continuum with strong Balmer emission, of widths $\sim$550 km s^{-1} FWHM; the spectral appearance is similar to what was seen for SN 1997bs. Blueshifted He I absorption indicated additional high-velocity (3000–5000 km s^{-1}) gas. Smith et al. [112] interpret this as two shock regimes, fast and slow, arising from the same eruption, analogous to the velocities observed for the Homunculus around η Car. The spectrum of UGC 2773-OT had quite a different appearance, more similar to S Dor in outburst, with a cooler (7000–8000 K) continuum, relatively weaker Balmer emission lines, and a forest of absorption lines. The emission line widths and blueshifts of absorption features indicate wind speeds of $\sim$350 km s^{-1}.

The erratic, "flickering" of both objects prior to outburst would imply that both stars are related to LBVs or giant eruption LBVs. SN 2009ip, after reaching maximum, swiftly declined in brightness by 3.2 mag, only to recover shortly thereafter to its former brightness interpreted as interpret this as the sudden ejection of an optically thick shell [112]. Foley et al. [41] interpret both the high-velocity ejecta in SN 2009ip, as well as this rapid decrease and return of the object's flux, to be the result of multiple explosions in quick succession.

The precursor for SN 2009ip was an extremely luminous (an astounding $M_V \approx -10$ mag) star of initial mass $\sim$50–80 $M_\odot$ [112], while the UGC 2773-OT precursor was estimated to be less luminous, $M_V \approx -7.8$ mag, and less massive, $\sim$20 $M_\odot$ ($\gtrsim$25 $M_\odot$ [41]).

UGC 2773-OT exhibited evidence of an infrared excess, which Smith et al. [112] and Foley et al. [41] both ascribe to pre-existing circumstellar dust. They also interpret the presence of strong Ca II absorption, as well as [Ca II] and Ca II emission, in the object's spectrum has been as being due to destruction of the pre-existing dust. However, it is more likely that the infrared excess arises from free-free emission. Similar to the yellow hypergiant IRC + 10420 [51], the Ca II emission likely results in the star's wind via radiative de-excitation from the strong Ca II H and K absorption upper levels, and the [Ca II] emission arises in a region beyond the denser wind.

Whereas SN 2009ip occurred in relative isolation, as seen in the pre-outburst *HST* images, UGC 2773-OT appears to be part of a putative stellar cluster; Foley et al. [41] used the properties of these neighboring stars to place further constraints on the precursor's initial mass. Smith et al. [112] suggested that both SN 2009ip and UGC 2773-OT are in between η Car and SN 1954J/V12 in terms of their observed properties. Both objects could still be in an outburst phase as of this writing. Indeed, SN 2009ip has just recently been reported to have experienced two additional outbursts since July 2010 after subsequent fading [27].

11.3.9 SN 2003hy: Probably Not an Impostor

Matheson [71] discussed the interesting case of SN 2003hy in IC 5145. Its early spectra resembled those of the SN impostors discussed above, with narrow emission features atop broader Balmer emission, albeit with distinct P Cygni profiles. However, its absolute magnitude ($M \approx -18$ mag) was significantly brighter than what we would expect for the impostors, and its spectra taken at later times more closely resembled a core-collapse SN (see [71], his Fig. 2). The moral of the SN 2003hy story is that continued, (semi-)periodic monitoring of these events (in fact, of any normal SNe as well) is required, since now several objects, true SNe or otherwise, have been "chameleons," transforming their spectral appearance and revealing possible surprises in their nature. However, it is intriguing to speculate whether the early spectra of SN 2003hy were providing evidence for a high-mass progenitor, possibly one that had gone through an LBV phase prior to explosion. The SN ejecta had clearly interacted at early times with a dense shell of matter, which was overrun at later times. In the next section we discuss recent SNe which more distinctly show evidence for very massive progenitors that have η Car or LBV-type eruptions prior to the terminal explosion.

11.4 The SN Impostor/SN Connection

Evidence is pointing to LBVs, or stars just evolving from the LBV phase, as progenitors of some SNe, especially the SNe IIn. This naturally implies that SN impostors may be the progenitors of real SNe. As highly massive stars, we would, of course, expect the impostors to inevitably explode. The extreme luminosity of the SNe IIn 2006gy in NGC 1260 [83, 109] and 2006tf in an anonymous galaxy [110] may well arise from the interaction of the SN shock with dense CSM immediately surrounding the progenitor, resulting from extreme mass loss at $\sim 0.1\,M_{\odot}$ year^{-1} or more, with a slow wind speed shedding ~ 1–$2\,M_{\odot}$ in the decades just prior to the SN explosion. The CSM for these SNe may be initially so dense that it is opaque to much of the radiation generated from the interaction [106, 110]. Such conditions are very reminiscent of the giant eruptions of very massive, evolved stars, such as η Car and the impostors.

Not all of these extremely luminous SNe are IIn by definition. The next most luminous SN 2005ap is an unusual SN II [94], without the usual spectral signatures of circumstellar interaction seen for SNe IIn. Another very bright SN, 2005gj, has been considered one of the few SN Ia/IIn "hybrids" (e.g., [1]); however, evidence has also been presented that it arose from the explosion of an evolved, hot star [120]. Not all of the SNe possibly connected to LBV-like progenitors are necessarily especially luminous. Due the similarity of the famous rings around underluminous SN 1987A in the LMC to rings seen around the Galactic LBV HD 168625, Smith [102] speculated the progenitor had an LBV-like eruption prior to explosion. It was proabably a post-red supergiant like the less luminous LBVs [48].

N enhancement, relative to, e.g., O and S, in the CSM around impostors, such as SN 2002kg/V37 [130], is consistent with gas rich in the products of CNO processing, which is lost during the LBV phase (e.g, [61]), although N enriched atmospheres are common in many O-type stars. Such N enhancements have also been seen in the ejecta and CSM associated with the SNe IIn 1995N [42] and 1998S [43]. Furthermore, it has been argued that both SNe IIn 1994W [20] and 1995G [18] have CSM ejected in a "violent event" only years before explosion.

Is there more direct evidence linking LBVs, especially the impostors, and SN explosions? Quite possibly. Two very striking examples now exist; SN 2006jc in UGC 4904 and SN 2005gl in NGC 266. Although SN 2006jc was not a SN IIn, but instead a peculiar SN Ib stripped of H and He (e.g., the immediate CSM with which the SN was interacting soon after explosion is He-rich [40]). What appears to have been a powerful outburst (with $M_R^0 \approx -14$ mag) was identified at the location of the SN just $\sim$2 years prior to the explosion [40, 86]! Since we expect SNe Ib/c to arise from single WR stars (in the case of SN 2006jc, most likely in the carbon-rich WC stage), and LBV-like outbursts are not observed from WR stars, this may imply that the SN 2006jc progenitor evolved *very rapidly* from the LBV phase, through the WN phase, to become a WC star at explosion. This further implies that the star must have been very massive. Gal-Yam et al. [45] nominally identified the progenitor of SN IIn 2005gl as a luminous ($M_V^0 = -10.3$ mag) point source seen in pre-explosion *HST* images, although it was not absolutely clear whether or not it was a single star. However, if it were, the luminosity would indicate that it could be an LBV at maximum. Based on late-time *HST* imaging of the SN, Gal-Yam and Leonard [44] showed that the point source present before the explosion has since *vanished*, and that spectra of the SN clearly show signatures of a pre-SN LBV outburst. This is the first definitive case linking SNe (specifically, SNe IIn) to LBVs and SN impostors.

The SN IIn subclass itself spans a very broad range of properties (e.g., [35]). Many SNe IIn, such as 1988Z, 1995N, and 1998S, are almost certainly associated with the collapse of the core in high-mass stars. These show clear signatures of the interaction of the SN shock with a pre-existing, dense CSM. However, one SN IIn, the very bright ($M_V \approx -19.5$ mag) 1994W, assumed to be the explosion (albeit with a surprisingly low ^{56}Ni yield [113]) of a star with an extended envelope into a more extended, dense CSM [20], may not have been a core-collapse event at all—it may merely be the energy released from the interaction of two massive gas shells which were ejected 1.5 year apart from a very massive star [26]. The energetics required appear to be very similar to that of the Great Eruption from η Car and is distinctly reminscent of the SN impostors (although none of these events has come close to -19.5 mag in brightness!). In particular, the star should survive the multiple shell ejection/interaction [26]. The impostors themselves have been mistyped as (confused for?) SNe IIn. The spectral characteristics of narrow emission atop intermediate- or broad-width features for the Balmer lines, in particular, may result from a variety of phenomena: stars undergoing eruptive mass ejections, a post-eruption phase where dense shells collide, and, finally, the explosion of the star and interaction with the pre-SN ejected matter. All of these phenomena may involve the end times for very massive stars.

If LBVs, and therefore, SN impostors, become SNe, Smith [103] has suggested that the LBV phase be of a longer duration ($\sim 10^5$ year) than expected from previous estimates [48]. The longer timescale would allow the star to evolve beyond core H burning for the explosion to occur. This also means that the LBV phase must include periods of extended quiescence. The progenitor star must emerge from quiescence and violently erupt not long, i.e., only years, before terminal explosion, so that the SN ejecta encounter dense CSM early, as observed in SNe IIn, SNe Ibn (e.g., SNe 2006jc, 2000er, 2002ao [87]), and transitional cases (e.g., SN 2005la [88]). This would imply that some of the impostors could well be within years to decades of exploding. It has already been over five decades since the outburst of SN 1954J/V12! Continued periodic monitoring of these events is therefore clearly warranted. Such monitoring will most likely happen, with the various dedicated SN search programs continually observing the various nearby host galaxies. SN impostors are particularly interesting as rare, nearby examples of very massive stars, especially in this exciting, final phase of their lives.

Acknowledgements We acknowledge the generosity of R P Kirshner for making spectra of SNe 1999bw and 2001ac from the CfA Supernova group available to us. Based on observations obtained at the Hale Telescope, Palomar Observatory as part of a continuing collaboration between Caltech, NASA/JPL, and Cornell University. This work is based in part on observations made with the Spitzer Space Telescope, which is operated by the Jet Propulsion Laboratory, Caltech under a contract with NASA.

References

1. G. Aldering, et al., Nearby supernova factory observations of SN 2005gj: another type Ia supernova in a massive circumstellar envelope. ApJ. **650**, 510–527 (2006)
2. I. Aretxaga, et al., SN 1988Z: spectro-photometric catalogue and energy estimates. MNRAS. **309**, 343–354 (1999)

3. R. Barbon, F. Ciatti, L. Rosino, Photometric properties of type II supernovae. A&A. **72**, 287–292 (1979)
4. R. Barbon, et al., SN 1993J in M 81: one year of observations at Asiago. A&AS. **110**, 513–519 (1995)
5. F. Bertola, The supernovae in NGC 1073 and NGC 1058. Ann. Ap. **27**, 319–326 (1964)
6. F. Bertola, Contr. Asiago Obs. No. 171, 3 (1965)
7. M. Blaylock, et al., Mount Wilson and Palomar photographic supernova spectra. PASP. **112**, 1439–1445 (2000)
8. M.T. Botticella, et al., SN 2008S: an electron-capture SN from a super-AGB progenitor? MNRAS. **398**, 1041–1068 (2009)
9. D. Branch, J.J. Cowan, Radio emission from the site of supernova 1961v in NGC 1058. ApJ. **297**, L33–36 (1985)
10. D. Branch, et al., In search of the progenitors of type Ia supernovae. PASP. **107**, 1019–1029 (1995)
11. E. Cappallero, et al., The rate of supernovae from the combined sample of five searches. A&A. **322**, 431–441 (1997)
12. R.M. Catchpole, et al., Spectroscopic and photometric observations of SN 1987a. II - Days 51 to 134. MNRAS. **229**, 15P–25P (1987)
13. R.A. Chevalier, The circumstellar interaction model for the radio emission from a Type I supernova. ApJ. **285**, L63–L66 (1984)
14. R.A. Chevalier, Circumstellar interaction around supernovae, in *From Twilight to Highlight: The Physics of Supernovae*, ed. by W. Hillebrandt, B. Leibundgut (Springer, Berlin/Heidelberg, 2003)
15. R.A. Chevalier, C. Fransson, Emission from circumstellar interaction in normal Type II supernovae. ApJ. **420**, 268–285 (1994)
16. R.A. Chevalier, C. Fransson, T.K. Nymark, Radio and X-ray emission as probes of type IIP supernovae and red supergiant mass loss. ApJ. **641**, 1029–1038 (2006)
17. Y.-H. Chu, et al., The nature of SN 1961V. AJ. **127**, 2850–2855 (2004)
18. N.N. Chugai, I.J. Danziger, A massive circumstellar envelope around the Type-IIn supernova SN 1995G. AstL. **29**, 649–657 (2003)
19. N.N. Chugai, R.A. Chevalier, P. Lundqvist, Circumstellar interaction of the type Ia supernova 2002ic. MNRAS. **355**, 627–637 (2004)
20. N.N. Chugai, et al., The Type IIn supernova 1994W: evidence for the explosive ejection of a circumstellar envelope. MNRAS. **352**, 1213–1231 (2004)
21. J.J. Cowan, R.B.C. Henry, D. Branch, Radio and optical studies of supernova 1961V in NGC 1058. ApJ. **329**, 116–121 (1988)
22. P. Cox, et al., Millimeter emission of eta Carinae and its surroundings. A&A. **297**, 168–174 (1995)
23. P.A. Crowther, Physical properties of Wolf-Rayet stars. ARA&A. **45**, 177–219 (2007)
24. K. Davidson, The relation between apparent temperature and mass-loss rate in hypergiant eruptions. ApJ. **317**, 760–764 (1987)
25. K. Davidson, R.M. Humphreys, Eta Carinae and its environment. ARA&A. **35**, 1–32 (1997)
26. L. Dessart, D.J. Hillier, S. Gezari, S. Basa, T. Matheson, SN 1994W: an interacting supernova or two interacting shells? MNRAS. **394**, 21–37 (2009)
27. A.J. Drake, et al., New luminous outbursts of supernova impostor SN 2009ip. The Astronomer's Telegram, ATEL 2897 (2010)
28. L. Drissen, J.-R. Roy, C. Robert, A new luminous blue variable in the giant extragalactic H II region NGC 2363. ApJ. **474**, L35–L38 (1997)
29. L. Drissen, et al., Physical parameters of erupting luminous blue variables: NGC 2363-V1 caught in the Act. ApJ. **546**, 484–495 (2001)
30. G. Duszanowicz, S. Nakano, K. Itagaki, Variable star near NGC 3432. CBET. 1534 (2008)
31. N. Elias-Rosa, et al., The massive progenitor of the Type II-linear supernova 2009kr. ApJ. **714**, L254–L259 (2010)
32. A. Elmhamdi, Hydrogen and Helium traces in Type Ib-c supernovae. A&A. **450**, 305–330(2006)

33. A. Fassia, et al., Optical and infrared photometry of the Type IIn SN 1998S: days 11-146. MNRAS. **318**, 1093–1104 (2000)
34. R.A. Fesen, Recent optical images of the site of supernova 1961v in NGC 1058. ApJ. **297**, L29–L31 (1985)
35. A.V. Filippenko, Optical spectra of supernovae. ARAA. **35**, 309–355 (1997)
36. A.V. Filippenko, T. Matheson, L.C. Ho, The "Type IIb" supernova 1993J in M81: a close relative of Type Ib supernovae. ApJ. **415**, L103–L106 (1993)
37. A.V. Filippenko, A.J. Barth, G.C. Bower, L.C. Ho, G.S. Stringfellow, R.W. Goodrich, A.C. Porter, Was Fritz Zwicky's "Type V" SN 1961V a genuine supernova? AJ. **110**, 2261–2273 (1995) [Erratum: AJ. **112**, 806 (1996)]
38. A.V. Filippenko, W.D. Li, M. Modjaz, Supernova 1999bw in NGC 3198. IAUC. 7152 (1999)
39. A.V. Filippenko, W.D. Li, R.R. Treffers, M. Modjaz, The lick observatory supernova search with the Katzman automatic imaging telescope, in *Small-Telescope Astronomy on Global Scales*, ed. by W.P. Chen, C. Lemme, B. Paczyński (ASP, San Francisco, 2000)
40. R.J. Foley, et al., SN 2006jc: a Wolf-Rayet star exploding in a dense He-rich circumstellar medium. ApJ. **657**, L105–L108 (2007)
41. R.J. Foley, et al., The diversity of massive star outbursts I: observations of SN 2009ip, UGC 2773 OT2009-1, and their progenitors. ApJ. **732**, 32 (2010)
42. C. Fransson, et al., Optical and ultraviolet spectroscopy of SN 1995N: evidence for strong circumstellar interaction. ApJ. **572**, 350–370 (2002)
43. C. Fransson, et al., Hubble space telescope and ground-based observations of SN 1993J and SN 1998S: CNO processing in the progenitors. ApJ. **622**, 991–1007 (2005)
44. A. Gal-Yam, D.C. Leonard, A massive hypergiant star as the progenitor of the supernova SN 2005gl. Nature **458**, 865–867 (2009)
45. A. Gal-Yam, et al., On the progenitor of SN 2005gl and the nature of Type IIn supernovae. AJ. **656**, 372–381 (2007)
46. R.W. Goodrich, G.S. Stringfellow, G.D. Penrod, A.V. Filippenko, SN 1961V: an extragalactic Eta Carinae analog? ApJ. **342**, 908–916 (1989)
47. W.C.G. Ho, et al., BVRI photometry of supernovae. PASP. **113**, 1349–1364 (2001)
48. R.M. Humphreys, K. Davidson, The luminous blue variables: astrophysical geysers. PASP. **106**, 1025–1051 (1994)
49. R.M. Humphreys, D.B. McElroy, The initial mass function for massive stars in the galaxy and the magellanic clouds. ApJ. **284**, 565–577 (1984)
50. R. Humphreys, K. Davidson, N. Smith, Eta Carinae's second eruption and the light curves of the eta Carinae variables. PASP. **111**, 124–1131 (1999)
51. R. Humphreys, K. Davidson, N. Smith, Crossing the yellow void: spatially resolved spectroscopy of the post-red supergiant IRC +10420 and its circumstellar ejecta. ApJ. **124**, 1026–1044 (2002)
52. R.M. Humphreys, L.A. Helton, T.J. Jones, The three-dimensional morphology of VY Canis Majoris. I. The kinematics of the ejecta. AJ. **133**, 2716–2729 (2007)
53. R. Humphreys, K. Davidson, M. Koppelman, The early spectra of Eta Carinae 1892 to 1941 and the onset of its high excitation emission spectrum. AJ. **135**, 1249–1263 (2008)
54. R.M. Humphreys, et al., SN 2010U: a luminous Nova in NGC 4214. ApJ. **718**, L43–L47 (2010)
55. K. Iwamoto, et al., Theoretical light curves for the Type Ic supernova SN 1994I. ApJ. **437**, L115–L118 (1994)
56. D.D. Kelson, et al., The Hubble space telescope key project on the extragalactic distance scale. XIX. The discovery of cepheids in and a new distance to NGC 3198. ApJ. **514**, 614–636 (1999)
57. A.R. Klemola, Precise positions for two supernovae. PASP. **98**, 464–466 (1986)
58. C.S. Kochanek, D.M. Szczygiel, K.Z. Stanek, The supernova impostor impostor SN 1961V: Spitzer shows that Zwicky was right (again). ApJ. **737**, 76 (2011)
59. C.T. Kowal, W.L.W. Sargent, Supernovae discovered since 1885. AJ. **76**, 756–764 (1971)
60. C.T. Kowal, et al., The 1971 palomar supernova search. PASP. **84**, 844–849 (1972)

61. N. Langer, et al., Towards an understanding of very massive stars. A new evolutionary scenario relating O stars, LBVs and Wolf-Rayet stars. A&A. **290**, 819–833 (1994)
62. S.S. Larsen, T. Richtler, Young massive star clusters in nearby galaxies . I. Identification and general properties of the cluster systems. A&A. **345**, 59–72 (1999)
63. T. Lejeune, D. Schaerer, Database of Geneva stellar evolution tracks and isochrones for (UBV)J(RI)C JHKLL'M, HST-WFPC2, Geneva and Washington photometric systems. A&A. **366**, 538–546 (2001)
64. D.C. Leonard, et al., The distance to SN 1999em in NGC 1637 from the expanding photosphere method. PASP. **114**, 3564 (2001)
65. W. Li, et al., A Hubble space telescope snapshot survey of nearby supernovae. PASP. **114**, 403–415 (2002)
66. W. Li, et al., Nearby supernova rates from the Lick observatory supernova search. II. The observed luminosity functions and fractions of supernovae in a complete sample. MNRAS. **412**, 1473–1507 (2011)
67. K.S. Long, P.F. Winkler, W.P. Blair, A rapid decline in the optical emission from SN 1957D in M83. ApJ. **395**, 632–636 (1992)
68. K. Maeda, et al., Asphericity in supernova explosions from late-time spectroscopy. Sci. **319**, 1220–1223 (2008)
69. J.C. Martin, M.D. Koppelman, η Carinae's brightness variations since 1998: Hubble space telescope observations of the central star. AJ. **127**, 2352–2361 (2004)
70. P. Massey, An unprecedented change in the spectrum of S Doradus: as cool as it gets. PASP. **112**, 144–147 (2000)
71. T. Matheson, Supernova impostors in the center for astrophysics SN database, in *The Fate of the Most Massive Stars*, ed. by R. Humphreys, K. Stanek (ASP, San Francisco, 2005)
72. T. Matheson, M. Calkins, Supernova 2001ac in NGC 3504. IAUC. 7597 (2001)
73. T. Matheson, et al., Optical spectroscopy of supernova 1993J during its first 2500 Days. AJ. **120**, 1487–1498 (2000)
74. T. Matheson, et al., Optical spectroscopy of Type Ib/c supernovae. AJ. **121**, 1648–1675 (2001)
75. J.R. Maund, et al., The massive binary companion star to the progenitor of supernova 1993J. Nature **427**, 129–131 (2004)
76. J.R. Maund, et al., Faint supernovae and supernova impostors: case studies of SN 2002kg/NGC 2403-V37 and SN 2003gm. MNRAS. **369**, 390–406 (2006)
77. P.A. Mazzali, et al., An asymmetric energetic Type Ic supernova viewed off-axis, and a link to gamma ray bursts. Sci. **308**, 1284–1287 (2005)
78. P.A. Mazzali, et al., Models for the Type Ic hypernova SN 2003lw associated with GRB 031203. ApJ. **645**, 1323–1330 (2006)
79. G. Meynet, A. Maeder, Stellar evolution with rotation. X. Wolf-Rayet star populations at solar metallicity. A&A. **404**, 975–990 (2003)
80. R. Minkowski, Spectra of supernovae. PASP. **53**, 224–225 (1941)
81. M.J. Montes, et al., Radio observations of SN 1979C: evidence for rapid presupernova evolution. ApJ. **532**, 1124–1131 (2000)
82. K. Nomoto, et al., Type Ia supernovae: progenitors and diversities, in *From Twilight to Highlight: The Physics of Supernovae*, ed. by W. Hillebrandt, B. Leibundgut (Springer, Berlin, 2003)
83. E.O. Ofek, et al., SN 2006gy: an extremely luminous supernova in the galaxy NGC 1260. ApJ. **659**, L13–L16 (2007)
84. J.P. Ostriker, J.E. Gunn, Do pulsars make supernovae? ApJ. **164**, L95–L104 (1971)
85. M. Papenkova, W.D. Li, Variable star in field of NGC 3432. IAUC. 7415 (2000)
86. A. Pastorello, et al., A giant outburst two years before the core-collapse of a massive star. Nature **447**, 829 (2007)
87. A. Pastorello, et al., Massive stars exploding in a He-rich circumstellar medium. I. Type Ibn (SN 2006jc-like) events. MNRAS. **389**, 113–130 (2008a)
88. A. Pastorello, et al., Massive stars exploding in a He-rich circumstellar medium. II. The transitional case of SN 2005la. MNRAS. **389**, 131–140 (2008b)

89. A. Pastorello, et al., Multiple major outbursts from a restless luminous blue variable in NGC 3432. MNRAS. **408**, 181–198 (2010)
90. F. Patat, A. Pastorello, J. Aceituno, 2003gm in NGC 5334. IAUC. 8167 (2003)
91. V. Petit, L. Drissen, P.A. Crowther, Spectral evolution of the luminous blue variable NGC 2363-V1. I. observations and qualitative analysis of the ongoing giant eruption. ApJ. **132**, 1756–1762 (2006)
92. J.L. Prieto, et al., Discovery of the dust-enshrouded progenitor of SN 2008S with spitzer. ApJ. **681**, L9–L12 (2008)
93. J.L. Prieto, et al., The SN 2008S progenitor star: gone or again self-obscured? ApJL. (2010) Submitted (arXiv:1007.0011)
94. R.M. Quimby, et al., SN 2005ap: a most brilliant explosion. ApJ. **668**, L99–L102 (2007)
95. M.W. Richmond, et al., UBVRI photometry of the Type IC SN 1994I in M51. AJ. **111**, 327–339 (1996)
96. A.G. Riess, et al., BVRI light curves for 22 Type Ia supernovae. AJ. **117**, 707–724 (1999)
97. E.M. Schlegel, A new subclass of Type II supernovae? MNRAS. **244**, 269–271 (1990)
98. D.J. Schlegel, D.P. Finkbeiner, M. Davis, Maps of dust infrared emission for use in estimation of reddening and cosmic microwave background radiation foregrounds. ApJ. **500**, 525–553 (1998)
99. N.A. Silbermann, et al., The Hubble space telescope key project on the extragalactic distance scale. VI. The cepheids in NGC 925. ApJ. **470**, 1–37 (1996)
100. S.J. Smartt, J.J. Eldridge, R.M. Crockett, J.R. Maund, The death of massive stars — I. Observational constraints on the progenitors of Type II-P supernovae. MNRAS. **395**, 1409–1437 (2009)
101. N. Smith, The dusty homunculus nebula of Eta Carinae, in *The Fate of the Most Massive Stars*, ed. by R. Humphreys, K. Stanek (ASP, San Francisco, 2005)
102. N. Smith, Discovery of a nearby twin of SN 1987A's nebula around the luminous blue variable HD 168625: was Sk -69 202 an LBV? AJ. **133**, 1034–1040 (2007)
103. N. Smith, Episodic mass loss and pre-SN circumstellar envelopes, in *Massive Stars as Cosmic Engines*, ed. by F. Bresolin P.A. Crowther J. Puls (Cambridge University Press, Cambridge, 2008)
104. N. Smith, P.S. Conti, On the role of the WNH phase in the evolution of very massive stars: enabling the LBV instability with feedback. ApJ. **679**, 1467–1477 (2008)
105. N. Smith, G.J. Ferland, The structure of the Homunculus. II. modeling the physical conditions in η Carinae's molecular shell. ApJ. **655**, 911–919 (2007)
106. N. Smith, R. McCray, Shell-shocked diffusion model for the light curve of SN 2006gy. ApJ. **671**, L17–L20 (2007)
107. N. Smith, S.P. Owocki, On the role of continuum-driven eruptions in the evolution of very massive stars and population III stars. ApJ. **645**, L45–L48 (2006)
108. N. Smith, R.M. Humphreys, R.D. Gehrz, Post-eruption detection of variable 12 in NGC 2403 (SN 1954j): another η Carinae variable. PASP. **113**, 692–626 (2001)
109. N. Smith, et al., SN 2006gy: discovery of the most luminous supernova ever recorded, powered by the death of an extremely massive star like η Carinae. ApJ. **666**, 1116–1128 (2007)
110. N. Smith, et al., SN 2006tf: precursor eruptions and the optically thick regime of extremely luminous Type IIn supernovae. ApJ. **686**, 467–484 (2008)
111. N. Smith, et al., SN 2008S: a cool super-eddington wind in a supernova impostor. ApJ. **697**, L49–L53 (2009)
112. N. Smith, et al., Discovery of precursor luminous blue variable outbursts in two recent optical transients: the fitfully variable missing links UGC 2773-OT and SN 2009ip. AJ. **139**, 1451–1467 (2010)
113. J. Sollerman, R.J. Cumming, P. Lundqvist, A very low mass of ^{56}Ni in the ejecta of SN 1994W. ApJ. **493**, 933–939 (1998)
114. C.J. Stockdale, et al., The fading radio emission from SN 1961V: evidence for a Type II peculiar supernova? AJ. **122**, 283–287 (2001)

115. R.B. Stothers, C.-W. Chin, Evolution of massive stars into luminous blue variables and Wolf-Rayet stars for a range of metallicities: theory versus observation. ApJ. **468**, 842–850 (1996)
116. B. Sugerman, M. Meixner, J. Fabbri, M. Barlow, Supernova 1999bw in NGC 3198. IAUC. 8442 (2004)
117. N.B. Suntzeff, et al., SN 1987A in the LMC. II - Optical photometry at Cerro Tololo. AJ. **96**, 1864–1873 (1988)
118. G.A. Tammann, A. Sandage, The stellar content and distance of the galaxy NGC 2403 IN the M81 group. ApJ. **151**, 825–860 (1968)
119. T.A. Thompson, et al., A new class of luminous transients and a first census of their massive stellar progenitors. ApJ. **705**, 1364–1384 (2009)
120. C. Trundle, R. Kotak, J.S. Vink, W.P.S. Meikle, SN 2005gj: evidence for LBV supernovae progenitors? A&A. **483**, L47–L50 (2008)
121. M. Turatto, Classification of supernovae, in *Supernovae and Gamma-Ray Bursters*, ed. by K. Weiler (Springer, Berlin, 2003)
122. M. Turatto, et al., The Type II supernova 1988Z in MCG + 03-28-022 - Increasing evidence of interaction of supernova ejecta with a circumstellar wind. MNRAS. **262**, 128–140 (1993)
123. V.P. Utrobin, Supernova SN 1961v - an explosion of a very massive star. Ap. Space Sci. **98**, 115–147 (1984)
124. S.D. Van Dyk, The η Carinae analogs, in *The Fate of the Most Massive Stars*, ed. by R. Humphreys K. Stanek (ASP, San Francisco, 2005)
125. S.D. Van Dyk, et al., Type "IIn" supernovae: a search for radio emission. AJ. **111**, 1271–1277 (1996)
126. S.D. Van Dyk, C.Y. Peng, A.J. Barth, A.V. Filippenko, The environments of supernovae in post-refurbishment Hubble space telescope images. AJ. **118**, 2331–2349 (1999)
127. S.D. Van Dyk, C.Y. Peng, J.Y. King, A.V. Filippenko, R.R. Treffers, W. Li, M.W. Richmond, SN 1997bs in M66: another extragalactic η Carinae analog? PASP. **112**, 1532–1541 (2000)
128. S.D. Van Dyk, A.V. Filippenko, W. Li, Possible recovery of SN 1961V in Hubble space telescope archival images. PASP. **114**, 700–707 (2002)
129. S.D. Van Dyk, W. Li, A.V. Filippenko, Supernova 1954J (variable 12) in NGC 2403 unmasked. PASP. **117**, 553–562 (2005)
130. S.D. Van Dyk, et al., The Type IIn supernova 2002kg: the outburst of a luminous blue variable star in NGC 2403. arXiv:astro-ph/0603025 (2006)
131. S.D. Van Dyk, T. Matheson, It's Alive! The Supernova Impostor 1961V. ApJ. in press (2012)
132. F. Viallefond, W.M. Goss, H II regions in M33. III - Physical properties. A&A. **154**, 357–369 (1986)
133. R.M. Wagner, et al., Discovery and evolution of an unusual luminous variable star in NGC 3432 (supernova 2000ch). PASP. **116**, 326–336 (2004)
134. K.W. Weiler, et al., Long-term radio monitoring of SN 1993J. ApJ. **671**, 1959–1980 (2007)
135. K. Weis, D.J. Bomans, SN 2002kg - the brightening of LBV V37 in NGC 2403. A&A. **429**, L13–L16 (2005)
136. C.L. Williams, et al., Radio emission from SN 1988Z and very massive star evolution. ApJ. **581**, 396–403 (2002)
137. B. Wolf, Empirical amplitude-luminosity relation of S Doradus variables and extragalactic distances. A&A. **217**, 87–91 (1989)
138. S.E. Woosley, N. Langer, T.A. Weaver, The evolution of massive stars including mass loss: presupernova models and explosion. ApJ. **411**, 823–839 (1993)
139. F. Zwicky, NGC 1058 and its supernova 1961. ApJ. **139**, 514–519 (1964)
140. F. Zwicky, Supernovae, in *Stars and Stellar Systems*, vol. VIII, ed. by L.H. Aller D.B. McLaughlin (University of Chicago Press, Chicago, 1965)

Chapter 12
Instability & Mass Loss near the Eddington Limit

S.P. Owocki and N.J. Shaviv

Abstract We review the physics of continuum-driven mass loss and its likely role in η Carinae and LBVs. Unlike a line-driven wind, which is inherently limited by self-shadowing, continuum driving can in principle lead to mass-loss rates up to the "photon-tiring" limit, for which the entire luminosity is expended in lifting the outflow. We discuss how instabilities near the Eddington limit give rise to a clumped atmosphere, and how the associated "porosity" can regulate a continuum-driven flow. We also summarize recent time-dependent simulations in which a mass flow stagnates because it exceeds the tiring limit, leading to complex time-dependent inflow and outflow regions. Porosity-regulated continuum driving in super-Eddington epochs can probably explain the large, near tiring-limit mass loss inferred for LBV giant eruptions. However, while these extreme flows can persist over dynamically long periods, they cannot be sustained for an evolutionary timescale; so ultimately it is stellar structure and evolution that sets the overall mass loss.

12.1 Introduction

LBV giant eruptions or "supernova imposters" are characterized by strongly enhanced luminosity and substantial mass ejection. An extreme example is η Carinae, which during its giant eruption of 1840–1860 ejected $10 M_{\odot}$ or more at speeds of 300–800 $\mathrm{km\,s^{-1}}$, implying a kinetic energy rate comparable to the peak radiative

S.P. Owocki (✉)
Bartol Research Institute, Department of Physics & Astronomy, University of Delaware, Newark, DE 19716 USA
e-mail: owocki@bartol.udel.edu

N.J. Shaviv
Racah Institute of Physics, Hebrew University, GivUat Ram, Jerusalem 91904 Israel
e-mail: shaviv@phys.huji.ac.il

K. Davidson and R.M. Humphreys (eds.), *Eta Carinae and the Supernova Impostors*, Astrophysics and Space Science Library 384, DOI 10.1007/978-1-4614-2275-4_12,

luminosity $L \approx 20 \times 10^6 L_\odot$ seen during the eruption [12, 25, 43]. The total energy release is only a few percent of the 10^{51} erg non-neutrino output in a typical supernova, but represents a substantial fraction of the binding energy of the layers outside the star's core. Such extreme mass loss is orders of magnitude greater than one expects in a standard CAK model for radiative driving by scattering in metal-ion lines [7].

The association of η Car and other LBVs with the empirical upper limit of luminosity for observed stars [23] has led to a general view that such strong episodes of mass loss may result from the star's approach to the classical Eddington limit, at which the continuum radiative force due to Thomson scattering equals the inward force of gravity [1, 10, 24, 25]. Normally, if the radiative flux approaches the Eddington limit in a star's interior, convection arises and carries enough of the energy flux to keep the interior sub-Eddington, as we explain in Sect. 12.2.3 below. On the other hand, no fundamental constraint prohibits the existence of a quasi-steady state with very high radiation fluxes in the outer layers of a star. As we show in Sect. 12.2.4, a super-Eddington luminosity may persist for a substantial length of time once we consider that high fluxes also give rise to instabilities.

This can lead to a situation analogous to classical novae, wherein the sudden onset of shell burning induces a super-Eddington brightness that lasts for several months; e.g., Nova LMC 1988 #1 was super-Eddington for about 50 days [36]. Since this is much longer than any dynamical timescale in the nova system, it can result in a quasi-steady wind mass loss, driven by continuum rather than line opacity. In fact, it was realized that optically thick continuum-driven winds explain many characteristics of these objects long before it was understood how super-Eddington luminosities could arise [3]. Unlike LBVs with poorly determined masses, novae have the Chandrasekhar mass as a strict upper limit, so there is no doubt that they are indeed super-Eddington [40].

Whatever mechanism may trigger such a super-Eddington brightening in LBVs, a key issue is how the continuum driving is regulated to keep the stellar interior gravitationally bound, while allowing a sustainable mass loss from near the surface. As discussed below, it seems that interior convection, flow stagnation, and the 'porosity' of a clumped medium may all play a role. A likely result is a mass loss rate that approaches the "photon-tiring limit", associated with the finite energy available to lift material out of the star's gravitational potential. This is several orders of magnitude higher than can be achieved by line-driving, and is indeed comparable to the mass loss rate, $\sim 1 M_\odot$ year^{-1}, inferred for the giant eruption of η Carinae [33].

However, the associated mass loss time scale, $t_{\dot{M}} = M/\dot{M} \sim 100$ year, is so short that this state can only be sustained for a few years before inducing fundamental readjustments in the stellar structure. After the super-Eddington condition has thus been quenched, a much longer thermal timescale probably elapses before another outburst [25,29]. But even with a limited duty cycle, such eruptions can dramatically reduce the star's mass within its evolutionary timescale. Such LBV mass loss may play a central role in the evolution and final fate of the most massive stars, and may be a key factor in setting the stellar upper mass limit. Moreover, unlike CAK line-driving, these continuum-driven processes do not depend directly on metallicity. Thus they may play a similar role in the first generation of massive stars ("Population III"), which are thought to have helped to reionize the universe following the recombination epoch of the Big Bang.

Below we elaborate on this role of radiation forces for both the interior structure (Sect. 12.2) and mass loss (Sect. 12.3) of massive stars. A concluding discussion (Sect. 12.4) considers general issues related to the energy source of the giant eruptions and various implications for massive star evolution near the Eddington limit.

12.2 Radiation Pressure in Massive Stars

The key to understanding the internal structure of massive stars is to recognize the effects that arise in the presence of very strong radiative flux. In Sect. 12.2.1 we begin by discussing the radiative force and the Eddington limit. In Sect. 12.2.2, we demonstrate how the radiative pressure changes the mass luminosity relation. In Sect. 12.2.3 we show that as the Eddington luminosity is approached, convection is necessarily excited, implying that the interior of a star never becomes super-Eddington. In Sect. 12.2.4 we discuss the instabilities that porosify the outer atmosphere. The combined effects allow us to build a coherent picture, in Sect. 12.2.5, for *quasi-static* structure in super-Eddington stars, such as η Car during its giant eruption.

12.2.1 Radiative Force and the Eddington Limit

Let us begin by considering the general form for the radiative acceleration $\mathbf{g}_{rad}$ associated with opacity κ_ν (a.k.a. the mass absorption coefficient, with units cm^2 g^{-1}) in stellar material with a radiative flux $\mathbf{F}_\nu$ at photon frequency ν:

$$\mathbf{g}_{rad} = \int_0^\infty \frac{\kappa_\nu \mathbf{F}_\nu}{c} d\nu \,. \tag{12.1}$$

Here κ_ν includes continuum processes – Thomson scattering by electrons plus bound-free and free-free absorption – and bound-bound transitions.

In a static interior or atmosphere, where saturation of spectral lines keeps the associated line-force small, $\mathbf{g}_{rad}$ is dominated by continuum processes, mainly Thomson scattering in the cases discussed here. Because Thomson opacity is gray (frequency-independent), it can be pulled outside the frequency integration in (12.1). In an idealized, spherically symmetric, radiative envelope, the bolometric flux $F = \int_0^\infty F_\nu d\nu$ is purely radial, $F = L/4\pi r^2$ where L is the bolometric luminosity. The radiative acceleration associated with a gray opacity κ is thus

$$g_{rad} = \frac{\kappa F}{c} = \frac{\kappa L}{4\pi r^2 c} \equiv \Gamma g \,. \tag{12.2}$$

Here Γ is the Eddington parameter L/L_{Edd}, where $L_{\mathrm{Edd}} \equiv 4\pi GMc/\kappa$ and M is the stellar mass. This is the ratio of the radiative acceleration to the local gravitational acceleration $g = GM/r^2$. Since both gravity and radiative flux have the same r^{-2} dependence, Γ is almost independent of radius in regions where the gradients of κ, L, and M are small.[1] For the classical case of pure Thomson scattering and "normal" chemical composition, the Eddington parameter is quantitatively

$$\Gamma_{\mathrm{e}} \approx 2.6 \times 10^{-5} \frac{(L/L_{\odot})}{(M/M_{\odot})}. \tag{12.3}$$

As discussed in the next subsection, stellar luminosity scales with a high power of the stellar mass, $L \propto M^3$, so hot massive stars with $M > 10M_{\odot}$ generally have $\Gamma_{\mathrm{e}} > 0.1$, possibly even approaching unity. Indeed $\Gamma_{\mathrm{e}} = 1$ defines the *classical Eddington limit* for the L/M ratio where an idealized star would become unbound.

It should be emphasized, however, that this does not represent an appropriate condition for normal steady mass loss, which requires an outwardly increasing radiative force that goes from being *less* than gravity inside the star to *exceeding* gravity in the outflowing wind. Section 12.3 summarizes how the necessary force modulation can occur through line-desaturation for line driving, and through porosity of spatial structure for continuum driving.

12.2.2 *Stellar Structure Scaling for Luminosity vs. Mass*

The interior structure of a star is set by the dual requirements for hydrostatic balance and energy transport. As Eddington found [14], we can use these to derive a simple scaling relation for stellar luminosity vs. mass. Let us begin with the classic formula for radiative flux F_{rad} via outward diffusion of radiation energy density U_{rad},

$$F_{rad} = -\frac{c}{3\kappa\rho}\frac{dU_{rad}}{dr}. \tag{12.4}$$

Since $P_{rad} = U_{rad}/3$, (12.4) can be rearranged to show that the gradient of radiation pressure is a radiative force per unit volume,

$$\frac{dP_{rad}}{dr} = -\frac{\kappa F_{rad}}{c}\rho = -g_{rad}\rho = -\Gamma g\rho = -\Gamma\frac{GM}{r^2}\rho. \tag{12.5}$$

Subtracting this from the hydrostatic equation that sets the gradient of total pressure $P = P_{gas} + P_{rad}$, we find that *gas* pressure in a static model satisfies

$$\frac{dP_{gas}}{dr} = -(1-\Gamma)\frac{GM}{r^2}\rho. \tag{12.6}$$

[1] As discussed below, there are various circumstances in which this is not the case.

Combined with mass conservation and the ideal gas law, (12.4) and (12.6) form the basis for Eddington's $n = 3$ polytrope model for stars with radiative interiors [14]. In such a model the Eddington parameter has a simple scaling with stellar mass,

$$\frac{\Gamma}{(1-\Gamma)^4} = \mu^4 \left(\frac{M}{18.3 M_\odot}\right)^2 \approx \left(\frac{M}{48 M_\odot}\right)^2 , \tag{12.7}$$

where the molecular weight μ is approximately $4/(5X+3)$ in a fully ionized mix with hydrogen mass fraction X. The second equality in (12.7) applies for hydrogen abundance $X = 0.7$, $\mu \approx 0.62$.

This scaling can be understood from the average gradients in (12.4) and (12.6) in terms of stellar mass M and radius R. Using the ideal gas law $P_{gas} \sim \rho T$ with $\rho \sim M/R^3$, (12.6) implies that the characteristic interior temperature scales as

$$T \sim (1-\Gamma)\,\frac{M}{R} . \tag{12.8}$$

With the further proportionalities $F_{rad} \sim L/R^2$ and $U_{rad} \sim T^4$, (12.4) gives

$$L \sim \frac{R^4 T^4}{M} . \tag{12.9}$$

Combining (12.8) and (12.9), we can eliminate both R and T to find

$$L \sim (1-\Gamma)^4 M^3 \quad \text{or} \quad \frac{\Gamma}{(1-\Gamma)^4} \sim M^2 , \tag{12.10}$$

which agrees with (12.7). Note that this result does not depend on the nature of energy generation in the stellar core.[2]

Figure 12.1 shows a log-log plot of the resulting variation of luminosity vs. mass. For $M \lesssim 20 M_\odot$ it implies a strong $L \sim M^3$ scaling, but for higher masses the $1-\Gamma$ term acts as a strong repeller away from the Eddington limit, causing a bend toward a linear asymptotic scaling $L \sim M$.

Formally this scaling suggests it is possible to have stars with arbitrarily large mass, approaching arbitrarily close to the Eddington limit. But surveys of dense young clusters are providing evidence for a cutoff in the stellar mass distribution at about $M \approx 150-200\,M_\odot$ [15,28,31,52].[3]

---

[2]Of course these simple relations have to be modified to accomodate gradients in the molecular weight as a star evolves from the zero-age main sequence. Equation 12.10 fails entirely for cool giant and supergiant stars where convection dominates the energy transport, but such objects are beyond the scope of this review.

[3][Editors' comment:] This has long been suspected on other grounds. For instance, if one extrapolates the empirical upper limit in the HR diagram to high temperatures, it approaches the ZAMS around $200\,M_\odot$ [23–25]. Various clues seem to implicate the same mass range.

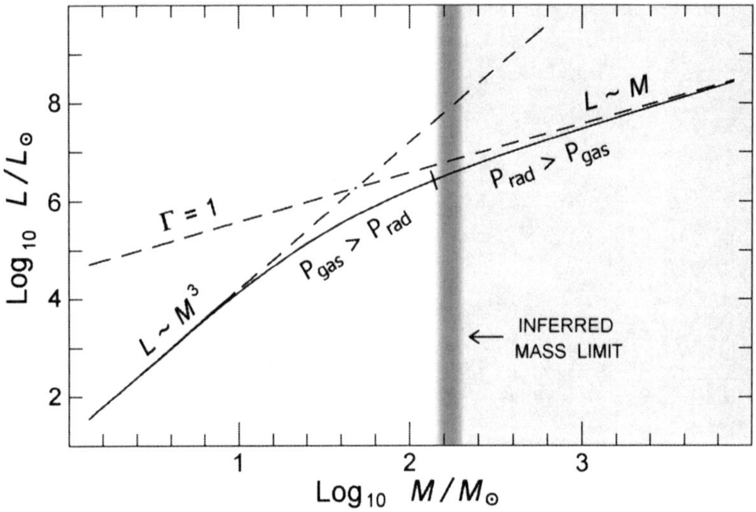

Fig. 12.1 Schematic log-log plot of stellar luminosity vs. mass for idealized $n = 3$ polytrope models, (12.7). For real stars the shape and normalization of the curve are somewhat altered by convection, variations in opacity and molecular weight, and other effects. The "inferred mass limit" at 150–200 $M_{\odot}$ is based mainly on observations, not theory

In Fig. 12.1, this inferred upper mass limit is slightly to the right of the $\Gamma = 0.5$ point; therefore $P_{rad} > P_{gas}$ in the most massive stars (This is particularly true for η Car [12]). Dominance by radiation pressure is somewhat analogous to having a heavier fluid support a lighter one, and, as discussed in Sect. 12.2.4, such a configuration may be subject to various kinds of instabilities, leading to spatial clumping and/or LBV-like brightness variations [37–39, 48]. Resulting mass loss may play a key role in setting the stellar upper mass limit.

12.2.3 Convective Instability of a Super-Eddington Stellar Interior

If the energy flux locally exceeds the Eddington limit in a stellar interior, this usually does *not* initiate mass loss. Instead $\Gamma \gtrsim 1$ induces convective instability [27]. In the deep interior the gas density is large enough to carry most of the heat flux via convection, thus lowering the radiative flux so Γ_{rad} remains safely below unity.

This suggests that a super-Eddington outflow should originate in near-surface layers where convection becomes inefficient. Within "mixing length" formalism [22], the convective energy flux is roughly

$$F_{conv} \approx v_{conv}\, l\, \frac{dU_{gas}}{dr} \lesssim v_s H \frac{dP_{gas}}{dr} \approx \rho v_s^3 , \qquad (12.11)$$

apart from factors of order unity. Here v_{conv}, l, and U_{gas} are the convective velocity, mixing length, and gas thermal energy density. In an ideal gas $U \sim P_{gas} = \rho v_s^2$, where ρ and v_s are the gas density and sound speed. If we limit the convective velocity to the sound speed and the mixing length to the gas pressure scale height $H = P_{gas}/(dP_{gas}/dr)$, then the inequality in (12.11) sets an upper bound on the convective energy flux. Below the layers where this limit is breached, convection can carry nearly the full energy flux, $F_{conv} \approx F = L/4\pi r^2$. But above that radius the convective flux is limited by the decline in ρv_s^3. This shortfall demands a compensating increase in the radiative flux and thus in the radiative Eddington parameter:

$$\Gamma_{rad} = \frac{\kappa F_{rad}}{gc} = \frac{\kappa(F - F_{conv})}{gc} \approx \frac{L - 4\pi r^2 \rho v_s^3}{L_{\rm Edd}} . \qquad (12.12)$$

Consider now the critical radius R_c where convective inefficiency leads to $\Gamma_{rad} > 1$. If we put the sonic radius for a wind outflow [29] at this location, then (12.12) implies an associated mass loss rate[4]

$$\dot{M}_{conv} \approx 4\pi R_c^2 \rho_c v_s \approx \frac{L - L_{\rm Edd}}{v_s^2} \approx (4\, M_\odot\, {\rm year}^{-1}) \left(\frac{\Gamma - 1}{\Gamma}\right) \frac{L_6}{T_5} , \qquad (12.13)$$

where L_6 and T_5 denote $L/10^6 L_\odot$ and $T/10^5$K. In the near-surface layers where convection becomes inefficient, T_5 is of order unity and $v_s \sim 40\,{\rm km\,s^{-1}}$.

For luminosities well above the Eddington limit, (12.13) gives an unsustainably high mass loss rate. The "photon tiring limit," set by the energy required to lift material out of the star's gravitational potential [32, 33], is

$$\dot{M}_{tir} \approx \frac{L}{v_{esc}^2/2} \approx \frac{L}{GM/R_c} \approx (0.032\, M_\odot\, {\rm year}^{-1}) \frac{(R_c/R_\odot)}{(M/M_\odot)} L_6 . \qquad (12.14)$$

If $\Gamma \gtrsim 1.5$, $T_5 \sim 1$, and $M/R \sim M_\odot/R_\odot$, (12.13) predicts a mass loss rate that exceeds the tiring limit by a factor of order $(v_{esc}/v_s)^2 \sim 100$. This means that L cannot sustain a mass loss initiated near R_c; the flow would stagnate at about $r \approx 1.01\, R_c$ with the above parameters. In other words no consistent super-Eddington outflow can originate at the top of the region of efficient convection [38].

12.2.4 Radiative Instability and Porosification of Atmospheres

If we are to explain the super-Eddington episode in η-Car and other objects in terms of sustained outflow, the above analysis shows that the sonic radius R_s, where the net radial force vanishes, must be shifted outward to a layer whose mass density is much

---

[4] [Clarification:] In this review as in most discussions of stellar mass loss, $\dot{M}$ denotes $|dM/dt|$ rather than the negative quantity dM/dt.

smaller than at the inefficient convection radius R_c. Thus the region $R_c < r < R_s$ must be quasi-hydrostatic with a declining density and pressure, even though $\Gamma_{rad} > 1$ traditionally implies a net outward force there!

The resolution to this paradox has two aspects. First, as an atmosphere approaches the Eddington luminosity, many possible instabilities can give rise to inhomogeneities. Second, optically thick inhomogeneities ("clumps"), reduce the effective opacity, thus allowing larger fluxes without increasing the radiative force.

Dating back to early work by Spiegel [46, 47], there have been speculations that atmospheres supported by radiation pressure would exhibit instabilities not unlike that of Rayleigh-Taylor, associated with the support of a heavy fluid by a lighter one, leading to formation of "photon bubbles". Quantitative stability analyses [39, 48] show that even a simple case of a pure "Thomson atmosphere" – i.e., supported by Thomson scattering of radiation by free electrons – develops lateral inhomogeneities. The analysis by Shaviv [39] suggests that these instabilities share many of the properties of "strange mode" pulsations [18,34]. They are favored when radiation pressure dominates over gas pressure, and occur when the temperature perturbation term becomes non-local. In strange mode instabilities, the relevant term arises because temperature in the diffusion limit depends on the radial gradient of the opacity perturbations. In the pure-Thompson lateral instability, the term depends on lateral radiative flux, which involves non-radial structure comparable in size to the scale height. More instabilities exist if the model is not a pure Thomson atmosphere; absorption opacity can induce the aforementioned strange modes, and magnetic fields imply additional phenomena [2,4,6,16].[5] The specific physical causes of these instabilities, however, are not fundamental to the discussion here. The two essential points are: (1) As atmospheres approach the Eddington limit, non-radial instabilities make them inhomogeneous on a horizontal scale comparable to the vertical scale height; and (2) Such inhomogeneities change the ratio between the radiative *flux* and the radiative *force*.

In the presence of inhomogeneities in density ρ, the outward radiation flux can be written as a volume average $\langle F \rangle_V$, while the force per unit volume that it exerts is $\langle F\kappa\rho \rangle_V /c$. Thus an "effective" opacity for the force can be defined as [37],

$$\kappa_{\text{eff}} \equiv \frac{\langle F\kappa\rho \rangle_V}{\langle F \rangle_V \langle \rho \rangle_V}. \tag{12.15}$$

(This is analogous to opacity variation in *frequency* space in non-gray atmospheres, wherein radiative force depends on a flux-weighted mean integrated over frequency instead of volume. The "Rosseland mean" defined in textbooks is pertinent.)

---

[5][Editors' comment:] One should also recall the "modified Eddington limit" often proposed in the 1980s for LBVs and η Car; see [1, 10, 24] and especially Sect. 5 in [25]. This is a hypothetical instability, based on temperature dependence of absorption opacity added to Thomson scattering in an 8,000–30,000 K atmosphere with $0.7 < \Gamma_e < 1$. Whether correct or not (which so far as we know has never been carefully tested), the MEL was a conceptual predecessor of some ideas outlined in this chapter; for instance it would obviously entail 3-D spatial inhomogeneities.

For some opacity laws, spatial inhomogeneity can enhance the effective opacity defined above. More generally, however, as in the special case of Thomson scattering, $\kappa_{\rm eff}$ is reduced. A calculable example is the limit of small isotropic perturbations in an optically thick Thomson-scattering atmosphere with negligible gas heat capacity, such that $\nabla \cdot \mathbf{F} = 0$. This case approximates a hot atmosphere, but deep enough for the inhomogeneities to remain opaque. Then one finds [37]

$$\kappa_{\rm eff} = \left[1 - \left(\frac{\mathscr{D} - 1}{\mathscr{D}}\right)\sigma^2\right]\kappa_0, \tag{12.16}$$

where σ is the normalized standard deviation of ρ and $\mathscr{D}$ is the number of relevant spatial dimensions. This result shows that inhomogeneities tend to reduce the effective opacity, but not in a one-dimensional system. In other words the porosity effect is intrinsically *non-radial* and tends to *decrease* the effective opacity.

Because the "porous" state depends on non-linear behavior of relevant instabilities, and, in particular, on the mechanism saturating them, there are currently no ab initio calculations predicting characteristics such as $\kappa_{\rm eff}(\Gamma)$ in the nonlinear state. This will require elaborate radiative hydrodynamic simulations.

12.2.5 Quasi-Static Structure of Super-Eddington Stars

The effects described in the previous subsections combine to give an overall picture for how stars can surpass the Eddington limit but remain, in a sense, stable (Fig. 12.2).

Region (**A**): As explained in Sect. 12.2.3, deep inside the star where the density is sufficiently high, excess flux above the Eddington luminosity is carried by convection. Thus we have a bound interior with $L_{\rm rad} < L_{\rm Edd} < L_{\rm tot}$.

Region (**B**): Farther out where convection is inefficient, radiative instabilities cause the atmosphere to become inhomogeneous. This reduces the effective opacity and therefore increases the effective Eddington luminosity $L_{\rm Edd}^{\rm eff}$. This layer is bound, not because the radiation flux is lowered as in the convective region, but because the effective opacity is reduced: $L_{\rm Edd} < L_{\rm rad} < L_{\rm Edd}^{\rm eff}$.

Region (**C**): Farther out, each clump becomes transparent and the effective opacity returns to the microscopic value: $L_{\rm Edd}^{\rm eff} \approx L_{\rm Edd}$. The resulting mass outflow has its sonic point where $L \approx L_{\rm Edd}^{\rm eff} \gtrsim L_{\rm Edd}$. More about this in Sect. 12.3.

Region (**D**): Since the mass loss rate is large enough to be opaque, the photosphere resides in the wind itself [8, 25].

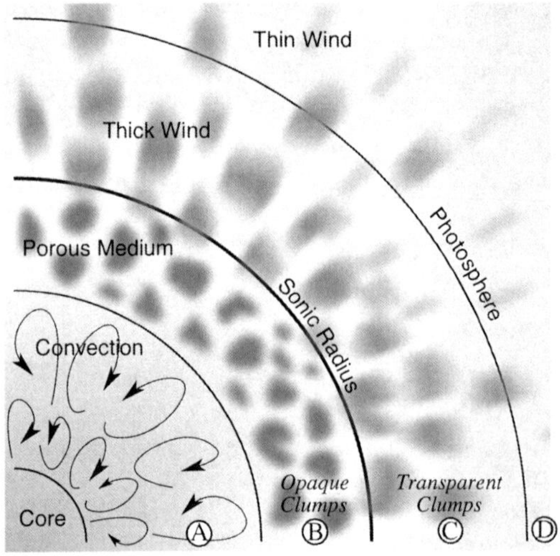

Fig. 12.2 The structure of a super-Eddington star from [41], see text (Sect. 12.2.5)

12.3 Radiatively Driven Mass Loss

12.3.1 Line-Driven Stellar Winds

As a basis for developing a model for continuum-driven mass loss from super-Eddington phases of LBV stars, let us review the more well-established theory for steady line-driven winds (For an account of stellar winds in general, see [29]).

The resonant nature of absorption in spectral lines leads to an opacity that is inherently much stronger than for free electrons. Comparing a bound electron to a free one, the enhancement in frequency-integrated cross-section can be represented as the "quality" value of the resonance, essentially the ratio of frequency to the damping frequency γ [17]. If we use a classical approximation for the latter,

$$Q \approx \frac{\nu}{\gamma} \approx \frac{3}{8\pi^2} \frac{\lambda}{r_e} \sim 10^7 , \qquad (12.17)$$

where λ is photon wavelength and $r_e = e^2/m_e c^2$ is the classical radius of the electron. Thus a bound electron typically scatters photons ten million times more effectively than a free electron does! Since only about 0.01% of the electrons in relevant layers of a hot star with typical metallicity are in bound states, the net enhancement is of the order of $\bar{Q} \sim 1{,}000$. In the idealized optically thin limit, the total spectral line force would exceed the Thomson-scattering force by a comparable factor [17]. For very massive stars one one might therefore expect an enormous outward acceleration, $\bar{Q}\, \Gamma_e g \sim 1000\, g$.

In reality, of course, self-absorption in the lines prevents this from occurring. The effective optical depth in a mass flow is related to the local velocity gradient (the

Sobolev approximation [29, 45]). For a single line with frequency-averaged opacity $\kappa_q = q\kappa_e$, the acceleration factor $q\Gamma_e$ is reduced to

$$\Gamma_{\text{line}} \approx \left[\frac{1-\exp(-qt)}{qt}\right] q\Gamma_e\,, \tag{12.18}$$

where $t = \kappa_e \rho c/(dv/dr)$ is the Sobolev optical depth of a line with unit strength $q = 1$ [7, 45]. In CAK line-driven wind theory, the number distribution of spectral lines vs. line-strength is approximated as a power law, $dN/dq \sim (q/\bar{Q})^{\alpha-1}$ with index α in the range 0.5–0.7. The associated total radiation force factor is then

$$\Gamma_{\text{lines}} = \frac{\bar{Q}\Gamma_e}{(1-\alpha)(\bar{Q}t)^{\alpha}} \propto \left(\frac{1}{\rho}\frac{dv}{dr}\right)^{\alpha}. \tag{12.19}$$

The latter proportionality emphasizes the scaling of the line-force with the velocity gradient and the *inverse* of the mass density. This keeps the line acceleration less than gravity in the dense, nearly static atmosphere, but also allows its outward increase to drive the outflowing wind.

The CAK mass loss rate is set by the associated critical density at which the outward line acceleration is just sufficient to overcome the gravity reduced by Thomson scattering, i.e., $\Gamma_{\text{lines}} \approx 1 - \Gamma_e$:

$$\dot{M}_{\text{CAK}} = \left(\frac{\alpha}{1-\alpha}\right)\left[\frac{\bar{Q}\,\Gamma_e}{1-\Gamma_e}\right]^{-1+1/\alpha}\frac{L}{c^2}\,. \tag{12.20}$$

Here we have used the fact that $v\,dv/dr \approx (1-\Gamma_e)g$ in this type of solution. The same property implies the CAK velocity law $v(r) \approx (1-R/r)^{1/2}v_\infty$, with the wind terminal speed being proportional to the effective surface escape speed,

$$v_\infty \propto v_{\text{esc}} \approx \left[(1-\Gamma_e)\frac{GM}{R}\right]^{1/2}. \tag{12.21}$$

As a star approaches the classical Eddington limit $\Gamma_e \to 1$, these standard CAK scalings formally predict the mass loss rate to diverge as $\dot{M} \propto 1/(1-\Gamma_e)^{(1-\alpha)/\alpha}$, but with a vanishing terminal flow speed $v_\infty \propto \sqrt{1-\Gamma_e}$. The former might appear to provide an explanation for the large mass losses inferred in LBV's, but the latter fails to explain the moderately high inferred ejection speeds, e.g. the 500–800 km/s kinematic expansion inferred for the Homunculus nebula of η Carinae [12, 13, 43, 44].

So one essential point is that line-driving cannot explain the extremely large mass loss rates needed to explain the Homunculus nebula. To maintain the moderately high terminal speeds, the $\Gamma_e/(1-\Gamma_e)$ factor would have to be of order unity. Then, with optimal realistic values $\alpha = 1/2$ and $\bar{Q} \approx 2{,}000$ for the line opacity parameters [17], the maximum mass loss due to line driving is [42]

$$\dot{M}_{\text{max,lines}} \approx 1.4 \times 10^{-4} L_6\, M_\odot\ \text{year}^{-1}\,, \tag{12.22}$$

where $L_6 = L/10^6 L_\odot$. Even for peak luminosities of a few times $10^7 L_\odot$ during η Car's great eruption, this limit is several orders of magnitude below the rate needed to form the Homunculus. Therefore, if mass loss in that event occurred via a wind, it must have been driven chiefly by continuum radiation force rather than lines [5, 35].

12.3.2 Continuum-Driven Winds Regulated by Porous Opacity

As discussed in Sect. 12.2.4, stars that approach or exceed the Eddington limit are expected to have complex spatial structure, and this has led to a new "porosity" paradigm [37, 38] for quasi-steady continuum driving from super-Eddington stars. As we now describe, the basic formalism can be cast in terms that draw heavily on the above theory for wind driving by line-opacity [33].

To begin, consider a medium in which material has coagulated into discrete clumps of individual optical thickness $\tau_{cl} = \kappa \rho_b \ell$, where ℓ is the clump size, and the clump density is enhanced compared to the mean density of the medium by a volume filling factor $\zeta = \rho_b/\rho$. The effective opacity of this medium can then be approximated by a form like the scaling of force in a single line (cf. 12.18),

$$\kappa_{\rm eff} \approx \left[\frac{1 - \exp(-\tau_{cl})}{\tau_{cl}} \right] \kappa \tag{12.23}$$

For optically thin clumps ($\tau_{cl} \ll 1$) this gives the usual microscopic opacity, $\kappa_{\rm eff} \approx \kappa$; but in the optically thick limit ($\tau_{cl} \gg 1$) $\kappa_{\rm eff}$ is reduced by a factor of $1/\tau_{cl}$. In that case the overall effective opacity is roughly the clump cross section divided by the clump mass: $\kappa_{\rm eff} \approx \kappa/\tau_{cl} \approx \ell^2/m_{cl}$. The critical mean density at which each clump becomes optically thin is $\rho_o \approx 1/\kappa h$, where $h = \ell/\zeta$ is a characteristic "porosity length" parameter. The overall radiative acceleration would likewise be reduced by a factor that depends on the mean density.

More realistically, it seems likely that structure should occur with a range of compression strengths and length scales. Noting the similarity of the single-scale and single-line correction factors (12.18) and (12.23), let us draw upon an analogy with the power-law distribution of line-opacity in the standard CAK model of line-driven winds, and thereby consider a *power-law-porosity* model wherein the associated structure has a broad range of porosity length h. As detailed by [33], this leads to an effective Eddington parameter that scales as

$$\Gamma_{\rm eff} \approx \left(\frac{\rho_o}{\rho} \right)^{\alpha_p} \Gamma \; ; \quad \rho > \rho_o \, , \tag{12.24}$$

where α_p is a porosity power index analogous to the CAK line-distribution power index α, and $\rho_o = 1/\kappa h_o$, with h_o now the porosity-length associated with the strongest (i.e. most optically thick) clump.

In rough analogy with the convective mixing length [22] discussed in Sect. 12.2.3, let us further assume that h_o scales with gravitational scale height $H = v_s^2/g$. Then the requirement that $\Gamma_{\rm eff} = 1$ at the wind sonic point yields a mass loss rate scaling with luminosity. For the canonical case $\alpha_p = 1/2$, this takes the form [33]

$$\dot{M}_{\rm por} \approx 4\,(\Gamma - 1)\left(\frac{L}{v_s c}\right)\left(\frac{H}{h_o}\right) \tag{12.25}$$

$$\approx 0.004\,(\Gamma - 1)\,\frac{(L/10^6 L_\odot)}{(v_s/20\,{\rm km\,s^{-1}})}\left(\frac{H}{h}\right) M_\odot\ {\rm year}^{-1}. \tag{12.26}$$

Comparision with the CAK scaling (12.20) for a line-driven wind shows that the mass loss can be substantially higher from a super-Eddington star with porosity-regulated, continuum driving. Applying the extreme luminosity $L \approx 2 \times 10^7 L_\odot$ estimated for the η Car's 1840–1860 outburst, which implies an Eddington parameter $\Gamma \sim 5$, the derived mass loss rate for a canonical porosity length $h_o = H$ is $\dot{M}_{\rm por} \approx 0.3 M_\odot$ year^{-1}, quite comparable to the likely average during that epoch.

Overall, it seems that, together with the ability to drive fast outflow speeds (of order the surface escape speed), the porosity formalism offers a basis for self-consistent dynamical modeling of even the most extreme LBV outbursts, which, like the giant eruption of η Carinae, approach the photon tiring limit.

12.3.3 Photon Tiring in a Simple Super-Eddington Wind Model

Before discussing simulations of porosity models with base mass flux above the tiring limit, let us first examine analytic models of photon tiring for continuum-driven winds in which the Eddington factor is assumed to have an explicit spatial dependence $\Gamma(r)$. Specifically, let us assume that it increases outward from $\Gamma(r) < 1$ in a static interior, crossing unity at some radius R_s which represents the sonic point of a mass outflow. The density ρ_s and sound speed v_s at this point set the mass loss rate $\dot{M} = 4\pi R_s^2 \rho_s v_s$, but otherwise gas pressure has negligible effect on the further supersonic acceleration. The steady-state equation of motion thus reduces to

$$v\frac{dv}{dr} \approx -\frac{[1 - \Gamma(r)]\,GM}{r^2}\ ;\quad r \gtrsim R_s. \tag{12.27}$$

Unlike the porosity models above, this equation of motion has no explicit dependence on density; so the resulting velocity law $v(r)$ would not depend on the amount of mass accelerated. More realistically, a given radiative luminosity can only produce a limited mass loss rate before the energy expended in accelerating the outflow against gravity requires a notable reduction in the radiative energy flux itself. To take account of this "photon tiring," we reduce the radiative luminosity according to the kinetic and potential energy of the flow:

$$L_{\text{reduced}}(r) = L_* - \left[\frac{v^2}{2} + \frac{GM}{R_s} - \frac{GM}{r}\right]\dot{M}. \tag{12.28}$$

Here L_* and M are practically the star's total energy outflow and mass. (The assumed r-dependence of Γ is due to a gradient in $\kappa_{\text{eff}}\,L_{\text{reduced}}$). Defining scaled variables

$$w = \frac{v^2}{(2GM/R_s)} \quad \text{and} \quad x = 1 - \frac{R_s}{r}, \tag{12.29}$$

we find a dimensionless equation of motion with photon tiring:

$$\frac{dw}{dx} = -1 + [1 - m(w+x)]\,\Gamma(x). \tag{12.30}$$

Parameter m is the "tiring number,"

$$m = \frac{GM\dot{M}}{R_s L_*}, \tag{12.31}$$

which is the fraction of radiative energy expended in lifting the wind out of the stellar gravitational potential. Using integrating factors, one can obtain an explicit solution in terms of the integral quantity $\Lambda(x) = \int_0^x dx'\Gamma(x')$:

$$w(x) = -x + \frac{1}{m}\left[1 - e^{-m\Lambda(x)}\right] + w(0)\,e^{-m\Lambda(x)}, \tag{12.32}$$

where for typical hot-star atmospheres the sonic point boundary value is very small, $w(0) = v_s^2 R_s/2GM < 10^{-3}$.

As a simple example, consider the form $\Gamma(x) = 1 + 0.1\sqrt{x}$. Figure 12.3 plots solutions $w(x)$ vs. x for various m. If m is small, the flow reaches a finite speed at large radii ($x \to 1$); but for larger m it eventually decelerates, stopping at some finite stagnation point x_{stag} where $w(x_{\text{stag}}) = 0$. The latter solutions represent flows in which the mass loss rate is too high for the given luminosity to lift the material to full escape at large radii. By considering the critical case where $w = 0$ at $x = 1$, we can define a maximum mass loss rate, given according to (12.32) by the relation

$$m_{\text{max}} = 1 - \exp[-m_{\text{max}}\Lambda(1)]. \tag{12.33}$$

Note that $\Lambda(1)$ is the average of $\Gamma(x)$ in the range $0 \le x \le 1$. A simple approximation accurate to about 10% in this range is $m_{\text{max}} \approx 1 - \exp[2 - 2\Lambda(1)]$.

Regardless of how large $\Lambda(1)$ becomes, it is always true that $m_{\text{max}} < 1$. By comparison, the maximum mass loss allowed by convective inefficiency (12.13)

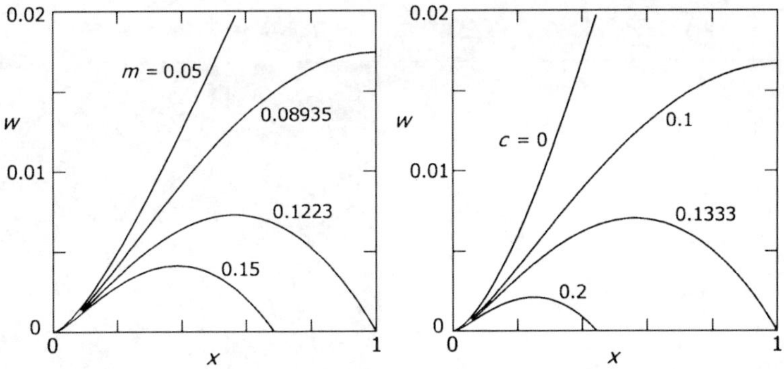

Fig. 12.3 Scaled kinetic energy w vs. radial parameter x for simple continuum-driven wind models, illustrating flow stagnation due to photon tiring (*left*) or to a post-peak decline of opacity (*right*). The curves are labeled with the photon tiring number m or opacity peak parameter c, see text

would correspond to a tiring number of order $m_{conv} \approx GM/R_s v_s^2 \approx 2v_{esc}^2/v_s^2$. Since typically $m_{conv} \gg 1$, we again see that any super-Eddington outflow initiated near the layer where convection becomes inefficient would stagnate by photon tiring.

In the limit of negligible tiring $m \ll 1$, solution (12.32) becomes $w(x) \approx \Lambda(x) - x$. Then the critical case of marginal escape, $w(1) = 0$, requires $\Lambda(1) = 1$. The right-hand part of Fig. 12.3 illustrates solutions for the specific nonmonotonic example $\Gamma(x) = 1 + 0.1\sqrt{x} - cx$, where c represents a tendency for $\kappa_{\rm eff}$ to decrease at large radii. For all $\Lambda(1) < 1$ (i.e., $c > 0.1333$), stagnation occurs at the radius where $\Lambda(x) = x$, precluding a steady flow. In a time-dependent model, material might accumulate there and then fall back to the star. This represents another way in which, instead of steady outflow, a super-Eddington region may give rise to an extended envelope with a 3-dimensional mass circulation and possibly a density inversion.[6]

12.3.4 *Simulation of Stagnation and Fallback Above the Tiring Limit*

For porosity models in which the mass flux exceeds the photon tiring limit, recent numerical simulations have explored the resulting complex pattern of infall and outflow [50, 51]. Despite the likely 3-D nature of such flow patterns, in order to keep the computation tractable this initial exploration assumes 1-D spherical symmetry, though allowing time-dependent density and flow speed. The total rate

[6] [Editors' comment:] In somewhat cooler hypergiant winds where other processes supplement radiative driving, phase changes from ionized to non-ionized hydrogen may produce interesting effects in unsteady outflows that resemble the case mentioned here. See, e.g., [26].

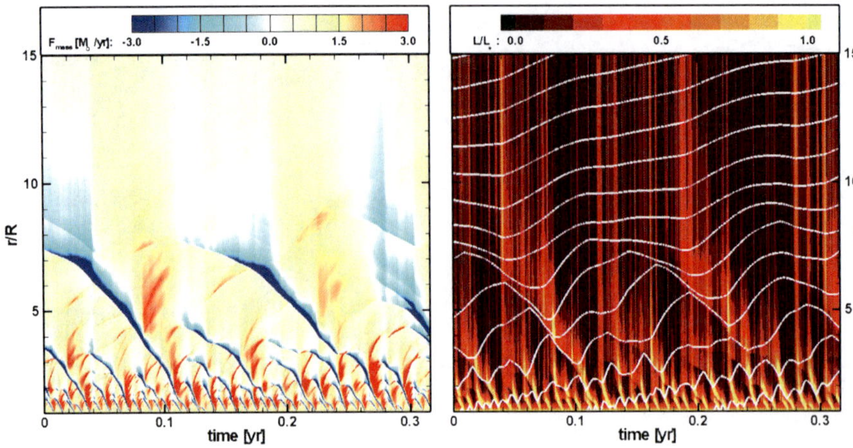

Fig. 12.4 Grayscale plot of radius and time variation of mass flux (*left*) and luminosity (*right*) in a time-dependent 1-D simulation of a super-Eddington wind with a porosity-mediated base mass flux which exceeds the photon tiring limit. The white contours on the right trace the height progression of fixed mass shells

of work done by the radiation on the outflow (or vice versa in regions of inflow) is again accounted for by a radial dependence of the radiative luminosity,

$$\frac{dL}{dr} = -\dot{m}\, g_{\rm rad} = -\frac{\kappa_{\rm eff}\, \rho v L}{c}, \qquad (12.34)$$

where $\dot{m}(r,t) = 4\pi\rho v r^2$ is the varying local mass flux. The second equality follows from the definition (12.1) of radiative acceleration $g_{\rm rad}$ with a gray opacity $\kappa_{\rm eff}$, set here by porosity-modified Thomson scattering. At each time step, (12.34) is integrated from an assumed lower boundary luminosity $L(R)$ to give the local radiative luminosity $L(r)$ at all radii $r > R$. Using this to compute $g_{\rm rad}(r,t)$, the time-dependent equations for mass and momentum conservation are evolved forward to obtain the variations of density $\rho(r,t)$ and flow speed $v(r,t)$. (For simplicity, the temperature is fixed at the stellar effective temperature.) The base Eddington parameter is assumed to be $\Gamma = 10$, and the analytic porosity mass flux is 2.3 times the tiring limit.

Figure 12.4 illustrates the flow structure as a function of radius and time long after the initial condition. The left panel shows the local mass flux, ranging from $3\, M_\odot$ year^{-1} inward (black) to $3\, M_\odot$ year^{-1} outward (white). In the right panel, shading represents the local luminosity in units of the base value $L(r)/L(R)$, ranging from zero (black) to one (white); in addition, superposed lines represent the radius and time variations of selected mass shells.

Both panels show the remarkably complex nature of the flow, with positive mass flux from the base overtaken by a hierarchy of infall structures from stagnated flow above. Re-energization of the radiative flux by this infall gives the overlying region an outward impulse. The shell tracks show that, once material reaches a radius $r \approx 5R$, its infall episodes become increasingly longer and eventually it drifts outward. The overall result is a net time-averaged mass loss through the outer boundary. Its rate is very close to the photon-tiring limit, and the terminal flow speed $v_\infty \approx 50\,\mathrm{km\,s^{-1}}$ is substantially below the gravitational escape speed $v_{esc} \approx 600\,\mathrm{km\,s^{-1}}$.

Of course the structure in a more realistic 3-D model is likely to be even more complex, and may lead to a highly porous medium. But it seems that one property of a super-Eddington star may be a mass loss rate comparable to the photon tiring limit.

12.4 Discussion

12.4.1 LBV Eruptions: Enhanced Winds or Explosions?

A key theme of this book is that extreme LBV eruptions can be "supernova imposters", characterized by substantial brightening and large mass ejections. In this chapter we have modeled this mass loss in terms of a quasi-steady, continuum-driven wind that results when luminosity exceeds the Eddington limit. But an alternative paradigm is that such eruptions might be interior explosions that did not have sufficient energy to completely disrupt the star.

The overpressure from an explosion propagates through the star on a very short dynamical time scale, of order R/v_s where v_s is now the sound speed in very high-temperature gas heated by energy deposition of the explosion. In supernovae this sound speed is on the order of the mass ejection speed, $\sim$10,000 $\mathrm{km\,s^{-1}}$. Even in a "failed" explosion it would be on the order of the surface escape speed, $\sim$ a few hundred $\mathrm{km\,s^{-1}}$, implying a dynamical time like the free fall time: a few days or less. The initial release of radiative energy results from expansion and peaks on a somewhat longer timescale of days or weeks for supernovae. (SN radioactive beta-decay becomes important in the later stages.) It is difficult to see how such a direct pressure-driven explosion can persist for years as seen in LBV eruptions.

This then is perhaps the key argument for a radiation-driven model. If energy is released in the deep interior, its *radiative* signature can take up to a much longer diffusion time to reach the surface.[7] This can be long as a few years.

In contrast to the explosive disruption of supernovae, the total energy in an LBV eruption is well below the total stellar binding energy. Thus even if this energy were released suddenly in the deep interior, the initial dynamical response would quickly stagnate, leaving radiative diffusion as the substitute transport process. The associated excess luminosity may push the object over the Eddington limit, leading

---

[7] Since the luminous stars are likely to be mostly convective (Sect. 12.2.3), the limiting time scale is that of the convective diffusion mixing length time in the stellar core, which due to the high density is much longer than the dynamical time scale.

to strong, *radiatively driven* mass loss. Because the diffusion time scale is much longer than any dynamical time, the essential processes can be modeled as a quasi-steady continuum-driven wind as described above.

12.4.2 Trigger & Energy Source for Super-Eddington Luminosity

In a supernova the energy source is core-collapse to a neutron star or black hole. But in a giant LBV eruption, the post-eruption survival of an intact star, and indications that some LBVs can undergo multiple events of this type, show that the energy source cannot be a one-time singular event. Some other mechanism must provide the energy.

The total energy associated with an extreme eruption, mainly radiative luminosity and mechanical (i.e., kinetic and potential) energy in the wind, is a few times 10^{49} erg. For an eruption lasting a few decades, this corresponds to a power output of order $10^7\,L_\odot$, which is only a few times the Eddington luminosity. This suggests that energy for the eruption can be supplied by the nuclear burning itself. If during quiescence the star shines at nearly the Eddington luminosity, a 10% increase in the central temperature would allow the CNO cycle or other reactions to supply the extra energy. Two implications follow.

First, the change in the binding energy required to get the increased temperature is only 10% of the binding energy of an $n = 3$ polytrope. Taking a mass $M \sim 100\,M_\odot$ and radius $R \sim 100\,R_\odot$, the energy change is $\sim 0.1\,\Gamma GM^2/(2R) \sim 10^{49}$ ergs. In other words, the energy associated with the eruption is comparable to the change in binding energy needed in order to supply the extra energy through nuclear reactions. This implies that the whole star can participate in the eruption.

Second, because the energy for the eruption can be continuously generated, many different scenarios can be envisioned for the triggering mechanism. Moreover, an instability originating in the outer layers, which then forces the interior to progressively adjust downward in the manner of a geyser [9,25], may be as favorable as a mechanism that originates in the core. Thus quite a few mechanisms have been suggested to explain LBV eruptions. Here are several examples.

In a linear analysis of radial modes using standard opacity tables, Glatzel and Kiriakidis found mode-coupling instabilties in the outer layers of massive stars [19]. Guzik et al. then showed how such an instability can propagate inward as convection arises [21]. Since convection takes a finite time to adjust, a super-Eddington luminosity can arise, with consequent ejection of mass.

Other instabilities originating in stellar envelopes were found by Stothers and Chin [49], who conjectured that η Car is repeatedly encountering ionization-induced dynamical instability, and by Maeder [30], who found that super-Eddington layers may occur somewhat deeper in.[8]

---

[8][Editors' comment:] An older idea also seems pertinent here, see footnote in Sect. 12.2.4.

As a contrast, other instabilities could originate at the core. Guzik [20], for example, suggested that nonradial gravity mode oscillations grow slowly to an amplitude sufficient to cause an episode of mixing of hydrogen-rich material downward into hotter denser layers, which would generate a burst of nuclear energy release. Most recently, Woosley et al. [53] have emphasized the potential role of pulsational pair instability in the stellar core for powering either supernovae or giant LBV eruptions.

Because luminous stars have large radiative pressure support, their outer regions tend to be loosely bound and so exhibit various instabilities. This explains the plethora of suggested mechanisms. Unfortunately, however, none seems markedly better than the others.

It is yet unclear why LBVs erupt, and what sets the eruption time scale, the total ejected mass, or what determines the eruption repetition rate. *There is currently no model which predicts these numbers.*

12.4.3 Evolution near the Eddington Limit

The evolution of extremely massive stars is not yet understood. In particular, the potentially key evolutionary role of continuum-driven mass loss has only recently begun to be fully appreciated, and the super-Eddington conditions driving this mass loss depend on instabilities that are still quite poorly understood.

Nevertheless, because the mass loss rates associated with super-Eddington states are so extremely large, it is possible to discuss the average state of these objects without relying on details. Such huge mass loss rates simply cannot be sustained over the evolutionary time scales. Thus, once a star reaches a near-Eddington state, it must evolve without crossing the Eddington limit for too long, since the large mass loss serves as a negative feedback to reduce the luminosity of the star.

To see this behavior more quantitatively, we can again apply the Eddington $n = 3$ polytrope model discussed in Sect. 12.2.2, particularly the scaling for Γ in terms of stellar mass M and "molecular weight" μ. Recall (12.7):

$$\frac{\Gamma}{(1-\Gamma)^4} = \mu^4 \left(\frac{M}{18.3 M_\odot}\right)^2,$$

where $\mu = 1/(2 - 5Y/4)$ for a mix of ionized hydrogen and helium with helium mass fraction Y. As the star evolves, its μ increases. If the mass loss rates are small (e.g., in a line-driven wind), M remains nearly constant so the increase in μ forces an increase in Γ. Without the effect of porosity, this would cause the star to approach the Eddington luminosity without actually reaching it.

However, once Γ exceeds some critical value, instabilities can make the atmosphere porous and allow a super-Eddington flux which implies a continuum-driven wind. According to (12.26), this mass loss would evaporate the star on a time scale much shorter than its several-million-year lifetime, unless $(\Gamma - 1) \ll 0.1$.

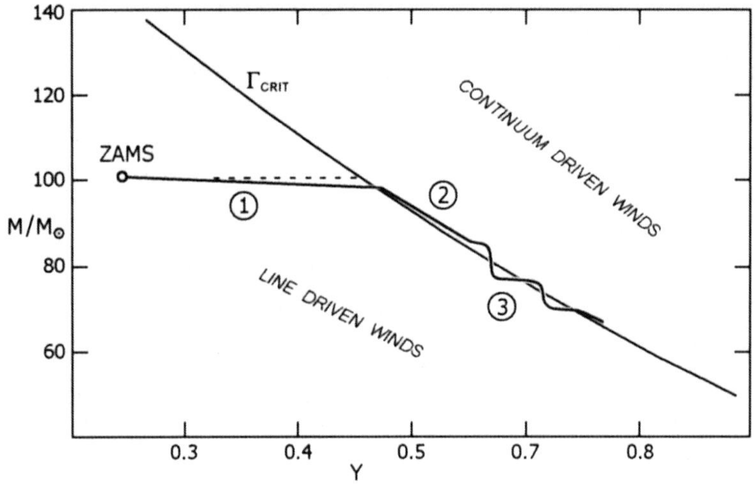

Fig. 12.5 Evolution of a very massive star: mass vs. helium fraction. (1) As long as a star is well below the Eddington limit, it will suffer only minor line driven mass-loss. (2) Once μ is large enough, the star exceeds the Eddington luminosity and loses mass rapidly. This forces the evolution to follow the Eddington luminosity from above. (3) Due to a yet unknown instability, excursions to high luminosities and mass-loss rates may occur with quiescent periods between. The *average* mass loss rate allows the star to follow L_{Edd}. Here we assume that $\Gamma_{\mathrm{crit}} = 0.5$, but this number is actually unknown as it depends on the instabilities that cause atmospheres to become porous

But (12.7) also shows that a decrease in M tends to reduce the equilibrium luminosity; so we can reasonably expect $(\Gamma - 1)$ automatically to remain small. As μ continues to increase, the star moves downward along evolutionary stage 2 in Fig. 12.5, remaining near the Eddington limit. Eventually, unknown instabilities may cause the track to become unsteady like stage 3 in the figure. The example of η Car suggests that this may indeed be the case.

The evolution shown in Fig. 12.5 can continue as long as hydrogen is burning in the core, and very likely also after hydrogen is depleted. This is because similar constraints apply to the remaining WR star. The difference is that typical masses are smaller for helium stars, because μ is larger. Towards the end of hydrogen burning, the remaining mass will be typically much less than the ZAMS value.

12.5 Summary

The large mass loss rates observed in some LBVs, and especially in η-Carinae, strongly suggest that continuum driven winds play an important role. This is because continuum driving is the only mechanism known to give a steady state wind with a mass loss rate much larger than line driving can achieve.

A continuum-driven wind, however, can only operate in a system above the Eddington luminosity. This implies that such a star should have a rather unique structure. Its "static" part should be composed of a large convective region, encompassing most of the star, and a *porous* layer where inhomogeneities are responsible for a lowered opacity. At the radius where opacity cannot be sufficiently reduced by porosity, there is a transition between the gravitationally bound layer below and the radiatively accelerated wind above. Because the mass loss rate can be very large, the wind itself is optically thick so the photosphere resides within the wind.

Another peculiar aspect of continuum driven winds is that because the mass loss rate is determined from conditions at the "static" surface of the star, the wind conditions are blind to whether the available radiative luminosity is sufficient to actually drive the wind to infinity. If the potential well is too deep, the result is a *photon-tired wind*, whose flow stagnates and falls back.

Current 1-D simulation of photon-tired winds suggest that a flow of this type develops a layer with a hierarchical structure of shocks and sonically moving gas, which allows a large kinetic "luminosity" without driving mass above the photon-tired limit. It is not unreasonable that η Carinae actually reached this state, since the ejected mass divided by the 20 years of the giant eruption is, to within the rather large observational uncertainty, comparable to the photon-tired mass loss rate.

Although a super-Eddington state with a continuum-driven wind can be stationary on a dynamical timescale, the large mass loss rate should substantially alter the stellar structure on an evolutionary timescale. Stellar structure and evolution should effectively control the overall level of mass loss, implying that mass loss and evolution are inextricably linked. We thus expect very massive stars that reach the Eddington limit (which they do once the hydrogen mass fraction decreases sufficiently) to evolve with a luminosity that is kept slightly above the Eddington value, with a relatively modest continuum driven wind.

Perhaps the key unresolved questions concern the origin of variability in LBVs and their giant mass loss episodes. While there is a general understanding of how a super-Eddington state with the large mass loss can exist, there is yet no theory that predicts, for example, the two-decade-long giant eruption of η Car.

Acknowledgements S.P.O. acknowledges partial support from NSF and N.J.S. the support of ISF grant 1325/06. We thank J. MacDonald, N. Smith, R. Townsend, and A.J. van Marle for many helpful discussions.

References

1. I. Appenzeller, The role of radiation pressure in LBV atmospheres, in *Physics of Luminous Blue Variables*, ed. by K. Davidson et al. (Kluwer, Dordrecht, 1989), p. 195
2. J. Arons, Photon bubbles – overstability in a magnetized atmosphere. ApJ. **388**, 561 (1992)
3. G.T. Bath, G. Shaviv, Classical novae – a steady state, constant luminosity, continuous ejection model. MNRAS. **175**, 305 (1976)

4. M.C. Begelman, Super-eddington fluxes from thin accretion disks? ApJ. **568**, L97 (2002)
5. A.A. Belyanin, Optically thick super-Eddington winds in galactic superluminal sources. A&A. **344**, 199 (1999)
6. O. Blaes, A. Socrates, Local radiative hydrodynamic and magnetohydrodynamic instabilities in optically thick media. ApJ. **596**, 509 (2003)
7. J. Castor, D. Abbott, R. Klein, Radiation-driven winds in of stars. ApJ. **195**, 157 (1975)
8. K. Davidson, The relation between apparent temperature and mass-loss rate in Hypergiant Eruptions. ApJ. **317**, 760–764 (1987)
9. K. Davidson, Plinian Eruptions a la Eta Carinae, in *Physics of Luminous Blue Variables*, ed. by K. Davidson et al. (Kluwer, Dordrecht, 1989), p. 101
10. K. Davidson, in *Physics of Luminous Blue Variables*, ed. by K. Davidson et al. (Kluwer, Dordrecht, 1989), p. 203
11. K. Davidson, The physical nature of η Carinae, in *The Fate of the Most Massive Stars*, ed. by R. Humphreys, K. Stanek. ASP Conference Series, vol. 332 (Astronomical Society of the Pacific, San Francisco, 2005), pp. 101–110
12. K. Davidson, R.M. Humphreys, Eta Carinae and its environment. Annu. Rev. Astron. Astrophys. **35**, 1–32 (1997)
13. K. Davidson, N. Smith, T.R. Gull, K. Ishibashi, D.J. Hillier, The shape and orientation of the Homunculus nebula based on spectroscopic velocities. AJ. **121**, 1569–1577 (2001)
14. A.S. Eddington, *The Internal Constitution of the Stars,* Chapter 6. (Cambridge University Press, Cambridge, 1926)
15. D.F. Figer, An upper limit to the masses of stars. Nature **434**, 192 (2005)
16. C.F. Gammie, Photon bubbles in accretion discs. MNRAS. **297**, 929 (1998)
17. K. Gayley, An improved line-strength parameterization in hot-star winds. ApJ. **454**, 410 (1995)
18. W. Glatzel, On the origin of strange modes and the mechanism of related instabilities. MNRAS. **271**, 66 (1994)
19. W. Glatzel, M. Kiriakidis, Stability of massive stars and the Humphreys/Davidson limit. MNRAS. **263**, 375 (1993)
20. J.A. Guzik, Instability considerations for massive star Eruptions, in *The Fate of the Most Massive Stars*, ed. by R. Humphreys, K. Stanek. ASP Conference Series, vol. 332 (Astronomical Society of the Pacific, San Francisco, 2005), p. 204
21. J.A. Guzik, A.N. Cox, K.M. Despain, M.S. Soukup, A nonlinear study of LBVs and possible outbursts, in *Luminous Blue Variables: Massive Stars in Transition*, ed. by A. Nota, H. Lamers. ASP Conference Series, vol. 120 (Astronomical Society of the Pacific, San Francisco, 1997), p. 138
22. C.J. Hansen, S.D. Kawaler, V. Trimble, in *Stellar Interiors: Physical Principles, Structure, and Evolution*, 2nd edn., ed. by C.J. Hansen, S.D. Kawaler, V. Trimble (Springer-Verlag, New York, 2004)
23. R.M. Humphreys, K. Davidson, Studies of luminous stars in nearby galaxies. III – Comments on the evolution of the most massive stars in the milky way and the large Magellanic cloud. ApJ. **232**, 409 (1979)
24. R.M. Humphreys, K. Davidson, The most luminous stars. Science **223**, 243 (1984)
25. R.M. Humphreys, K. Davidson, The luminous blue variables: astrophysical geysers. PASP. **106**, 1025–1051 (1994)
26. R.M. Humphreys, K. Davidson, N. Smith, Crossing the yellow void: spatially resolved spectroscopy of the post-red supergiant IRC +10420 and its circumstellar ejecta. AJ. **124**, 1026 (2002)
27. P. Joss, E. Salpeter, J. Ostriker, On the "critical luminosity" in stellar interiors and stellar surface boundary conditions. ApJ. **181**, 429 (1973)
28. S.S. Kim, D.F. Figer, R.P. Kudritzki, F. Najarro, The arches cluster mass function. ApJL. **653**, L113 (2006)
29. H.J.G.L.M. Lamers, J.P. Cassinelli, *Introduction to Stellar Winds* (Cambridge University Press, Cambridgem, 1999)

30. A. Maeder, On the evolutionary status and instability mechanism of the luminous blue variables, in *Physics of Luminous Blue Variables*, ed. by K. Davidson et al. (Kluwer, Dordrecht, 1989), p. 15
31. M.S. Oey, C.J. Clarke, Statistical confirmation of a stellar upper mass limit. ApJL. **620**, L43 (2005)
32. S. Owocki, K. Gayley, The physics of stellar winds near the Eddington limit, in *Luminous Blue Variables: Massive Stars in Transition*, ed. by A. Nota, H. Lamers. ASP Conference Series, vol. 120 (Astronomical Society of the Pacific, San Francisco, 1997), p. 121
33. S. Owocki, K. Gayley, N. Shaviv, A porosity-length formalism for photon-tiring-limited mass loss from stars above the Eddington limit. ApJ. **558**, 802 (2004)
34. J. Papaloizou, F. Alberts, J.E. Pringle, G.J. Savonije, On the nature of strange modes in massive stars. MNRAS. **284**, 821 (1997)
35. T. Quinn, B. Paczynski, Stellar winds driven by super-Eddington luminosities. ApJ. **289**, 634 (1985)
36. G.J. Schwarz, P.H. Hauschildt, S. Starrfield, P.A. Whitelock, E. Baron, G. Sonneborn, A multiwavelength study of the early evolution of the classical Nova LMC 1988 1. MNRAS. **300**, 931 (1998)
37. N. Shaviv, The Eddington luminosity limit for multiphased media. ApJ. **494**, L193 (1998)
38. N. Shaviv, The porous atmosphere of N7 Carinae. ApJ. **532**, L137 (2000)
39. N. Shaviv, The nature of the radiative hydrodynamic instabilities in radiatively supported Thomson atmospheres. ApJ. **549**, 1093 (2001)
40. N. Shaviv, The theory of steady-state super-Eddington winds and its application to Novae. MNRAS. **326**, 126 (2001)
41. N. Shaviv, Exceeding the Eddington limit, in *The Fate of the Most Massive Stars*, ed. by R.M. Humphreys, K.Z. Stanek. ASP Conference Series, vol. 332 (Astronomical Society of the Pacific, San Francisco, 2005) pp. 180–188
42. N. Smith, S. Owocki, On the role of continuum-driven Eruptions in the evolution of very massive stars and population III stars. ApJL. **645**, 45 (2006)
43. N. Smith, R.D. Gehrz, P.M. Hinz, W.F. Hoffmann, J.L. Hora, E.E. Mamajek, M.R. Meyer, Mass and kinetic energy of the Homunculus Nebula around η Carinae. AJ. **125**, 1458–1466 (2003)
44. N. Smith, K. Davidson, T.R. Gull, K. Ishibashi, D.J. Hillier, Latitude-dependent effects in the stellar wind of eta Carinae. ApJ. **586**, 432–450 (2003)
45. V.V. Sobolev, *Moving Envelopes of Stars* (Harvard University Press, Cambridge, 1960)
46. E. Spiegel, Photohydrodynamic instabilities of hot stellar atmospheres, in *Physique des Mouvement dans les Atmospheres Stellaires*, ed. by R. Cayrel, M. Steinberg (CNRS, Paris, 1976) p. 19
47. E. Spiegel, Photoconvection, in *Problems in Stellar Convection*, ed. by E. Spiegel, J.-P. Zahn (Springer, Berlin, 1977), 267
48. E. Spiegel, L. Tao, Photofluid instabilities of hot stellar envelopes. Phys. Rep. **311**, 163 (1999)
49. R.B. Stothers, C.-W. Chin, On two theories of the cyclical outbursts of eta Carinae. ApJ. **489**, 319 (1997)
50. A.J. van Marle, S.P. Owocki, N. Shaviv, Continuum-driven winds from super-Eddington stars: a tale of two limits, in *First Stars III*, ed. by B.W. O'Shea, A. Heger, T. Abel, AIP Conference Proceedings, vol. 990 (American Institute of Physics, Melville, 2008), pp. 250–253
51. A.J. van Marle, S.P. Owocki, N. Shaviv, Numerical simulations of continuum-driven winds of super-Eddington stars. MNRAS. **389**, 1353 (2008)
52. C. Weidner, P. Kroupa, Evidence for a fundamental stellar upper mass limit from clustered star formation. MNRAS. **348**, 187 (2004)
53. S.E. Woosley, S. Blinnikov, A. Heger, Pulsational pair instability as an explanation for the most luminous supernovae. Nature **450**, 390 (2007)

Chapter 13
The Final Stages of Massive Star Evolution and Their Supernovae

Alexander Heger

Abstract In this chapter I discuss the final stages in the evolution of massive stars – stars that are massive enough to burn nuclear fuel all the way to iron group elements in their core. The core eventually collapses to form a neutron star or a black hole when electron captures and photo-disintegration reduce the pressure support to an extent that it no longer can hold up against gravity. The late burning stages of massive stars are a rich subject by themselves, and in them many of the heavy elements in the universe are first generated. The late evolution of massive stars strongly depends on their mass, and hence can be significantly effected by mass loss due to stellar winds and episodic mass loss events – a critical ingredient that we do not know as well as we would like. If the star loses all the hydrogen envelope, a Type I supernova results, if it does not, a Type II supernova is observed. Whether the star makes neutron star or a black hole, or a neutron star at first and a black hole later, and how fast they spin largely affects the energetics and asymmetry of the observed supernova explosion. Beyond photon-based astronomy, other than the sun, a supernova (SN 1987) has been the only object in the sky we ever observed in neutrinos, and supernovae may also be the first thing we will ever see in gravitational wave detectors like LIGO. I conclude this chapter reviewing the deaths of the most massive stars and of Population III stars.

13.1 Introduction

Stellar evolution depends on several initial parameters. The initial mass of the star is generally considered to be the most important and the key parameter for stellar evolution. We also classify stars usually as "massive" or "very massive" based on their initial mass and discuss the final fate of stars in terms of their initial mass.

A. Heger (✉)
School of Physics and Astronomy, University of Minnesota, Minneapolis, MN 55455, USA
e-mail: alex@physics.umn.edu

K. Davidson and R.M. Humphreys (eds.), *Eta Carinae and the Supernova Impostors*, Astrophysics and Space Science Library 384, DOI 10.1007/978-1-4614-2275-4_13,

Table 13.1 Mass ranges of massive stars

Mass range ($M_\odot$)	Name
10 – 100	Massive stars
100 – 1,000	Very massive stars
1,000 – 10,000	Extremely massive stars
10,000 – 100,000	Ultra massive stars
100,000 – 1,000,000	Hyper massive stars
1,000,000 – 10,000,000	Mega massive stars
$\gtrsim$150,000	supermassive stars[a]

[a]No hydrostatic hydrogen burning possible. The limit given is valid only for primordial stars.

Table 13.2 Metallicity ranges of stars[a]

[Fe/H] ("metallicity")	Name
$\lesssim -1$	Metal-poor stars
$\lesssim -2$	Very metal-poor stars
$\lesssim -3$	Extremely metal-poor stars
$\lesssim -4$	Ultra metal-poor stars
$\lesssim -5$	Hyper metal-poor stars
$\lesssim -6$	Mega metal-poor stars
Zero metallicity	Primordial stars (Population III stars)

[a]Adopted from presentations by Tim Beers.

We will also adopt this approach in the present chapter of the book (Table 13.1). For the purpose of this chapter I use the term "*massive stars*" for those stars that are massive enough to make a core collapse supernova; this excludes stars that leave a white dwarf, even if they form a thermonuclear Type Ia supernova later.[1]

The initial composition of the star, usually expressed in terms of the *metallicity* (Table 13.2) or "population." is the second most important parameter. In our own Galaxy, we distinguish between stars that have formed relatively recently (Population I) and "old" stars (Population II) that typically have low metallicity and are found in the halo or other old structures of the Galaxy. But there also must have been an elusive "first" generation of stars that formed from the primordial gas left over by the Big Bang. Such "primordial stars" are usually referred to as "Population III". There is some ongoing discussion in the community about how to exactly name stars at the transition from Population III to Population II in terms of Population; as an example, we refer the reader to [62]. No Population III star has ever been observed, but if our theory of Big Bang nucleosynthesis is correct, they must have existed. We will come back to this in Sect. 13.5.

From observations, we know that not all stars with the same metallicity and same composition necessarily evolve the same. It has been shown that the effects of rotation can also be quite strong [31, 34, 37, 53, 56], altering the mass limits for different final stages by as much as a factor 2, and making some evolutionary

[1]Not all stars that make supernovae are "*massive*" stars; the process for making Type Ia supernovae is much more complicated and most scenarios in the literature involve typically two stars.

outcomes like gamma-ray bursters (GRBs) possible [38, 89, 98]. Some current models also include the effect of dynamo action and magnetic fields [36,54,78,79].

Another important complication in stellar evolution is due to binary star evolution, in particular for "close" binary stars where here "closeness" means that the orbit is tight enough that the stars will interact in the form of mass exchange, mass ejection from the binary system, or even merger of the stars during their lifetime. About half of all massive stars are found in close binary systems. Binary stars can also interact by tidal forces affecting their spin and internal mixing. Lastly, the initial properties like rotation rate may be effected by binarity. Binary star evolution is a very rich topic as it has many more initial physical parameters than for a single star. A detailed discussion of binary star evolution, however, is beyond the scope of this article; for a reference see, e.g., [64]. Here I focus on the evolution of single stars only.

Usually stellar evolution is considered to be just a function of the parameters mentioned above [84]. However, stars may also have peculiar features such as "fossil" magnetic fields like AmP stars [8], or other features that were preserved as the star reached the main sequence. These are observable and may also affect evolution, however, our knowledge of their frequency and magnitude is limited; they are less systematically studied than the other parameters, and also are not often included in current theoretical models.

Finally, we should emphasize that there is an implication from stars having hydrodynamic instabilities – like convection – in their interior. These instabilities may vary in timing and magnitude from star to star, even for the same initial mass, rotation, composition, ... In most cases, of course, we would expect that they "average out" in the long run, but if there is insufficient negative feedback there could be a "random walk" process that may lead to a somewhat different evolution from star to star, maybe sometimes even push the evolution beyond some branching points. *Stars may be subject to "weather"* like on earth where a minor change can lead to big hurricanes instead of calm sea.

13.2 Final Stages of Massive Stars

As an over-all general picture we can describe the evolution of a massive star as a process of ongoing contraction to increasingly higher central density and temperature (Fig. 13.1). Even the formation of stars, crudely, starts from the contraction of a cold molecular gas cloud to hot spherical object that balances heat loss at the surface by energy generation in the center.

The contraction phases come to a halt when temperature and density become large enough to transform nuclei (Table 13.3 and labels in Fig. 13.1) of low binding energy (low mass excess) to more tightly bound nuclei, and the rate of energy release in this process is large enough to balance the energy loss at the surface of the stars ("luminosity") or in its interior (due to neutrinos). This theoretical picture of the basics of massive star evolution, their late evolution stages, and their explosions, has been known for decades (e.g. [3, 13, 17, 59, 93]).

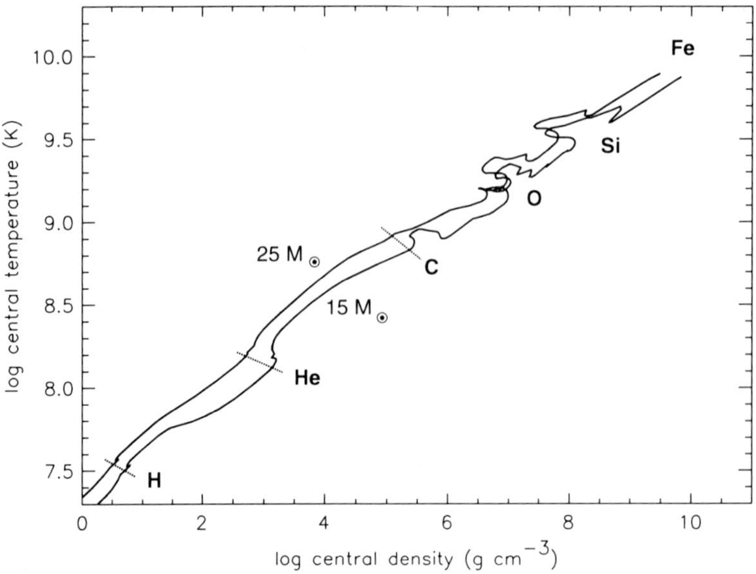

Fig. 13.1 Evolution of the central temperature and density in stars of 15 and 25 $M_\odot$ from birth as hydrogen burning stars until iron core collapse (Table 13.1). In general, the trajectories follow a line of $\rho \propto T^3$, but with some deviation downwards (towards higher ρ at a given T) due to the decreasing entropy of the core. Non-monotonic behavior is observed when nuclear fuels are ignited and this is exacerbated in the 15 $M_\odot$ model by partial degeneracy of the gas (From [95])

Table 13.3 Hydrostatic nuclear burning stages in massive and very massive stars. The table gives burning stages, main products (ashes), typical temperatures and burning time scales for a 20 $M_\odot$ Population I star (solar composition), and for a 200 $M_\odot$ Population III star (primordial composition)

Burning stage		20 $M_\odot$, Pop I		200 $M_\odot$, Pop III	
Fuel	Main product	T (10^9 K)	Time (year)	T (10^9 K)	Time (year)
H	He	0.02	10^7	0.15	210^6
He	O, C	0.2	10^6	0.3	210^5
C	Ne, Mg	0.8	10^3	1.2	10
Ne	O, Mg	1.5	3	2.5	310^{-6}
O	Si, S	2.0	0.8	3.0	210^{-6}
Si	Fe	3.5	0.02	4.5	310^{-7}

13.2.1 Hydrogen Burning

The star spends the vast majority of its evolution, about 90%, just fusing hydrogen to helium. This is generally referred to as the "main-sequence evolution" as stars during that phase of evolution typically obey a specific relation between surface temperature and luminosity, the "main sequence" (though some stars, e.g., when rotating rapidly, may leave the main sequence while still burning hydrogen in the center). In massive stars this burning occurs by the CNO cycle. The typical burning temperature is about 2–3 $\times 10^7$ K, and about 7–8% of the energy is carried away by neutrinos due to weak decays (to fuse four protons to a helium, two of them need to be converted to neutrons by a β^+ decay or electron capture, EC). The neutrino loss from thermal processes is negligible. Since the proton capture on ^{14}N is the slowest reaction in the CNO cycle, most of the initial CNO isotopes in the burning region of massive stars are converted to that isotope, and hence this is what we find as a typical signature of material that has been exposed to hydrogen burning in massive stars.

For Population III stars no initial CNO elements are present, so in this case the CNO cycle cannot work – at first. The pp-chains(s) for hydrogen burning do not provide a high enough nuclear energy generation rate for massive stars to balance the energy loss at the surface (its luminosity). In this case, the star continues to contract until a central temperature of 1.2–1.5 $\times$ 10^8 K is reached, higher the larger the mass of the star. At that point it starts to fuse helium to carbon – and makes the carbon needed for the CNO cycle (recall: that hydrogen is still around as it could not burn fast enough). We find that the star rebounds to slightly lower central temperature, usually around 10^8 K, but still much higher than in the case of later generations of stars. The amount of carbon mass needed to run the CNO cycle is remarkably small: only about 10^{-9}! Carbon production continues at some level due to the high temperature, and just before central hydrogen depletion a central mass fraction of 10^{-7} of CNO isotopes is reached. As hydrogen completely depletes in the center of the star, the final CNO mass fraction (mostly in the form of ^{14}N) may become as high as 10^{-5}.

13.2.2 Helium Burning

The next burning phase of massive stars is central helium burning.[2] At the start of helium burning, helium is first fused into carbon (triple alpha reaction). Then, as the carbon mass fraction becomes larger enough, the star starts to fuse carbon with another helium nucleus to oxygen (^{12}C$(\alpha,\gamma)^{16}$O). Generally, the more massive the star, the larger will be the ratio of oxygen to carbon at the end of central helium burning. Typical numerical results from current calculations for the final carbon mass fraction are around 10–20%. Some of the most severe uncertainties in nuclear

[2]There is another burning phase in between that varies in time scale and importance as a function of metallicity that is usually omitted but worth while mentioning for completeness: **nitrogen burning**. After hydrogen burning is completed, most of the initial metallicity of the star is converted to ^{14}N. For solar metallicity, the ^{14}N mass fraction in the helium core hence will be about 1–2%; for very metal-rich stars it may be as high as 10%. ^{14}N burns by two alpha captures and a β^+ decay to ^{22}Ne; the reaction usually is ^{14}N$(\alpha,\gamma)^{18}$F$(e^+\nu_e)^{18}$O$(\alpha,\gamma)^{22}$Ne. For metal-rich stars this burning phase can comprise a few percent of the helium burning lifetime and even become a distinct convective burning episode by itself; for lower metallicity stars is it usually negligible. This reaction can also be a significant source of ^{18}O where the second alpha capture is not occurring.

reaction rates for stellar evolution are for this phase of evolution, affecting both the triple alpha and $^{12}C + ^{4}He$ reaction [81].

The typical burning temperature for helium burning is around $2–3 \times 10^8$ K. Neutrino losses from thermal processes are still negligible, weak processes do not play a major role. Usually the core helium burning is also paired with a hydrogen-burning shell just outside the nucleus or core.

Central helium burning in massive stars usually takes around 10% of the star's lifetime; the luminosity during this phase is comparable to that during hydrogen burning,[3] withing a factor 2, however, the amount of energy released by helium burning is only about 10% of that released by hydrogen burning. A significant fraction of the luminosity of the star during this phase is also supplied by the hydrogen-burning shell. As a side note for stars of primordial composition: In this case we typically find a CNO mass fraction of $\times 10^{-7}$ as the hydrogen shell burning is hotter than central hydrogen burning and hence more ^{12}C is being made by the triple alpha process.

13.2.3 Carbon Burning

After helium is depleted in the center, the star contracts toward the next burning phase, carbon burning. During this contraction phase, however, a remarkable transition occurs: The star gets hot enough to produce pairs of neutrinos and anti-neutrinos that can leave the star without essentially any further interaction. These neutrinos also carry away energy directly from the stellar interior, without photons having to diffuse (or being transported by radiation) to the surface of the star. This energy loss steeply increases with the temperature of the gas. The star becomes a "neutrino star" – more luminous in neutrinos than in electromagnetic radiation (light from the surface). That is, during the post-helium burning contraction phase, the evolution continuously accelerates, neutrino losses lead to contraction which leads to higher temperature, which leads to more neutrino losses. A typical time-scale for this contraction phase is about 10–20 kyr, shorter the more massive the star. This phase is long enough for the star to change its position in the HR-Diagram – the time scale of the contraction is typically longer than the thermal (Kelvin-Helmholtz) time scale of the envelope. For example, the progenitor star of SN 1987A is thought to have made a transition from a red to a blue supergiant on such a time scale.

Eventually, the temperature becomes high enough to ignite carbon in the center of the star.[4] Carbon burning is by direct fusion of two ^{12}C nuclei. In this reaction only a little ^{24}Mg is made directly; in most cases $^{20}Ne + ^{4}He$, $^{23}Na + ^{1}H$, or $^{23}Mg + ^{1}n$ are produced. In particular, the lighter particles can react further, e.g., making ^{20}Ne

---

[3]It could become lower if the star experiences significant mass loss.

[4]For stars very close to the lower limit for making supernovae, carbon ignition may occur off-center as the gas in the center becomes electron-degenerate and hence the heating-contraction-cooling cycle is interrupted there.

by capture of ^{4}He on ^{16}O which is still present – dominant – from the preceding helium burning phase.

The typical time scale for central carbon burning is around 100–1000 years, and a typical burning temperature is around 8×10^8 K. Generally, it will proceed faster when it is radiative than when it is convective – there is less fuel and both the temperature and neutrino loss rate accelerate quickly as the burning moves outward. The energy balance in carbon burning is dominated by neutrino losses. At first, when a central burning (or a convective shell) ignites, some work will be done against gravity by expansion (pushing the overlaying layers outward), but when the burning comes into balance, energy generation is balanced by neutrino losses; the size of the convective regions is such that all the energy generated by burning is carried away by the neutrinos ("balanced powers"). When the burning extinguishes, the shell contracts again, and the gravitational energy first deposited is now radiated away in the form of neutrinos as well. Essentially none of the energy released by the burning reaches the surface in the form of radiation; whether the star does central carbon burning – or any later stage – is not observable at the surface. Note, that the acceleration of the burning by several orders of magnitude is not due to a lower energy release of carbon burning as compared to helium burning – in fact it is about comparable – but due to the onset of neutrino losses governing the energy loss from the stellar interior.

During carbon burning the outer structure of the star typically looks as follows: The helium-free carbon-oxygen core is surrounded by a, usually powerful hydrogen-free helium-burning shell. At the outer edge of the helium shell we find a hydrogen-burning shell. This hydrogen-burning shell may become very weak, depending on how strong the helium-burning shell is; the stronger the helium shell, the weaker the hydrogen shell (mirror principle).[5]

After central carbon burning, carbon fusion will continue in one or several shells further and further out in the mass coordinate (as measured from the center of the star) throughout the helium depleted core. The central burning phase can be radiative or convective; in more massive stars and for lower carbon abundance (which is also typical for more massive stars, see above) central carbon burning is "radiative", i.e., no central convection zone is present. An example for this is the 22 $M_\odot$ star shown in Fig. 13.2. For lower-mass stars (below $\sim 20\,M_\odot$) central carbon burning is still convective. As this depends on the carbon mass fraction, the exact limit depends on the uncertainty in the reaction rates that govern helium burning as well. The typical time scale for carbon shell burning is small compared to convective core carbon burning.

[5]This is because a powerful helium-burning shell will be radially more extended, pushing the hydrogen shell further out in radius where it becomes cooler and less dense and burns less efficiently.

13.2.4 Neon and Oxygen Burning

The next two burning phases, neon and oxygen burning, are closely related and usually occur in "pairs". Neon burning is different from the preceding phases in that two nuclei are not fused together directly, but rather one nucleus is disassembled while the other is being built up. Specifically, the first reaction $^{20}\mathsf{Ne}(\gamma,\alpha)^{16}\mathsf{O}$ is triggered by an energetic photon in the tail of the Planck spectrum and disassembles a $^{20}\mathsf{Ne}$ nucleus, an endoergic reaction. The $^{4}\mathsf{He}$ released can then be used to fuse with another $^{20}\mathsf{Ne}$ to make $^{24}\mathsf{Mg}$, an even more exoergic reaction, so that the resulting net reaction, $2\,^{20}\mathsf{Ne} \rightarrow {}^{16}\mathsf{O} + {}^{24}\mathsf{Mg}$ is exoergic. The neon burnt in this phase originates mostly from carbon burning, consequently the more carbon the star made, the more $^{20}\mathsf{Ne}$ will there be to burn.[6] The typical burning temperature is around 1.5×10^{9} K and the time scale is less than a year.

Oxygen burning is the heaviest burning stage that occurs by directly fusing two nuclei (of $^{16}\mathsf{O}$) together. Oxygen is the dominant species at the beginning of oxygen burning, around 80%. Typical burning temperatures are around $2\text{–}2.5\times10^{9}$ K. Central oxygen burning takes about 1 year or less; shell burning phases are faster. As with carbon burning, the excited $^{32}\mathsf{S}^{*}$ nucleus made by fusing the two oxygen nuclei dominantly de-excites by n, p, or $^{4}\mathsf{He}$ emission which do react further; $^{28}\mathsf{Si}$ usually is the dominant final product, with $^{32}\mathsf{S}$ a close second.

Oxygen burning follows shortly after a (central or shell) neon burning phase, and we find that a sequence of central and shell burning phases occurs in pairs of neon and oxygen burning shells. All of the energy produced by oxygen and neon burning is carried away by neutrinos; essentially none of it reaches the surface in the form of photons. In other words, the thermal time scale of the envelope, even the helium shell, is now much shorter than the evolution time scale in the core. We cannot tell whether a star is burning neon or oxygen in the center, or none of the above. Most of the stars we observe are not, because the phase is so brief compared to the lifetime of the star. Only when we see the supernova do we know that the star initiated oxygen burning a year before.

13.2.5 Silicon Burning

This is the last hydrostatic burning phase in the life of a massive star. More correctly it should be named "silicon and sulfur burning" as these two species contribute almost equally. The reaction sequence governing this burning phase is similar to that of neon burning: some nuclei are disassembled while others are being built up. Along the way, weak decays or electron captures convert some of the protons into neutrons. The final result is a core of isotopes of the iron group, an "iron core". No

[6] For the more massive stars a significant fraction of $^{20}\mathsf{Ne}$ can already be made at the end of central helium burning and survives through carbon burning.

more energy can be gained by fusing these nuclei further. The typical time scale for silicon burning is about 1 week, and the temperature is about 3.5×10^9 K.

After the central burning phase, silicon burning continues in shells until the iron core in the center collapses under it own weight. As it contracts, temperature rises, but instead of igniting another burning phase as it always occurred in the preceding episodes of stellar evolution, now there is no more fuel to burn. Instead, electron captures continue to decrease the pressure support from the degenerate electrons, and the iron nuclei start to disintegrate into lighter particles (mostly ^{4}He at first), a process that takes a lot of energy. Both processes accelerate the collapse further. At the end of this collapse phase, a neutron star or a black hole forms, and a supernova may result.

13.2.6 Summary of Nuclear Burning

Figure 13.2 shows all of the central and shell burning phases for a star of 22 $M_\odot$ *initial* mass and *initial* solar composition, and no rotation. Note that all burning phases except carbon start out as a convective (green hatching) central burning phase. The central burning is then followed by one or several shell burning phases. It is important to realize that these convective regions evolve essentially chemically homogeneously. When the burning phase ends, the fuel is depleted in the entire convective region. Hence the next burning shell of the same fuel will ignite somewhere toward the upper edge of that convection zone (and it will not necessarily always be convective itself, as, e.g., in the case of central carbon burning in the star shown here).

When we look at Fig. 13.2, we see that all of the nuclear fusion phases depend on the previous phases. Sometimes there are central and one or more shell burning phases before the next burning phase starts in the center. Depending on the mass of the core left behind, sometimes one more shell burning phase is needed before the next phase can ignite. If the core mass was just below the threshold limit, the ensuing evolution would be quite different than if it was just above. As a result, the late evolution phases of massive stars can be a non-linear function of the initial parameters for massive stars – it can even determine whether the star creates a neutron star or a black hole, and whether it ends its life as a bright supernova or just faintly disappears.

13.3 Supernovae from Massive Stars

Table 13.4 gives a comprehensive overview of the different types of supernovae. Supernovae resulting from the interaction of two stars, like coalescing white dwarfs or neutron stars, or a combination thereof are not included. The table is sorted by increasing initial stellar mass and includes the type of progenitor star at the time of explosions (see also Fig. 13.3), a rough mass scale, the physical model for

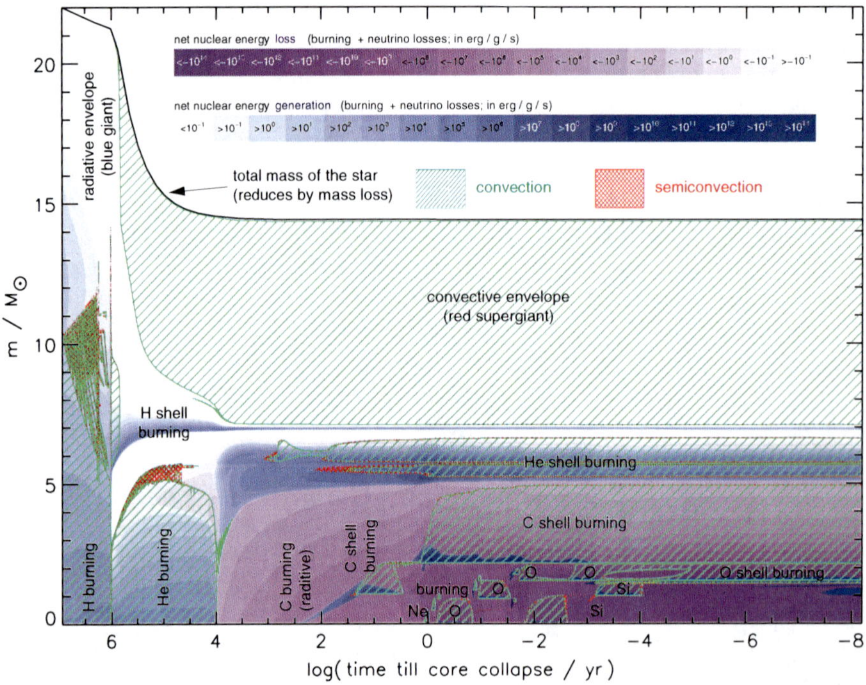

Fig. 13.2 Interior structure of a 22 M$_{\odot}$ star of solar composition as a function of time (logarithm of time until core collapse) and enclosed mass. The end of the x-axis corresponds to a time 0.25 s before core bounce. *Green hatching* and *red cross hatching* indicate convective and semi-convective regions. Convective regions are typically well mixed and evolve chemically homogeneously. *Blue shading* indicates energy generation and *pink shading* energy loss. Both take into account the sum of nuclear and neutrino loss contributions. Each shade of color corresponds to an order of magnitude in specific energy generation rate. The *thick black line* at the top indicates the total mass of the star. The total mass is reduced by mass loss due to stellar winds. Note that the mass loss rate actually increases at late times. The decreasing slope of the total mass of the star in the figure is due to the logarithmic scale chosen for the time axis (from [80])

the explosion mechanism, the typical type of supernova explosion, the explosion energy (in Bethe, 1 B $= 10^{51}$ erg), the physical energy source, and the Population (metallicity) of a typical progenitor. The mass ranges given are from a specific numerical stellar evolution model code (KEPLER [86]) and for commonly used assumptions on physical processes like mixing. For other stellar evolution codes the results may vary by 10% or more [65]. The reader is referred to the review papers [35, 76, 95].

13.3.1 Low-Mass Supernovae

For completeness, two types of supernovae (SNe) from "relatively low-mass stars" are also included: Type Ia supernovae and accretion-induced collapse (AIC)

Table 13.4 Mass ranges, progenitors, and supernovae

Progenitor	Mass ($M_\odot$)	Physical model	Supernova	Energy (B)	Energy source	Populations
CO white dwarf in binary	$\lesssim$6	Degenerate runaway	Type Ia	1	Nuclear	Any
SAGB star (AGB, then SN)	$\sim$6	Degenerate runaway	Type "I.5" (Type Ia inside RSG)	1	Nuclear	Any
MgNeO WD, accretion	6–8	AIC	Faint SN	$\sim$0.1	Gravity (NS)	Any
SAGB star (AGB, then SN)	8–10	EC SN	Faint Type II	<1	Gravity (NS)	[Z] <0
Massive star with H envelop	10–35	Fe core collapse	Type II	0.3–3	Gravity (NS)	Any
WR star	$\gtrsim$35	Fe core collapse	Type Ib/c	0.3–3	Gravity (NS/BH)	Pop I/II; rotation: III
Massive rotating star	$\gtrsim$25	Collapsar, GRB	Broad line Ib/a SN, "hypernova"	1–10	Gravity (BH)	Any
Very massive star	$\sim$100–150	Pulsational pair SN	Multiple, nested Type I/II SN	0.1–5	Nuclear, final: BH	Pop III, I?
Very massive star	$\sim$150–250	Pair SN	Broad, long Type I/II SN	5–100	Nuclear	Pop III, I?
Very massive star	$\gtrsim$250	Pair-collapsar	IMBH, SN, hard transient	0–100	Gravity (BH)	Pop III
Supermassive star	$\gtrsim$150,000	GR inst. + burning	Broad SN or SMBH	$\sim$100,000	Nuclear	Pop III?

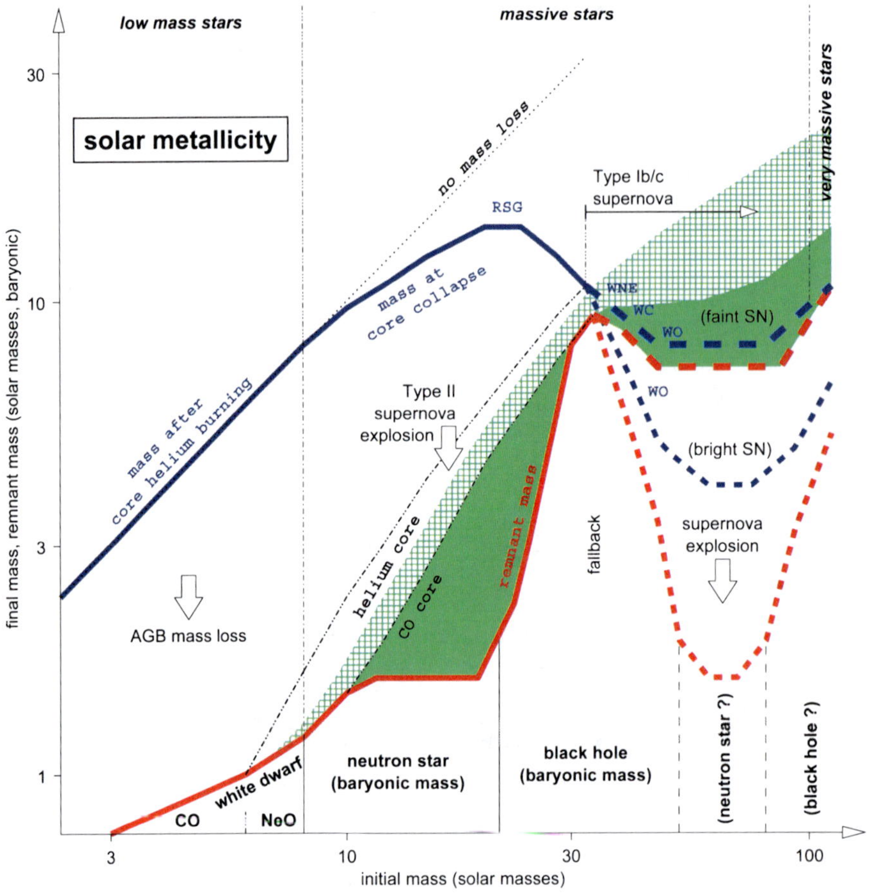

Fig. 13.3 Initial-final mass function (IFMF) of non-rotating stars of solar composition. Similar to Fig. 13.5. Mass loss reduces the mass of the envelope (*blue curve*) until, for a mass above ~33 M$_\odot$, the helium core is uncovered before the star reaches core collapse. At this point the star becomes a Wolf-Rayet (WR) star and the strong WR mass loss sets in. We give two scenarios for the uncertain strength of the WR mass loss rate: The *short dashed red and blue lines* are for a high mass loss rate. Here a "window" of initial masses may exist around 50 M$_\odot$ where neutron stars are still formed (bounded by higher and lower mass stars that make black holes). For a low WR mass loss rate (*long dashed red and blue lines*) the final mass at core collapse is higher and the "neutron star window" may not exist. Then only black holes are formed above ~21 M$_\odot$. "RSG" (red supergiant), "WE", "WC", and "WO" (Wolf-Rayet star spectral sub-types) indicate the last mass loss phase and the type star when it explodes. The heavy element production (green and green cross hatched) is given only for the low mass loss case (from [95])

supernovae (first two lines of Table 13.4). In both cases there is a binary star system in which one star has evolved to become a white dwarf star – presumably the initially more massive star, though more complicated scenarios have been discussed in the literature – that accretes material from a close companion star. Eventually the white dwarf star approaches the Chandrasekhar limit – the maximum mass for the white dwarf – and then the supernova is initiated. The result is very different depending on the kind of white dwarf star:

For carbon-oxygen (CO) white dwarf stars, due to the increasing density the carbon burning rate becomes fast enough that its energy generation overcomes the neutrino losses and center of the star becomes convective. This happens about 1,000 year before the supernova. The convection region slowly grows, heating an increasingly larger part of the star. The temperature continues to rise. Eventually a significant part of the white dwarf is convective, and the convection becomes increasingly more violent. Finally, the burning time-scale of fluid elements in the center becomes short compared to their cooling or mixing time-scale and thermonuclear runaway occurs [101]. Within about a second, the entire star is incinerated, blows up, and a Type Ia supernova occurs [45, 68]. For the progenitor, the limiting mass is such that it must result from a stellar core low enough in mass to not ignite central carbon burning. In commonly discussed scenarios, this implies that the CO white dwarf has no more than about $1.1\,\mathrm{M}_\odot$ when it is born; it will have to accrete about $0.3\,\mathrm{M}_\odot$ before it reaches the Chandrasekhar limit. Other scenarios for Type Ia SNe have also been discussed e.g., binary white dwarf coalescence and sub-Chandrasekhar mass models where a thick helium layer at the surface triggers the supernova [47, 70, 92].

For oxygen-neon-magnesium (ONeMg) white dwarfs ignition of oxygen requires much higher density than what was needed for carbon ignition in CO WDs. Instead, the Fermi energy of the degenerate electron gas becomes so high that the electron capture on nuclei ($^{23}\mathrm{Na}$, ...). This increases the mass per electron and the star contracts to even higher electron densities and electron Fermi energies, a runaway process. By the time oxygen could ignite, the binding energy of the star is larger than the energy that could be obtained from oxygen burning. Instead, the star continues to collapse to a neutron star [58]. This is called this an electron capture supernova (ECSN). The optical display from such a collapse is probably rather faint. It comes from a very compact object, a WD, and little mass is probably ejected. For a review and numerical simulations that also include rotation, see [1, 19].

Both of the above supernova scenarios include a binary star companion from which mass is accreted. Something similar, however, may occur inside the star itself if the core continues to grow during the AGB phase. Then, the competition between core growth in the center and mass loss from the surface determines whether the core can reach the Chandrasekhar limit or the envelope is lost before that and just a white dwarf star is left behind [65]. If the underlying core is a CO WD, the result is an object that often is referred to as "Type I.5" or "Type IIa" SN, essentially a Type Ia supernova exploding inside a hydrogen-rich red supergiant. The resulting supernova may be quite bright. In case the underlying core is a ONeMg WD, the resulting supernova may be somewhat less bright. Due to the steep density gradient at the edge of the core, less energy may be released and little enriched material is ejected [39,67, 85]. This entire discussion is based on theoretical models. Association with specific supernovae and observational evidence has to studied on a case-by-case basis.

13.3.2 Core Collapse Supernovae

The typical range of initial masses is about 10–100 $M_\odot$ for those stars that pass though all of the nuclear burning phases outlined in Sect. 13.2 (Table 13.3) and build up an iron core that finally collapses. A detailed discussion of such stars and their supernovae can also be found in [90].

As far as the ensuing supernova is concerned, there can be complications depending on whether a neutron star or a black is made and whether the hydrogen envelope was previously lost (or how much of it is left), or not, or even if the helium layer has been removed from the surface of the star. There can also be a minor complication depending whether the remaining hydrogen envelope is extended (red supergiant) or not (blue supergiant or WNL star), the former usually resulting in brighter supernovae.

Back to our Table 13.4. For stars of solar metallicity we estimate that for initial masses somewhere around $\sim$35 $M_\odot$ the hydrogen envelope is lost by the time the core collapses [22, 35] due to stellar winds and episodic mass loss. Stars of lower mass explode as Type II supernovae (hydrogen present in spectrum), whereas stars of higher mass are observed as Type Ib or Ic supernovae (not Type Ia) [35]. The detailed mass loss rates of stars as a function of initial mass, metallicity, and the different evolutionary stages, however are not known very accurately. This largely limits the accuracy of the above mass estimate. It may depend on some of the other factors for stellar models mentioned above. In particular, there may also be high mass loss episodes like eruptions that not well understood theoretically nor are the energies and mass lost well known from observations. Such eruptions may cause more extreme mass loss than assumed here. The exact time scale and sequence a star spends as red or blue supergiant is also not well known, and different mass loss rates are associated with these two stages.

For stars below $\lesssim$20 $M_\odot$ [26] a neutron star is made. Due mostly to neutrino energy deposition from the hot neutron star (in most cases), within about a second or less a shock is driven outward that causes the supernova explosion (e.g. [42,43,91]) with a characteristic explosion energy of about 1 B (10^{51} erg) [4, 93]. For a typical supernova, about $\gtrsim$99% of the total energy supply, $\sim$0.2 $M_\odot c^2$ is released in the form of neutrinos, less than 1% (but sill 1 B) goes into kinetic energy of the ejecta, and only $\sim 10^{49}$ erg go into observable light during the supernova explosion (some of the kinetic energy may thermalize later). Other mechanisms of powering bright supernovae have been suggested, e.g., the formation of a fast-spinning magnetized neutron star [46,88]. In this case rotational energy (some fraction of the gravitational binding energy) could be deposited in the form of electromagnetic work much more efficiently than in the case of neutrinos.

For initial stellar masses above $\gtrsim$20 $M_\odot$ we expect the star to form a black hole rather than a neutron star. The exact transition regime is not clear at present. We do not even know what the maximum mass of a neutron star is in the first place [18]. We also do not know whether there is large regime of initial masses in which first a massive hot neutron star is formed that then only later collapses to a black hole. In the latter case, those stars could still have a supernova explosions powered by the same mechanism that works for stars that make neutron stars, but leave behind a black hole instead. Observationally, we would observe a neutrino signal that suddenly shuts off (provided the SN is close enough we can measure the neutrinos). There certainly would also be a distinct signal for gravitational wave detectors should they ever see supernovae. Another complication is that despite the formation of a supernova shock, not all the material that moves outward at first may actually escape; some of it may fall back, and if the resulting total mass exceeds the maximum mass of a neutron star, a black hole would result [100].

On the other hand, even if a black hole was formed directly, for stars of high enough initial mass, there may still be an explosion. One possible mechanism for that is the collapsar model [87] initially conceived to explain gamma-ray bursts. For gamma-ray bursts (GRBs), the jet has to escape from the star on a time scale of the duration of the observed burst (a few to some 100 s), i.e., the travel time of the jet through the star should not be much bigger than the duration. This requires an early-type Wolf Rayet star, and excludes stars with hydrogen envelopes [52, 99]. But the mechanism may still work in other stars, giving a potentially powerful supernova explosion, even if there is no gamma-ray burst. On the other hand, powerful supernovae have been observed in association with gamma-ray bursts [94]. Such supernovae could be several times more energetic than normal supernovae, though they may not necessarily be correspondingly brighter as well. The gamma-ray burst itself, however, if pointed at us, can be much brighter than a regular supernova.

13.4 Very Massive Stars and Their Supernovae

Generally, we call stars of more than $100\,M_\odot$ "*very massive*." From the point of view of physics, this distinction *is* motivated by the onset of the pair instability near this initial mass. Until recently, it was thought that star of such high mass would form only rarely, if at all. A famous counter example is η Car, and if they did form, they would always loose so much mass before they die that their stellar deaths would essentially be indistinguishable from stars of lower initial mass.

Continuing our discussion of mass ranges as guided by Table 13.3, we will start by first discussing pair-instability supernovae (PSNe) and then the pulsational pair-instability supernovae (PPSNe), although the latter come from stars of lower mass.

13.4.1 Pair Instability Supernovae

Pair-Instability supernovae occur in helium cores of about 65–133 $M_\odot$ [32]. Without stellar mass loss, this corresponds to a range of initial masses of about 150–260 $M_\odot$. The prediction of this kind of supernovae, and the physical mechanism, as well as the nucleosynthesis produced have been known for some time [7, 20, 60, 73]. The physical mechanism is as follows: After central carbon burning (which is radiative in these massive stars), the entropy in these stars is high enough that they will produce electron-positron pairs (hence the name of the instability). This uses up some of the internal energy of the gas and converts it into the rest mass of electrons and positrons. The equation of state (EOS) of these stars, on the other hand, is dominated by radiation pressure, already very close to marginal stability. The pair creation now reduces the pressure just below stability, and the star becomes unstable. The increased gravity, when the star is compressed, is not outweighed by a sufficiently large increase in pressure[7] and a runaway collapse occurs. As the temperature rises, nuclear burning sets in and increases the entropy further. First neon and oxygen then silicon burn on a time scale of seconds as the star collapses.

Eventually, the temperature, due to compression and burning, can become high enough so that the electron-positron pairs created have enough energy to also contribute to the pressure and the pair-instability regime of the EOS is left. The nuclear burning in this regime produces enough energy to entirely disrupt the star; the kinetic energy can be as much as 100 times that of a normal supernova (Fig. 13.4).

At the low-mass end of the PSN regime the burning just barely goes through oxygen burning before the collapse is turned around. The explosion energy at this point is only about 4 B. As the mass increases, the burning runs through oxygen burning and silicon burning, making mostly ^{56}Ni in the center of the star. At the upper end of the mass regime a new kind of physics happens. The temperature becomes so high that ^{56}Ni is being photo-disintegrated back into alpha particles (^{4}He) – which costs a lot of energy. The burning further out, however, can provide enough energy to still turn the collapse around.[8] But for even higher mass, these alpha particles are also photo-disintegrated into free nucleons, sort of the reverse of hydrogen burning, which costs even more energy. If this point is reached, the collapse cannot be turned around; this determines the upper mass limit for the pair instability regime (Fig. 13.5). In the simulations this is a rather sharp transition from very powerful explosion to complete collapse; there are no partial explosions.

13.4.2 Pulsational Pair Instability Supernovae

For helium cores of $\lesssim$65 $M_\odot$ (initial stellar masses $\lesssim$150 $M_\odot$ for stars without mass loss or rotation), the energy release of the burning during the pair instability is not

[7]The adiabatic exponent of the gas drops below the critical adiabatic exponent needed for stability of the stars, which is 4/3; a completely radiation-dominated gas just has an adiabatic exponent of 4/3, the ideal ions gas has an adiabatic exponent of 5/3, so the sum of both usually gives something just slightly above 4/3 in such massive stars – until the pair-creation reduced the photon part to below 4/3.

[8]Strictly speaking, the stability of the star is determined by $\int_0^M \gamma(m)\frac{P(m)}{\rho(m)}\,\mathrm{d}m > \frac{4}{3}\int_0^M \frac{P(m)}{\rho(m)}\,\mathrm{d}m$.

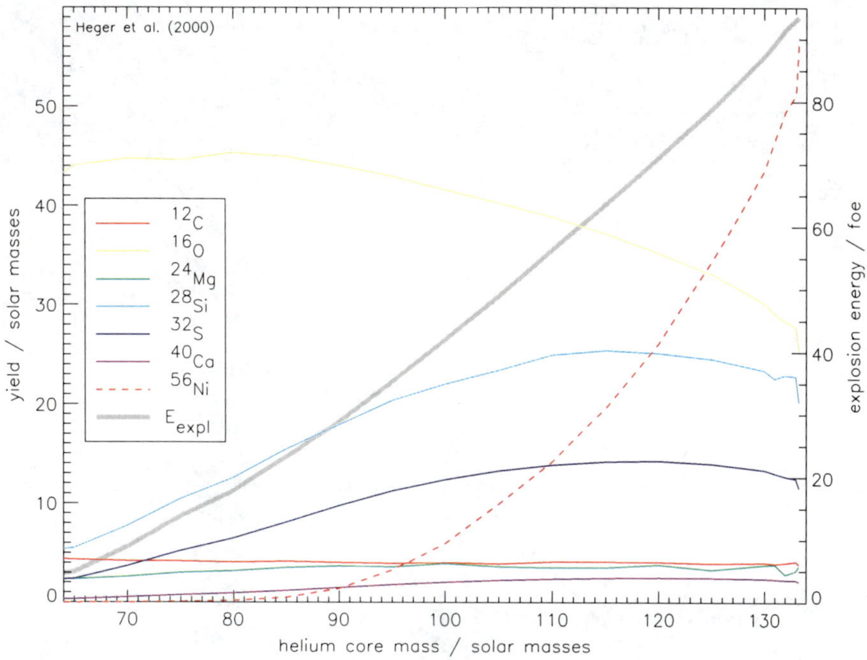

Fig. 13.4 Explosion energy and main nucleosynthesis yield of pair instability supernovae as a function of helium core mass. A 133 $M_\odot$ helium star explodes at about 100× the mass of a Type I supernova, 100× the explosion energy, and 100× the ^{56}Ni mass produced – 100 Type Ia supernovae in one spot. The energy unit on the right hand side, "foe", stands for 10^{51} erg ("ten to fify one erg".), a unit coined by Hans Bethe; in his honor now a new unit "B" (Bethe) as been introduced for this energy scale ($1\,\mathrm{B} = 1\times10^{51}$ erg) (from [95])

enough to entirely disrupt the star. Even though the total energy of the star may be positive when the collapse turns into an expansion, and in fact continues for some small time, much of the momentum is transferred into the outer layers which are ejected at high velocity, while the inner part is pulled back by gravity. The result is only a pulse with ejection of the outer layers, not a terminal explosion. We may call this a supernova "impostor", though in effect, in terms of the energy and mass ejected, it is more like a supernova than anything else. Potentially, some of them may even be as bright as a supernova, but only during the first pulse. Even normal core collapse supernova leave behind a neutron star or a black hole; in this case the massive star survives, so it would be still reasonable to call these pulses "supernovae" rather than supernova impostors, as the outbursts involve the entire star and an energy source from its center.

In the case discussed here, however, what is left behind after the pulse may still be massive enough to undergo further pulses. The pulses will continue until the object has shrunk below the limiting mass of $\lesssim 45\,M_\odot$ for the hydrogen-free core. This could be just a few, or dozens of pulses, depending on how energetic they are.

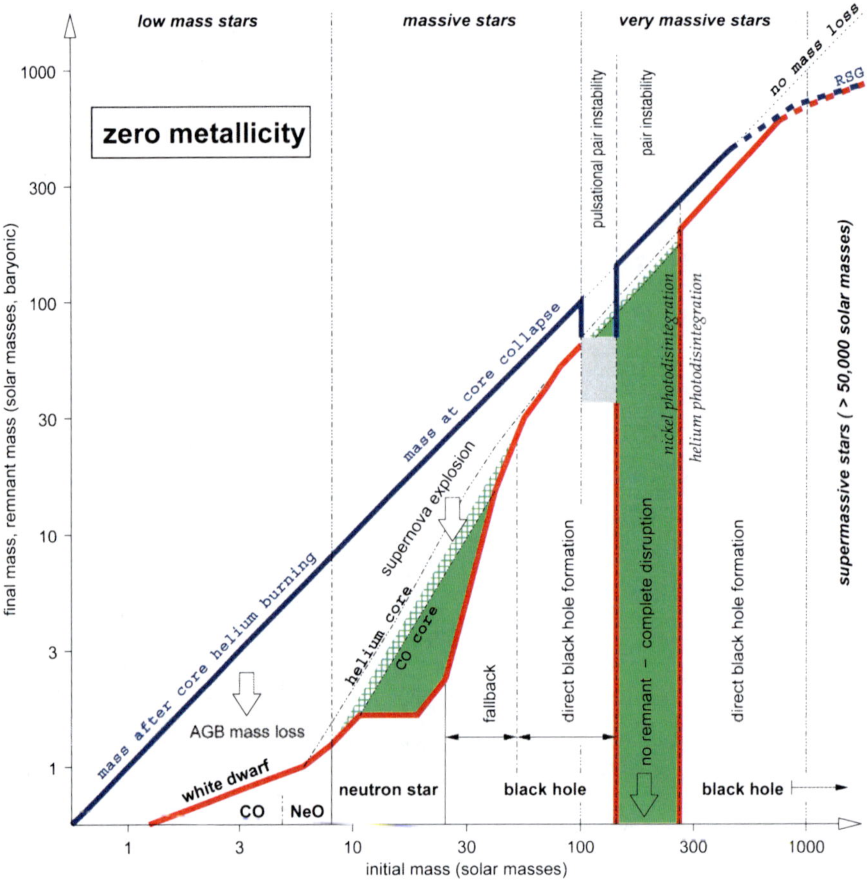

Fig. 13.5 Initial-final mass function (IFMF) of non-rotating primordial stars ($Z = 0$). The *x-axis* gives the initial stellar mass. The *y-axis* gives both the final mass of the collapsed remnant (*thick red curve*) and the mass of the star when the event begins that produces that remnant (e.g., mass loss in AGB stars, supernova explosion for those stars that make a neutron star, etc.; *thick blue curve*). Dark green indicates regions of heavy element ($Z > 2$) synthesis and cross-hatched green, regions of partial helium burning to carbon and oxygen. We distinguish four regimes of initial mass: *low mass stars* below $\sim 10\,M_\odot$ that form white dwarfs; *massive stars* between ~ 10 and $\sim 100\,M_\odot$; *very massive stars* between ~ 100 and $\sim 1{,}000\,M_\odot$; and *supermassive stars* (arbitrarily) above $\sim 1{,}000\,M_\odot$. Since no mass loss is expected for $Z = 0$ stars, the blue curve corresponds approximately to the line of no mass loss (*dotted*), except for ~ 100–$140\,M_\odot$ where the pulsational pair-instability ejects the outer layers of the star before it collapses, and above $\sim 500\,M_\odot$ where pulsational instabilities in red supergiants may lead to significant mass loss. Since the magnitude of the latter is uncertain, lines are *dashed*. In the low mass regime we assume, even in $Z = 0$ stars, that mass loss on the asymptotic giant branch (AGB) removes the envelope of the star leaving a CO or NeO white dwarf (though the mechanism and thus the resulting initial-final mass function may differ from solar composition stars). Massive stars are defined by stars that ignite carbon and oxygen burning non-degenerately and do not leave white dwarfs. The hydrogen-rich envelope and parts of the helium core (*dash-double-dotted curve*) are ejected in a supernova explosion. Below initial masses of $\sim 25\,M_\odot$ neutron stars are formed. Above that, black holes

When the mass has dropped below the limiting mass, we expect that the evolution will continue like that of any other massive star that has lost the hydrogen (and possibly helium) envelope due to some other mass loss mechanism, form a big iron core than eventually collapses and make a black hole. The terminal collapse could still be accompanied by a gamma-ray burst.

The detailed sequence of pulses can be very complicated and has not been studied in detail as a function of mass. This is because of the interaction of the neon, oxygen, and silicon shell burning during the pulses, and the feedback on the mass ejected. We expect a general relation between pulse energy and recurrence time which is also found in simulations. If the pulse is very energetic, the next pulse will take a longer time to occur than if the pulse was less energetic. If the entropy generated by the pulse was low, the temperature in the star when it settles back to hydrostatic equilibrium will be high and can cool efficiently by neutrinos, so the next pulse can ensue swiftly as the star contracts to pair-instability conditions. If, on the other hand, the pulse was energetic and a lot of entropy was generated, the star will be more extended and cooler when it gets back into hydrostatic equilibrium. In this case neutrino cooling will be less efficient or absent. In the extreme case, the star has to cool on the Kelvin-Helmholtz time scale for photon emission at the surface, some 10,000 year, whereas in the other extreme of a weak pulse it may be just weeks or less for the next pulse.

Can the eruptions of Eta Car be a pulsational pair instability? For the pulsational pair instability we would expect that even the first pulse would eject the hydrogen envelope. At least for red giant stars, the binding energy of the envelope is less than the energy of even a weak pulse of 10^{49} erg. It will be much higher for a blue star. But in order to have a delay between outbursts of $\sim$100 years we would expect that the energy of the pulse would have to be at least some 10^{51} erg (Eta Car's luminous energy in the great eruption was $\sim 10^{50}$ erg [40], although very high velocity gas recently detected in the outer ejecta suggests that its kinetic energy may have rivaled that of a SN[9]), and all the mass would be ejected in just one SN-like ejection, not as a continuous surface event lasting 20 years, as in the case of η Car.

Fig. 13.5 (continued) form, either in a delayed manner by fall back of the ejecta, or directly during iron core collapse (above $\sim$40 $M_\odot$). The defining characteristic of *very* massive stars is the electron-positron *pair instability* after carbon burning. This begins as a pulsational instability for helium cores of $\sim$40 $M_\odot$ ($M_{ZAMS} \sim 100\,M_\odot$). As the mass increases, the pulsations become more violent, ejecting any remaining hydrogen envelope and an increasing fraction of the helium core itself. An iron core can still form in hydrostatic equilibrium in such stars, but it collapses to a black hole. Above $M_{He} = 63\,M_\odot$ or about $M_{ZAMS} = 140\,M_\odot$, and on up to $M_{He} = 133\,M_\odot$ or about $M_{ZAMS} = 260\,M_\odot$ a single pulse disrupts the star. Above $260\,M_\odot$, the pair instability in non-rotating stars results in complete collapse to a black hole (from [95])

[9] See Chap. 7

It would also be accompanied by a shock hitting the surface of the star that would make a flash of X-rays or gamma-rays.[10]

Another interesting prediction from pulsational pair instabilities is that the ejected shells themselves could actually "interact". So not only could each pulse produce a supernova "impostor", but when the two shells collide, the kinetic energy could be thermalized and make a rather bright display because the shells would be optically thin, and the radiation not trapped as it is the case in normal core collapse supernova where only 1% of the energy can escape in the form of radiation, the rest is kinetic. Such an interaction has been suggested as a possible model for one of the brightest supernovae observed recently, SN 2006gy [96].

For both, PSN and PPSN to occur, the star must keep enough mass through the end of helium burning (the contraction toward carbon burning is comparably fast, though the star *does* undergo some structural changes during that phase which can affect mass loss). Hence the big question that we cannot answer from theory at the present is whether modern stars in the galaxy can retain that much mass or would always lose too much mass before reaching the final stages. And if so, what is the limiting metallicity, or the limiting mass as a function of metallicity to make pair instability supernovae?

13.4.3 Beyond Pair Instability

For helium cores more massive than about $\sim 133 M_\odot$ (initial masses $\gtrsim 260\,M_\odot$ without mass loss or rotation) the onset of photo-disintegration in the center will lead to collapse rather than explosion. For stars with rotation, we may still find some characteristic observable signatures, maybe a flash in the X-ray band (if the star was at high redshift) [27]. For stars up to about $\sim 150{,}000\,M_\odot$ we would expect to see just a collapse to a black hole during or at the end of the hydrostatic burning evolution. Stars of even higher mass cannot reach a state of hydrostatic and thermal equilibrium; in this case general relativity demands that the adiabatic exponent of the gas has to be *larger* than 4/3 for stability of the star[11] whereas the actual adiabatic exponent of the gas continues to decrease toward 4/3 as the stars become more radiation-dominated the more massive they are. If the initial composition of the star was sufficiently enriched in metals, a powerful explosion may be triggered due to rapid burning [28]. The question about these supermassive stars does bear some relevance to the question of making large seed black holes for the supermassive black holes present at the centers of galaxies, and appear to be

[10] Editor's note: Two recent supernovae have been confirmed to have had eruptions only a few years prior to the terminal explosion. See the Chaps. 10 and 11.

[11] Of course it always does, but it only becomes a large enough effect to be relevant for stars of that large mass.

present in quasars when the universe was just a billion years old [23,68,71]. Starting with a big seed would make it easier to explain these black holes, but we will have to explain how to make such big stars.

13.5 Primordial Stars

The first stars to form after the big bang may have been quite different from stars that form today. They formed from the primordial gas left over by the big bang – essentially just hydrogen and helium. This unique composition may have influenced both their formation and their evolution. Later generations of stars were polluted by the metals made by this first generation. Even if pockets of such gas were still around today [72], their formation environment would have changed [62], so one of the key properties we discuss below, the initial mass function, would have changed. Not being able to observe them or their supernovae directly, they appear somewhat elusive, but yet, a first generation surely must have existed.

13.5.1 Formation of the First Stars

Regions of star formation today are cold molecular clouds filled with dust and penetrated by magnetic fields made by the galaxy and stellar explosions. But, the environment of first star formation was quite different. Hot atomic hydrogen from cosmic recombination was pulled together by overdensities of dark matter. Cooling of that gas was inefficient due to lack of metals and there were probably no strong magnetic fields. Theoretical models predict that due to the higher gas temperature, the characteristic initial mass of primordial stars would be much higher [14, 51, 75]. The first detailed numerical simulations of the formation of primordial stars suggested that they may have been quite massive, around $100\,M_{\odot}$ with values in the literature ranging from lower limits around $30\,M_{\odot}$ to more than $300\,M_{\odot}$ (e.g. [2, 9–11, 57, 61]). A good review on the state of star formation theory 5 years ago can be found in [12].

Whereas the paradigm for a long time had been that primordial stars would always form as single stars from a monolithic collapse of the giant gas cloud, more recent simulations that allow much higher numerical resolution in the center of the star-forming region now indicate that a significant fraction of these, maybe most, may form as binaries (e.g. [30, 82]). Most of these calculations cannot yet follow the evolution of early binary star to final stars, so the outcome is not yet fully clear. This opens the possibility that the massive stars in Population III could have had an IMF similar to massive stars today.

On another avenue, supermassive star formation (Sect. 13.4.3), has been fueled by the idea of dark-matter annihilation in pre-galactic stars; a phenomenological particle physics model for dark matter. "Weakly interacting massive particles" (WIMPs), could self-annihilate inside stars in the early universe when the dark matter density was still much higher than it is today ([25, 41], and references therein). The first stars would actually form in the regions of largest overdensities for dark matter. Some optimistic models suggest that this could power stars for some time, hold up their contraction and evolution, and that way they could pick up 1,000 $M_\odot$ or more. Maybe even much more.

13.5.2 Evolution and Nucleosynthesis of the First Stars

Figure 13.5 shows the fate of massive Pop III stars, their supernovae and their remnants, as a function of initial mass. In contrast to Fig. 13.3 the main difference is that we assume no significant mass loss would happen before the star reaches the final supernova stage (only for the pulsational pair instability regime we indicate the loss of the outer layers of the star prior to core collapse).

The reduced – or vanishing – mass loss is one of the key ingredients for how primordial stars may behave differently. The changes to hydrogen burning have been discussed in Sect. 13.2.1. In the past it was been suggested that stars more massive than about 60 $M_\odot$ would be pulsationally unstable due to the "epsilon-mechanism" [74].[12] It has been shown, however, that this mechanism does not work for primordial stars due to the different way in which hydrogen burning is started [5]. If anything, this mechanism would only operate at higher metallicity, but in more metal-rich stars, opacity-driven oscillations at the outer layers would dominate [29].

Besides the traditional picture that Population III stars may hold on to most of their initial mass, those stars would also be more compact at the time they die; they would be blue supergiants, not red supergiants. Accordingly, they would have more fallback after the supernova explosion [33], making bigger remnants. Because of the more compact envelope, there would be less Rayleigh-Taylor instabilities after the passage of the SN shock inside the star, and accordingly less mixing [44]. Both of these seem to be consistent with observed abundance patterns of the most metal-poor stars in the galaxy [16,24,48]. Hence it is our hope that whereas we are not able to directly observe the first, observing their ashes, their nucleosynthesis fingerprints as preserved in ultra- and extremely metal-poor stars in the Galactic halo, we will be able to determine the IMF of the first generation of stars indirectly. Interestingly, preliminary results by [33] suggest that the average abundances observed by [15] do not appear to indicate the need for a deviation from the standard Salpeter IMF for massive Population III stars.

[12] nuclear-powered pulsations due to the high temperature sensitivity of the nuclear burning combined with the radiation-dominated equation of state of such massive stars.

13.5.3 Pair Instability Supernovae

This leaves the question whether there could have been a population of very massive stars early on during the epoch of primordial stars. Interestingly, a prediction about the nucleosynthesis yields of Population III does exist [32] (see also Fig. 13.4). The main point is that those stars would not make compact remnants, there is no neutron star and there would be no r-process. There are no initial metals, so there would be no s-process; in fact, these stars would make a very strong odd-even effect as a function of nuclear charge: elements of odd charge number would be about two orders of magnitude under-abundant relative to elements with even charge number, when compared to the solar abundance pattern. If we were to further assume that the ejecta of these early stars are incorporated in a next generation of metal-poor stars, we should find some of them that show such an abundance pattern. None have been found to date. Possible explanations are: (1) the second generation of stars that were enriched by the ejecta of pair-instability supernovae may also have been massive and therefore all of them have died by now, or (2) they enriched the next generation of stars to higher metallicities than just "extremely metal poor"; however, these stars are very rare and have not been found yet – most stars of higher metallicity were made later –, or (3) these stars formed in the regions of the highest overdensities that now are in the center of the Galaxy, not in the halo. So we just have not looked in the right place!

13.6 Summary and Outlook

The fundamental theory of the physics of the late stages of massive star evolution has been well known for a long time. In the last decades little has changed in our basic picture but some major uncertainties for real stars still exist [21, 55]. Uncertainties in the theory and modeling for mixing in stars remain, starting with the modeling of convection based on mixing-length theory [6, 66], the question of convective "overshooting" or penetration, and whether to use the Ledoux or the Schwarzschild criterion for determining convective boundaries. The formulation and efficiency of thermal transport and chemical mixing in semiconvective regions has been heavily debated in the literature for decades [49,50,77], but is still not solved. This is in addition to modeling rotation and magnetic fields in stars mentioned in the Introduction.

The other major source of uncertainty is mass loss. Red giant mass loss rates are largely unknown, though there have been some recent new approaches in modeling the final red supergiant phases of massive stars [97]. But for modeling the evolution of the most massive stars, an understanding of stellar eruptions and episodic mass loss as observed in η Car and similar very massive stars is essential. This is particularly true for the first generation of stars, assuming that they were very massive and may have behaved similarly. At the present, little about the detailed mechanism for the eruptions is known, though some suggestions can be found in the literature (e.g. [63, 83]). The pulsational pair-instability supernova (PPSN) mechanism does not seem a viable candidate (Sect. 13.4.2).

Hence a major challenge for stellar evolution theory will be to understand the observed massive ejections and that are not predicted by current theory. Only then will be able to extend our understanding of stellar evolution and stellar death to the high-mass end of the IMF and to the first generations of stars in the universe.

Acknowledgements I want to thank the editor, Roberta Humphreys, for her input, feedback, and patience with the completion of this article. Much of the scientific research on which this chapter is based has been done in collaboration with Stan Woosley, who I would like to thank for his continued support. Heger has been supported by the DOE Program for Scientific Discovery through Advanced Computing (SciDAC; DE-SC0002300/FC02-09ER41618), and by the US Department of Energy under grant DE-FG02-87ER40328. The early work on this chapter was carried out in part under the auspices of the National Nuclear Security Administration of the U.S. Department of Energy at Los Alamos National Laboratory and supported by Contract No. DE-AC52-06NA25396.

References

1. E.B. Abdikamalov, C.D. Ott, L. Rezzolla, L. Dessart, H. Dimmelmeier, A. Marek, H.-T. Janka, Axisymmetric general relativistic simulations of the accretion-induced collapse of white dwarfs. Phys. Rev. D **81**, 044012 (2010)
2. T. Abel, G.L. Bryan, M.L. Norman, The formation of the first star in the universe. Science **295**, 93–98 (2002)
3. D. Arnett, D. Arnett, D. Arnett, *Supernovae and Nucleosynthesis. An Investigation of the History of Matter, from the Big Bang to the Present*. Princeton Series in Astrophysics (Princeton University Press, Princeton, 1996)
4. W.D. Arnett, J.N. Bahcall, R.P. Kirshner, S.E. Woosley, Supernova 1987A. ARA&A. **27**, 629–700 (1989)
5. I. Baraffe, A. Heger, S.E. Woosley, On the stability of very massive primordial stars. ApJ. **550**, 890–896 (2001)
6. E. Böhm-Vitense, Über die Wasserstoffkonvektionszone in Sternen verschiedener Effektivtemperaturen und Leuchtkräfte. Mit 5 Textabbildungen. 'Zeitsch. für Astron. **46**, 108 (1958)
7. J.R. Bond, W.D. Arnett, B.J. Carr, The evolution and fate of very massive objects. ApJ. **280**, 825–847 (1984)
8. J. Braithwaite, H.C. Spruit, A fossil origin for the magnetic field in A stars and white dwarfs. Nature **431**, 819–821 (2004)
9. V. Bromm, P.S. Coppi, R.B. Larson, Forming the first stars in the universe: the fragmentation of primordial gas. ApJL. **527**, L5–L8 (1999)
10. V. Bromm, P.S. Coppi, R.B. Larson, The formation of the first stars. I. The primordial star-forming cloud. ApJ. **564**, 23–51 (2002)
11. V. Bromm, A. Ferrara, P.S. Coppi, R.B. Larson, The fragmentation of pre-enriched primordial objects. MNRAS. **328**, 969–976 (2001)
12. V. Bromm, R.B. Larson, The first stars. ARA&A. **42**, 79–118 (2004)
13. E.M. Burbidge, G.R. Burbidge, W.A. Fowler, F. Hoyle, Synthesis of the elements in stars. Rev. Mod. Phys. **29**, 547–650 (1957)
14. R.G. Carlberg, An estimate of the mass of zero metal stars. MNRAS. **197**, 1021–1029 (1981)

15. R. Cayrel, E. Depagne, M. Spite, V. Hill, F. Spite, P. François, B. Plez, T. Beers, F. Primas, J. Andersen, B. Barbuy, P. Bonifacio, P. Molaro, B. Nordström, First stars V – Abundance patterns from C to Zn and supernova yields in the early galaxy. A&A. **416**, 1117–1138 (2004)
16. N. Christlieb, B. Gustafsson, A.J. Korn, P.S. Barklem, T.C. Beers, M.S. Bessell, T. Karlsson, M. Mizuno-Wiedner, HE 0107-5240, a chemically ancient star. I. A detailed abundance analysis. ApJ. **603**, 708–728 (2004)
17. D.D. Clayton, *Principles of Stellar Evolution and Nucleosynthesis* (McGraw-Hill, New York, 1968)
18. P.B. Demorest, T. Pennucci, S.M. Ransom, M.S.E. Roberts, J.W.T. Hessels, A twosolar-mass neutron star measured using Shapiro delay. Nature **467**, 1081–1083 (2010)
19. L. Dessart, A. Burrows, C.D. Ott, E. Livne, S.-C. Yoon, N. Langer, Multidimensional simulations of the accretion-induced collapse of white dwarfs to neutron stars. ApJ. **644**, 1063–1084 (2006)
20. M.F. El Eid, K.J. Fricke, W.W. Ober, Evolution of massive pregalactic stars. I – Hydrogen and helium burning. II – Nucleosynthesis in pair creation supernovae and pregalactic enrichment. A&A. **119**, 54–68 (1983)
21. M.F. El Eid, L.-S. The, B.S. Meyer, Massive stars: input physics and stellar models. Space Sci. Rev. **147**, 1–29 (2009)
22. J.J. Eldridge, C.A. Tout, The progenitors of core-collapse supernovae. MNRAS. **353**, 87–97 (2004)
23. C. Filloux, F. Durier, J.A.F. Pacheco, J. Silk, Evolution of supermassive black holes from cosmological simulations. Int. J. Mod. Phys. D **19**, 1233–1240 (2010)
24. A. Frebel, W. Aoki, N. Christlieb, H. Ando, M. Asplund, P.S. Barklem, T.C. Beers, K. Eriksson, C. Fechner, M.Y. Fujimoto, S. Honda, T. Kajino, T. Minezaki, K. Nomoto, J.E. Norris, S.G. Ryan, M. Takada-Hidai, S. Tsangarides, Y. Yoshii, Nucleosynthetic signatures of the first stars. Nature **434**, 871–873 (2005)
25. K. Freese, C. Ilie, D. Spolyar, M. Valluri, P. Bodenheimer, Supermassive dark stars: Detectable in JWST. ApJ. **716**, 1397–1407 (2010)
26. C.L. Fryer, Mass limits for black hole formation. ApJ. **522**, 413–418 (1999)
27. C.L. Fryer, S.E. Woosley, A. Heger, Pair-instability supernovae, gravity waves, and gamma-ray transients. ApJ. **550**, 372–382 (2001)
28. G.M. Fuller, S.E. Woosley, T.A. Weaver, The evolution of radiation-dominated stars. I – Nonrotating supermassive stars. ApJ. **307**, 675–686 (1986)
29. W. Glatzel, M. Kiriakidis, The stability of massive main-sequence stars. MNRAS. **262**, 85–92 (1993)
30. S.C.O. Glover, P.C. Clark, R.S. Klessen, V. Bromm, Fragmentation in turbulent primordial gas. ArXiv e-prints, arXiv:1007.2763 (2010)
31. A. Heger, N. Langer, Presupernova evolution of rotating massive stars. II. Evolution of the surface properties. ApJ. **544**, 1016–1035 (2000)
32. A. Heger, S.E. Woosley, The nucleosynthetic signature of population III. ApJ. **567**, 532–543 (2002)
33. A. Heger, S.E. Woosley, Nucleosynthesis and evolution of massive metal-free stars. ArXiv e-prints, arXiv:0803.3161 (2008); ApJ. **724**, 341–373 (2010)
34. A. Heger, N. Langer, S.E. Woosley, Presupernova evolution of rotating massive stars. I. Numerical method and evolution of the internal stellar structure. ApJ. **528**, 368–396 (2000)
35. A. Heger, C.L. Fryer, S.E. Woosley, N. Langer, D.H. Hartmann, How massive single stars end their life. ApJ. **591**, 288–300 (2003)
36. A. Heger, S.E. Woosley, H.C. Spruit, Presupernova evolution of differentially rotating massive stars including magnetic fields. ApJ. **626**, 350–363 (2005)
37. R. Hirschi, G. Meynet, A. Maeder, Stellar evolution with rotation. XII. Pre-supernova models. A&A. **425**, 649–670 (2004)
38. R. Hirschi, G. Meynet, A. Maeder, Stellar evolution with rotation. XIII. Predicted GRB rates at various Z. A&A. **443**, 581–591 (2005)

39. L. Hüdepohl, B. Müller, H.-T. Janka, A. Marek, G.G. Raffelt, Neutrino signal of electron-capture supernovae from core collapse to cooling. PRL. **104**, 251101 (2010)
40. R.M. Humphreys, K. Davidson, N. Smith, Eta Carinae's second eruption and the light curves of the Eta Carinae variables. PASP. **111** (1999)
41. F. Iocco, A. Bressan, E. Ripamonti, R. Schneider, A. Ferrara, P. Marigo, Dark matter annihilation effects on the first stars. MNRAS. **390**, 1655–1669 (2008)
42. H.-T. Janka, Conditions for shock revival by neutrino heating in core-collapse supernovae. A&A. **368**, 527–560 (2001)
43. H.-T. Janka, K. Langanke, A. Marek, G. Martínez-Pinedo, B. Müller, Theory of core-collapse supernovae. PhR. **442**, 38–74 (2007)
44. C.C. Joggerst, S.E. Woosley, A. Heger, Mixing in zero- and solar-metallicity supernovae. ApJ. **693**, 1780–1802 (2009)
45. G.C. Jordan IV, R.T. Fisher, D.M. Townsley, A.C. Calder, C. Graziani, S. Asida, D.Q. Lamb, J.W. Truran, Three-dimensional simulations of the deflagration phase of the gravitationally confined detonation model of Type Ia supernovae. ApJ. **681**, 1448–1457 (2008)
46. D. Kasen, L. Bildsten, Supernova light curves powered by young magnetars. ApJ. **717**, 245–249 (2010)
47. M. Kromer, S.A. Sim, M. Fink, F.K. Röpke, I.R. Seitenzahl, W. Hillebrandt, Double-detonation sub-chandrasekhar supernovae: Synthetic observables for minimum helium shell mass models. ApJ. **719** 1067–1082 (2010)
48. D.K. Lai, M. Bolte, J.A. Johnson, S. Lucatello, A. Heger, S.E. Woosley, Detailed abundances for 28 metal-poor stars: Stellar relics in the milkyway. ApJ. **681**, 1524–1556 (2008)
49. N. Langer, M.F. El Eid, K.J. Fricke, Evolution of massive stars with semiconvective diffusion. A&A. **145**, 179–191 (1985)
50. N. Langer, K.J. Fricke, D. Sugimoto, Semiconvective diffusion and energy transport. A&A. **126**, 207 (1983)
51. R.B. Larson, Early star formation and the evolution of the stellar initial mass function in galaxies. MNRAS **301**, 569–581 (1998)
52. A.I. MacFadyen, S.E. Woosley, Collapsars: Gamma-ray bursts and explosions in "failed supernovae". ApJ. **524**, 262–289 (1999)
53. A. Maeder, G. Meynet, The evolution of rotating stars. ARA&A. **38**, 143–190 (2000)
54. A. Maeder, G. Meynet, Stellar evolution with rotation and magnetic fields. II. General equations for the transport by Tayler-Spruit dynamo. A&A. **422**, 225–237 (2004)
55. C. A. Meakin, T. Sukhbold, D. Arnett, Presupernova structure of massive stars. ArXive-prints, arXiv:1006.0513 (2010)
56. G. Meynet, A. Maeder, Stellar evolution with rotation. I. The computational method and the inhibiting effect of the μ-gradient. A&A. **321**, 465–476 (1997)
57. F. Nakamura, M. Umemura, On the mass of population III stars. ApJ. **515**, 239–248 (1999)
58. K. Nomoto, Evolution of 8–10 solar mass stars toward electron capture supernovae. II – Collapse of an O + NE + MG core. ApJ. **322**, 206–214 (1987)
59. K. Nomoto, M. Mashimoto, Presupernova evolution of massive stars. PhR. **163**, 13–36 (1988)
60. W.W. Ober, M.F. El Eid, K.J. Fricke, Evolution of massive pregalactic stars – Part two – Nucleosynthesis in pair creation supernovae and pregalactic enrichment. A&A. **119**, 61 (1983)
61. K. Omukai, R. Nishi, Formation of primordial protostars. ApJ. **508**, 141–150 (1998)
62. B. W. O'Shea, C.F. McKee, A. Heger, T. Abel, First Stars III Conference Summary, in *First Stars III* eds. by B.W. O'Shea, A. Heger. AIP Conf. Series. vol. 990, (AIP, Melville, 2008), p. D13
63. S.P. Owocki, K.G. Gayley, N.J. Shaviv, A porosity-length formalism for photon-Tiring-limited mass loss from stars above the eddington limit. ApJ. **616**, 525–541 (2004)
64. P. Podsiadlowski, P.C. Joss, J.J.L. Hsu, Presupernova evolution in massive interacting binaries. ApJ. **391**, 246–264 (1992)
65. A.J.T. Poelarends, F. Herwig, N. Langer, A. Heger, The supernova channel of super-AGB stars. ApJ. **675**, 614–625 (2008)

66. L. Prandtl, Z. Angew. Math. Mech. **5**, 136 (1925)
67. M.L. Pumo, M. Turatto, M.T. Botticella, A. Pastorello, S. Valenti, L. Zampieri, S. Benetti, E. Cappellaro, F. Patat, EC-SNe from super-asymptotic giant branch progenitors: theoretical models versus observations. ApJL. **705**, L138–L142 (2009)
68. M.J. Rees, Black hole models for active galactic nuclei. ARA&A. **22**, 471–506 (1984)
69. F.K. Röpke, W. Hillebrandt, W. Schmidt, J.C. Niemeyer, S.I. Blinnikov, P.A. Mazzali, A three-dimensional deflagration model for type Ia supernovae compared with observations. ApJ. **668**, 1132–1139 (2007)
70. A.J. Ruiter, K. Belczynski, S.A. Sim, W. Hillebrandt, M. Fink, M. Kromer, Type Ia supernovae and accretion induced collapse. ArXiv e-prints, arXiv:1009.3661 (2010)
71. S. Salviander, G.A. Shields, K. Gebhardt, E.W. Bonning, The black hole mass-galaxy bulge relationship for QSOs in the Sloan digital sky survey data release 3. ApJ. **662**, 131–144 (2007)
72. E. Scannapieco, P. Madau, S. Woosley, A. Heger, A. Ferrara, The detectability of pair-production supernovae at $z \lesssim 6$. ApJ. **633**, 1031–1041 (2005)
73. M. Schwarzschild, R. Härm, Evolution of very massive stars. ApJ. **128**, 348 (1958)
74. M. Schwarzschild, R. Härm, On the maximum mass of stable stars. ApJ. **129**, 637 (1959)
75. J. Silk, The first stars. MNRAS. **205**, 705–718 (1983)
76. S.J. Smartt, Progenitors of core-collapse supernovae. ARA&A. **47**, 63–106 (2009)
77. H.C. Spruit, The rate of mixing in semiconvective zones. A&A. **253**, 131–138 (1992)
78. H.C. Spruit, Differential rotation and magnetic fields in stellar interiors. A&A. **349**, 189–202 (1999)
79. H.C. Spruit, Dynamo action by differential rotation in a stably stratified stellar interior. A&A. **381**, 923–932 (2002)
80. J.W. Truran Jr., A. Heger, Origin of the elements, in *Meteorites, Comets and Planets: Treatise on Geochemistry*, ed. by A.M. Davis, vol. 1 (Elsevier B. V, Amsterdam, 2005), p. 1
81. C. Tur, A. Heger, S.M. Austin, On the sensitivity of massive star nucleosynthesis and evolution to solar abundances and to uncertainties in helium-burning reaction rates. ApJ. **671**, 821–827 (2007)
82. M.J. Turk, T. Abel, B. O'Shea, The formation of population III binaries from cosmological initial conditions. Science **325**, 601 (2009)
83. A.J. van Marle, S.P. Owocki, N.J. Shaviv, Numerical simulations of continuum-driven winds of super-Eddington stars. MNRAS. **389**, 1353–1359 (2008)
84. H. Vogt, Die Beziehung zwischen den Massen und den absoluten Leuchtkräften der Sterne. Astron. Nachr. **226**, 301 (1926)
85. S. Wanajo, K. Nomoto, H.-T. Janka, F.S. Kitaura, B. Müller, Nucleosynthesis in electron capture supernovae of asymptotic giant branch stars. ApJ.**695**, 208–220 (2009)
86. T.A. Weaver, G.B. Zimmerman, S.E. Woosley, Presupernova evolution of massive stars. ApJ. **225**, 1021–1029 (1978)
87. S.E. Woosley, Gamma-ray bursts from stellar mass accretion disks around black holes. ApJ. **405**, 273–277 (1993)
88. S.E. Woosley, Bright supernovae from magnetar birth. ApJL. **719**, L204–L207 (2010)
89. S.E. Woosley, A. Heger, The progenitor stars of gamma-ray bursts. ApJ. **637**, 914–921 (2006)
90. S.E. Woosley, A. Heger, Nucleosynthesis and remnants in massive stars of solar metallicity. PhR. **442**, 269–283 (2007)
91. S.E. Woosley, T.A. Weaver, The physics of supernova explosions. ARA&A. **24**, 205–253 (1986)
92. S.E. Woosley, T.A. Weaver, Sub-Chandrasekhar mass models for type IA supernovae. ApJ. **423**, 371–379 (1994)
93. S.E. Woosley, T.A. Weaver, The evolution and explosion of massive stars. II. Explosive hydrodynamics and nucleosynthesis. ApJS. **101**, 181 (1995)
94. S.E. Woosley, R.G. Eastman, B.P. Schmidt, Gamma-ray bursts and type IC supernova SN 1998BW. ApJ. **516**, 788–796 (1999)
95. S.E. Woosley, A. Heger, T.A. Weaver, The evolution and explosion of massive stars. Rev. Mod. Phys. **74**, 1015–1071 (2002)

96. S.E. Woosley, S. Blinnikov, A. Heger, Pulsational pair instability as an explanation for the most luminous supernovae. Nature **450**, 390–392 (2007)
97. S.-C. Yoon, M. Cantiello, Evolution of massive stars with pulsation-driven superwinds during the red supergiant phase. ApJL. **717**, L62–L65 (2010)
98. S.-C. Yoon, N. Langer, Evolution of rapidly rotating metal-poor massive stars towards gamma-ray bursts. A&A. **443**, 643–648 (2005)
99. W. Zhang, S.E. Woosley, A. Heger, The propagation and eruption of relativistic jets from the stellar progenitors of gamma-ray bursts. ApJ. **608**, 365–377 (2004)
100. W. Zhang, S.E. Woosley, A. Heger, Fallback and black hole production in massive stars. ApJ. **679**, 639–654 (2008)
101. M. Zingale, A.S. Almgren, J.B. Bell, A. Nonaka, S.E. Woosley, Low mach number modeling of type IA supernovae. IV. White Dwarf convection. ApJ. **704**, 196–210 (2009)

Index

K. Davidson and R.M. Humphreys (eds.), *Eta Carinae and the Supernova Impostors*, Astrophysics and Space Science Library 384, DOI 10.1007/978-1-4614-2275-4,

Printed by Publishers' Graphics LLC
SO20120503